Water Science and Technology Library

Volume 78

The aim of the Water Science and Technology Library is to provide a forum for dissemination of the state-of-the-art of topics of current interest in the area of water science and technology. This is accomplished through publication of reference books and monographs, authored or edited. Occasionally also proceedings volumes are accepted for publication in the series.

Water Science and Technology Library encompasses a wide range of topics dealing with science as well as socio-economic aspects of water, environment, and ecology. Both the water quantity and quality issues are relevant and are embraced by Water Science and Technology Library. The emphasis may be on either the scientific content, or techniques of solution, or both. There is increasing emphasis these days on processes and Water Science and Technology Library is committed to promoting this emphasis by publishing books emphasizing scientific discussions of physical, chemical, and/or biological aspects of water resources. Likewise, current or emerging solution techniques receive high priority. Interdisciplinary coverage is encouraged. Case studies contributing to our knowledge of water science and technology are also embraced by the series. Innovative ideas and novel techniques are of particular interest.

Comments or suggestions for future volumes are welcomed.

Vijay P. Singh, Department of Biological and Agricultural Engineering & Zachry Department of Civil Engineering, Texas A&M University, USA
Email: vsingh@tamu.edu

More information about this series at http://www.springer.com/series/6689

Vijay P. Singh · Shalini Yadav
Ram Narayan Yadava
Editors

Water Resources Management

Select Proceedings of ICWEES-2016

Springer

Editors
Vijay P. Singh
Department of Biological and Agricultural Engineering, and Zachry Department of Civil Engineering
Texas A&M University
College Station, TX
USA

Ram Narayan Yadava
AISECT University
Hazaribagh, Jharkhand
India

Shalini Yadav
Department of Civil Engineering
AISECT University
Bhopal, Madhya Pradesh
India

ISSN 0921-092X ISSN 1872-4663 (electronic)
Water Science and Technology Library
ISBN 978-981-10-5710-6 ISBN 978-981-10-5711-3 (eBook)
https://doi.org/10.1007/978-981-10-5711-3

Library of Congress Control Number: 2017946676

Printed on acid-free paper

This Springer imprint is published by Springer Nature
The registered company is Springer Nature Singapore Pte Ltd.
The registered company address is: 152 Beach Road, #21-01/04 Gateway East, Singapore 189721, Singapore

Preface

Fundamental to sustainable economic development, functioning of healthy ecosystems, reliable agricultural productivity, dependable power generation, maintenance of desirable environmental quality, continuing industrial growth, enjoyment of quality lifestyle, and renewal of land and air resources is water. With growing population, demands for water for agriculture and industry are skyrocketing. On the other hand, freshwater resources per capita are decreasing. There is therefore a need for effective water resources management strategies. These strategies must also consider the nexus between water, energy, environment, food, and society. With these considerations in mind, the International Conference on Water, Environment, Energy and Society (WEES-2016) was organized at AISECT University in Bhopal, MP, India, during March 15–18, 2016. The conference was fifth in the series and had several objectives.

The first objective was to provide a forum to not only engineers, scientists, and researchers, but also practitioners, planners, managers, administrators, and policy-makers from around the world for discussion of problems pertaining to water, environment, and energy that are vital for the sustenance and development of society.

Second, the Government of India has embarked upon two large projects: one on cleaning of River Ganga and the other on cleaning River Yamuna. Further, it is allocating large funds for irrigation projects with the aim to bring sufficient good-quality water to all farmers. These are huge ambitious projects and require consideration of all aspects of water, environment, and energy as well as society, including economics, culture, religion, politics, administration, law, and so on.

Third, when water resources projects are developed, it is important to ensure that these projects achieve their intended objectives without causing deleterious environmental consequences, such as water logging, salinization, loss of wetlands, sedimentation of reservoirs, loss of biodiversity, etc.

Fourth, the combination of rising demand for water and increasing concern for environmental quality compels that water resources projects are planned, designed, executed and managed, keeping changing conditions in mind, especially climate change and social and economic changes.

Fifth, water resources projects are investment intensive and it is therefore important to take a stock of how the built projects have fared and the lessons that can be learnt so that future projects are even better. This requires an open and frank discussion among all sectors and stakeholders.

Sixth, we wanted to reinforce that water, environment, energy, and society constitute a continuum and water is central to this continuum. Water resources projects are therefore inherently interdisciplinary and must be so dealt with.

Seventh, a conference like this offers an opportunity to renew old friendships and make new ones, exchange ideas and experiences, develop collaborations, and enrich ourselves both socially and intellectually. We have much to learn from each other.

Now the question may be: Why India and why Bhopal? India has had a long tradition of excellence spanning several millennia in the construction of water resources projects. Because of her vast size, high climatic variability encompassing six seasons, extreme landscape variability from flat plains to the highest mountains in the world, and large river systems, India offers a rich natural laboratory for water resources investigations.

India is a vast country, full of contrasts. She is diverse yet harmonious, mysterious yet charming, old yet beautiful, ancient yet modern. Nowhere we can find mountains as high as the snow-capped Himalayas in the north, the confluence of three seas and large temples in the south, long and fine sand beaches in the east as well as architectural gems in the west. The entire country is dotted with unsurpassable monuments, temples, mosques, palaces, and forts and fortresses that offer a glimpse of India's past and present.

Bhopal is located in almost the center of India and is situated between Narmada River and Betwa River. It is a capital of Madhya Pradesh and has a rich, several century-long history. It is a fascinating amalgam of scenic beauty, old historic city, and modern urban planning. All things considered, the venue of the conference could not have been better.

We received an overwhelming response to our call for papers. The number of abstracts received exceeded 450. Each abstract was reviewed and about two thirds of them, deemed appropriate to the theme of the conference, were selected. This led to the submission of about 300 full-length papers. The subject matter of the papers was divided into more than 40 topics, encompassing virtually all major aspects of water and environment as well energy. Each topic comprised a number of contributed papers and in some cases state-of-the-art papers. These papers provided a natural blend to reflect a coherent body of knowledge on that topic.

The papers contained in this volume, "Water Resources Management," represent one part of the conference proceedings. The other parts are embodied in six companion volumes entitled, "Hydrologic Modelling," "Groundwater," "Environmental Pollution," "Water Quality Management," "Climate Change Impacts," and "Energy and Environment." Arrangement of contributions in these seven books was a natural consequence of the diversity of papers presented at the conference and the topics covered. These books can be treated almost independently, although significant interconnectedness exists among them.

This volume contains two parts. Part I deals with some aspects of irrigation, encompassing farm irrigation systems, landscape gardening, energy assessment for drip irrigation, and micro-sprinklers. Part II is on water resources planning and management. It discusses water crisis, challenges in river health management, water supply systems, salt water intrusion, lake management, water supply demand assessment, integrated water resources management, among other topics.

The book will be of interest to researchers and practitioners in the field of water resources, hydrology, environmental resources, agricultural engineering, watershed management, earth sciences, as well as those engaged in natural resources planning and management. Graduate students and those wishing to conduct further research in water and environment and their development and management may find the book to be of value.

WEES-16 attracted a large number of nationally and internationally well-known people who have long been at the forefront of environmental and water resources education, research, teaching, planning, development, management, and practice. It is hoped that long and productive personal associations and friendships will be developed as a result of this conference.

College Station, USA Vijay P. Singh, Conference Chair
Bhopal, India Shalini Yadav, Conference Organizing Secretary
Hazaribagh, India Ram Narayan Yadava, Conference Co-Chair

Acknowledgements

We express our sincere gratitude to Shri Santosh Choubey, Chancellor, and Dr. V.K. Verma, Vice Chancellor, Board of Governing Body, and Board of Management of the AISECT University, Bhopal, India, for providing their continuous guidance and full organizational support in successfully organizing this international conference on Water, Environment, Energy and Society on the AISECT University campus in Bhopal, India.

We are also grateful to the Department of Biological and Agricultural Engineering, and Zachry Department of Civil Engineering, Texas A&M University, College Station, Texas, U.S.A., and International Centre of Excellence in Water Management (ICE WaRM), Australia, for their institutional cooperation and support in organizing the ICWEES-2016.

We wish to take this opportunity to express our sincere appreciation to all the members of the Local Organization Committee for helping with transportation, lodging, food, and a whole host of other logistics. We must express our appreciation to the Members of Advisory Committee, Members of the National and International Technical Committees for sharing their pearls of wisdom with us during the course of the Conference.

Numerous other people contributed to the conference in one way or another, and lack of space does not allow us to list all of them here. We are also immensely grateful to all the invited Keynote Speakers, and Directors/Heads of Institutions for supporting and permitting research scholars, scientists and faculty members from their organizations for delivering keynote lectures and participating in the conference, submitting and presenting technical papers. The success of the conference is the direct result of their collective efforts. The session chairmen and co-chairmen administered the sessions in a positive, constructive, and professional manner. We owe our deep gratitude to all of these individuals and their organizations.

We are thankful to Shri Amitabh Saxena, Pro-Vice Chancellor, Dr. Vijay Singh, Registrar, and Dr. Basant Singh, School of Engineering and Technology, AISECT University, who provided expertise that greatly helped with the conference organization. We are also thankful to all the Heads of other Schools, Faculty

Members and Staff of the AISECT University for the highly appreciable assistance in different organizing committees of the conference. We also express our sincere thanks to all the reviewers at national and international levels who reviewed and moderated the papers submitted to the conference. Their constructive evaluation and suggestions improved the manuscripts significantly.

Sponsors and Co-sponsors

The International Conference on Water, Environment, Energy and Society was Jointly organized by the AISECT University, Bhopal (MP), India and Texas A&M University, Texas, USA in association with ICE WaRM, Adelaide, Australia. It was partially supported by the International Atomic Energy Agency (IAEA), Vienna, Austria; AISECT University, Bhopal; M.P. Council of Science and Technology (MPCOST); Environmental Planning and Coordination Organization (EPCO), Government of Madhya Pradesh; National Bank for Agriculture and Rural Development (NABARD), Mumbai; Maulana Azad National Institute of Technology (MANIT), Bhopal; and National Thermal Power Corporation (NTPC), Noida, India. We are grateful to all these sponsors for their cooperation and providing partial financial support that led to the grand success to the ICWEES-2016.

Contents

About the Editors

Prof. Vijay P. Singh is University Distinguished Professor, Regents Professor, and the inaugural holder of the Caroline and William N. Lehrer Distinguished Chair in Water Engineering in the Department of Biological and Agricultural Engineering and Zachry Department of Civil Engineering at Texas A&M University. He received his B.S., M.S., Ph.D., and D.Sc. degrees in engineering. He is a registered professional engineer, a registered professional hydrologist, and an Honorary Diplomate of American Academy of Water Resources Engineers.

Professor Singh has extensively published the results of an extraordinary range of his scientific pursuits. He has published more than 900 journal articles; 25 textbooks; 60 edited reference books, including the massive Encyclopedia of Snow, Ice and Glaciers and Handbook of Applied Hydrology; 104 book chapters; 314 conference papers; and 72 technical reports in the areas of hydrology, ground water, hydraulics, irrigation engineering, environmental engineering, and water resources.

For his scientific contributions to the development and management of water resources and promoting the cause of their conservation and sustainable use, he has received more than 90 national and international awards and numerous honors, including the Arid Lands Hydraulic Engineering Award, Ven Te Chow Award, Richard R. Torrens Award, Norman Medal, and EWRI Lifetime Achievement Award, all given by American Society of Civil Engineers; Ray K. Linsley Award and Founder's Award, given by American Institute of Hydrology; Crystal Drop Award, given by International Water Resources Association; and Outstanding Distinguished Scientist Award given by Sigma Xi, among others. He has received three honorary doctorates. He is a Distinguished Member of ASCE, and a fellow of EWRI, AWRA, IWRS, ISAE, IASWC, and IE and holds membership in 16 additional professional associations. He is a fellow/member of 10 international science/engineering academies. He has served as President and Senior Vice President of the American Institute of Hydrology (AIH). Currently he is editor-in-chief of two book series and three journals and serves on editorial boards of 20 other journals.

Professor Singh has visited and delivered invited lectures in all most all parts of the world but just a sample: Switzerland, the Czech Republic, Hungary, Austria, India, Italy, France, England, China, Singapore, Brazil, and Australia.

Prof. Shalini Yadav is Professor and Head of the Department of Civil Engineering, AISECT University, Bhopal, India. Her research interests include solid and hazardous waste management, construction management, environmental quality and water resources. She has executed a variety of research projects/consultancy in Environmental and Water Science and Technology and has got rich experience in Planning, formulating, organizing, executing and management of R&D programs, seminars, and conferences at national and international levels. She has got to her credit guiding an appreciable number of M.Tech. and Ph.D. students. She has published more than 10 journal articles and 30 technical reports. Dr. Shalini has also visited and delivered invited lectures at different institutes/universities in India and abroad, such as Australia, South Korea, and Kenya.

Professor Shalini Yadav graduated with a B.Sc. in Science from the Bhopal University. She earned her M.Sc. in Applied Chemistry with a specialization in Environmental Science from Bhopal University and M.Tech. in Civil Engineering with a specialization in Environmental Engineering from Malaviya National Institute of Technology, Jaipur, India in 2000. Then she pursued the degree of Ph.D. in Civil Engineering from Rajiv Gandhi Technical University, Bhopal, India in 2011. Also, she is a recipient of national fellowships and awards. She is a reviewer for many international journals. She has been recognized for one and half decades of leadership in research, teaching, and service to the Environmental Engineering Profession.

Dr. Ram Narayan Yadava holds the position of Vice Chancellor of the AISECT University, Hazaribagh, Jharkhand. His research interests include solid mechanics, environmental quality and water resources, hydrologic modeling, environmental sciences and R&D planning and management. Yadava has executed a variety of research/consultancy projects in the area of water resources planning and management, environment, remote sensing, mathematical modeling, technology forecasting, etc.

He has got adequate experience in establishing institutes/organizations, planning, formulating, organizing, executing and management of R&D programs, seminars, symposia, conferences at national and international level. He has got to his credit guiding a number of M.Tech. and Ph.D. students in the area of mathematical sciences and Earth sciences. Dr. Yadava has visited and delivered invited lectures at different institutes/universities in India and abroad, such as USA, Canada, United Kingdom, Thailand, Germany, South Korea, Malaysia, Singapore, South Africa, Costa Rica, and Australia.

He earned an M.Sc. in Mathematics with a specialization in Special Functions and Relativity from Banaras Hindu University, India in 1970 and a Ph.D. in Mathematics with specialization in Fracture Mechanics from Indian Institute of Technology, Bombay, India, in 1975. Also, he is a recipient of Raman Research Fellowship and other awards. Dr. Yadava has been recognized for three and half decades of leadership in research and service to the hydrologic and water resources profession. Dr. Yadava's contribution to the state of the art has been significant in many different specialty areas, including water resources management, environmental sciences, irrigation science, soil and water conservation engineering, and mathematical modeling. He has published more than 90 journal articles; 4 textbooks; 7 edited reference books.

Part I
Irrigation

Identification of Farming Systems in Tribal Region of Zone IV-B of Rajasthan (India)

P. S. Rao and Hari Singh

Abstract The share of agriculture in gross domestic product has registered a steady decline from 36.4% in 1982–83 to 14.5% in 2013–14. Yet this sector continues to support more than half a billion people proving employment to 52% of the work force. It is also an important source of raw material and demand for many industrial products. Growth of agriculture over a period of time remained lower than the growth in non-agriculture sector. The gap began to widen since 1981–82 and more particularly since 1996–97, because of acceleration in the growth of industry and service sector. Survey on cropping system was carried out in tribal districts of Rajasthan. Using the stratified random sampling and probability proportion sampling, 450 farmers belonging to size groups, based on the size of holding were selected from five tehsils of tribal districts of Rajasthan. The secondary data has been taken from published records of state and central government. Data are interpreted as averages and percentages. The percentage of forest area to the total geographical area in Dungarpur and Banswara districts is more than double compared to forest area at the state level. It is evident that total cropped area in the state is 17,914,166 ha of which 55.76% (9,989,675 ha) kept under cereals and small millets, 24.84% (4,449,229 ha) under pulses, 18.82% (3,372,130 ha) under oilseeds crops and 0.58% (103,132 ha) under commercial crops. It is evident that of the total 577 lakh animals in the state, the population of goat is the highest (37.53%) followed by cattle (23.08%), buffalo (22.48%), and sheep (15.73%). In this study irrespective of the rain-fed and irrigated condition as well as location of the selected districts, four farming systems were observed. The study area is tribal dominating area where 50% tribal of the country are residing. An alternative farming system, which yields not only higher income but also utilizes family labor efficiency, needs to be evolved. Further, the system should help in restoration of ecological balance. The basic aim of integrated/sustainable farming system is to derive a set of resource development, management, and utilization practices that lead to a substantial and

P. S. Rao (✉)
Department of Soil and Water Engineering, CTAE, Udaipur, Rajasthan, India
e-mail: pratapsinghrao10@gmail.com

H. Singh
Department of Agricultural Economics, RCA, Udaipur, Rajasthan, India

V. P. Singh et al. (eds.), *Water Resources Management*, Water Science and Technology Library 78, https://doi.org/10.1007/978-981-10-5711-3_1

sustained increase in agricultural production. It is evident that out of 30 farmers, the maximum number of farmers (40%) adopted FS-II (Crop + Dairy) followed by FS-I (Crop + Vegetable) and FS-III (Crop + Dairy + Goat) and only 13.33% farmers followed FS-IV (Crop + Goat + Poultry). The highest cropping intensity was observed in FS-II (C+D), i.e., 94.44% followed by FS-I and FS-IV. The lowest cropping intensity was seen in FS-III (88.23%). In rain-fed area, cropping intensity was low due to non-availability of irrigation facilities and farmers were forced to take crops only in one season.

Keywords Socioeconomic · Farming systems · Tribal region · Cropping intensity

Introduction

Agriculture is mainstay of Indian Economy because of its high share in employment and livelihood creation not withstanding its reduced contribution to the notice GDP. The share of agriculture in gross domestic product has registered a steady decline from 36.4% in 1982–83 to 13.5% in 2012–13. Yet this sector continues to support more than half a billion people proving employment to 52% of the work force. It is also an important source of raw material and demand for many industrial products. Growth of agriculture over a period of time remained lower than the growth in non-agriculture sector. The gap began to widen since 1981–82 and more particularly since 1996–97, because of acceleration in the growth of industry and service sector. The study concluded by Poddar et al. (1995) in Karnataka has revealed that the optimization of resource use with the existing and improved technology led to cultivation of a few but more profitable crops in the new cropping patterns.

Rajasthan is the largest state of India with geographical area of 342.65 lakh hectares and occupies 10.41% land area of the country. About 46.30% area is under cultivation. Identification of efficient cropping systems with reference to productivity and sustainability has become imperative for different farming situations (Nanda et al. 1999). The whole state therefore, has been subdivides into 10 agro-climatic zones to cater the need-based and location-specific research demand. The zone IV-a&b Sub-humid Southern Plain and Aravali Hills covering Bhilwara, Udaipur, Rajsamand, Pratapgarh, Dungarpur, Banswara, and Sirohi district where maize, wheat, gram, paddy, and urad are the major crops grown in this area. This area is tribal dominating area where 50% tribal of the country are residing. An alternative farming system, which yields not only higher income but also utilizes family labor efficiency, needs to be evolved. Further, the system should help in restoration of ecological balance. The basic aim of integrated/sustainable farming system is to derive a set of resource development, management, and utilization practices that lead to a substantial and sustained increase in agricultural production. Since farming systems differ in different situations such studies should be location-specific (Singh 1998). Looking to the above factors, this study has been conducted with the objectives (i) to examine the resource characterization, (ii) to study the cropping system of the tribals, and (iii) analyze the farming system along with income and employment pattern in the study area. Keeping the above objectives in mind this study has been conducted with the following methodology.

Methodology

Survey on cropping system was carried out in two districts out of four districts of tribal abundance population (Banswara and Dungarpur) of Rajasthan, which falls under Zone IV-b (Humid Southern Plain). Stratified random sampling technique was used for selection of Tehsil, and villages. While farmers have been selected through probability proportion sampling, a total of 450 farmers were ultimately selected, based on the size of operational holding. The size of holding so selected was, small (1–2 ha), medium (2–4 ha), and large (above 4 ha). Selected farmers were from five tehsils of two tribal districts of Rajasthan of Zone IV-b (Humid Southern Plain). Two tehsils from Dungarpur and three tehsils from Banswara have been selected. From each tehsil, 15 farmers from rain-fed and 15 farmers from irrigated area in each category of small, medium, and large size holding have been selected. In this way, 45 farmers in rain-fed and 45 farmers in irrigated area have been selected from each selected tehsil. The data on socioeconomic parameters, like existing cropping system, and income and employment pattern were obtained in pre-tested schedule. The survey has been conducted by way of the questionnaires. The secondary data has been taken from published records of state and central government. Data were analyzed and interpreted by using averages and percentages.

Results

Land Utilization Pattern of the Study Area—Agriculture is the main occupation in both the selected districts. Land utilization pattern of Dungarpur and Banswara district (tribal) as well as of the Rajasthan State is presented in Table 1. Out of 343 lakh hectares geographical area of Rajasthan state, Dungarpur district contributed 3.85 lakh hectares whereas Banswara district contributed 4.54 lakh hectares. The percentage of forest area to the total geographical area in both the districts is more than double compared to forest area at the state average. The forest area of Dungarpur is, 16.00% while it is 20.12% in Banswara district compared to 8.02% of the state average.

Data further shows that area not available for cultivation is high in both the districts, Dungarpur (24.10%) and 13.80% in Banswara due to barren land area; which is again greater than state average (12.46%). Uncultivated land is more in Dungarpur district (14.68%) as compared to Banswara district (8.69%) which is due

Table 1 Land utilization pattern of the study area (2011–12) (area in ha)

S. No.	Particulars	Dungarpur district	Banswara district	Rajasthan state
A	Total geographical area	385,593 (100.00)	453,612 (100.00)	34,267,252 (100.00)
1	Area under forest	62,204 (16.13)	91,250 (20.12)	2,746,686 (8.02)
2	Land not available for cultivation	92,943 (24.10)	62,589 (13.80)	4,271,340 (12.46)
	(i) Non-agricultural uses	22,970 (5.96)	11,017 (2.43)	1,884,055 (5.50)
	(ii) Barren	69,973 (18.15)	15,172 (11.37)	2,387,285 (6.97)
3	Other uncultivated land	56,589 (14.68)	39,415 (8.69)	5,883,545 (17.18)
	(i) Cultivated waste	20,811 (5.40)	24,789 (5.46)	4,168,681 (12.17)
	(ii) Permanent pasture	34,539 (8.96)	11,509 (2.54)	1,693,790 (4.94)
	(iii) Trees and groove	1239 (0.03)	117 (0.03)	21,074 (0.001)
4	Fallow land	39,387 (10.21)	36,521 (8.05)	3,331,274 (9.72)
	(a) Current fallow	8373 (2.17)	4558 (1.00)	1,476,694 (4.31)
	(b) Other fallow	31,014 (8.04)	21,963 (4.84)	1,854,580 (5.41)
5	Net sown area	134,470 (34.87)	226,812 (50.00)	18,034,407 (52.63)
6	Area sown more than once	65,417 (8.04)	109,903 (24.23)	6,470,962 (18.88)
7	Total cropped area	199,887 (34.87)	336,742 (74.24)	24,505,369 (71.51)
B	Cropping intensity (%)	149	148	136
C	Irrigated area as percentage of net sown area (%)	34.26	44.19	39.49
D	Total net irrigated area	46,077 (100.00)	100,228 (100.00)	7,121,575 (100.00)
	(a) Area under canal irrigation	9295 (20.17)	59,019 (58.88)	1,843,797 (25.89)
	(b) Tanks	2489 (5.40)	5005 (4.99)	68,785 (9.65)
	(c) Wells and Tubewells	31,438 (68.22)	17,999 (17.96)	5,111,105 (71.77)
	(d) Others	2855 (6.20)	18,205 (18.16)	97,888 (1.87)

Source Statistical abstract, 2012. Directorate of economics and statistics Rajasthan, Jaipur (figures in parentheses denotes percentage to respective totals) to the permanent pasture, trees, and groove while it is 17.18% at state level. As far as fallow land is concerned Dungarpur districts have more area under fallow land (10.21%) and in Banswara it is (8.05%) as compared to state average (9.72%). Cropping intensity in Dungarpur (149%) and Banswara (148%) district is more than the state average (136%). Data show that net sown area in Dungarpur was (34.87%) which is less than both Banswara (50.00%) and the state (52.63%) as a whole. At the state level, area under canal and well irrigation accounted for 25.89 and 71.71% respectively. The irrigated area in Banswara district under canal (58.88%) and in Dungarpur under well and well tube (68.22%) was found highest.

It can be concluded that an alternative farming system should be adopted by the farmers, which yields not only higher income but also utilizes family labor efficiency. Further, the system should help in restoration of ecological balance. It is evident that out of 450 farmers, the maximum number of farmers (40%) adopted crop production along with dairy farming so that they will be mutually beneficial to each other where by-product of both the enterprises can be fully utilized by each other. Farming System-II (Crop + Dairy).

Area Under Various Crops in Study Area

There are number of crops grown in Rajasthan state as well as in the selected districts. The area put under different crops in Dungarpur and Banswara tribal districts and in the state is presented in Table 2.

It is evident from the table that total cropped area in the state is 17,914,166 ha of which 55.76% (9,989,675 ha) kept under cereals and small millets, 24.84% (4,449,229 ha) under pulses, 18.82% (3,372,130 ha) under oilseeds crops and 0.58% (103,132 ha) under commercial crops. In both the selected districts maize, wheat, gram, other kharif pulses and groundnut are the major important crops which yield a major share of fodder for dairy animals which support the farmers to withstand at the time of crop failure in adverse condition and survive on dairy through diversification.

Livestock Population in the Study Area

Almost all types of pet animals are reared in Rajasthan. The livestock population in the tribal districts and Rajasthan state is presented in Table 3. It is evident from the table that of the total 577 lakh animals in the state, the population of goat is the highest (37.53%) followed by cattle (23.08%), buffalo (22.48%) and sheep (15.73%).

Table 2 Area under different crops in the study area (2011–12) (area in ha)

S. No.	Crops	Banswara district	Dungarpur district	Rajasthan state
A	*Cereals*			
	Bajra	101 (0.00)	124 (0.0006)	5,027,993 (28.07)
	Sorghum	383 (0.12)	549 (0.003)	553,796 (3.09)
	Maize	145,482 (46.47)	82,205 (42.86)	1,045,591 (5.84)
	Wheat	87,177 (27.85)	45,085 (23.50)	2,935,341 (16.39)
	Barley	1,023 (0.33)	1100 (0.57)	278,016 (1.55)
	Rice	28,542 (9.12)	22,031 (11.49)	134,337 (0.70)
	Small millets	4,128 (1.32)	6,130 (3.20)	14,601 (0.01)
	Sub-total	266,836 (85.23)	157,224 (81.97)	9,989,675 (55.76)
B	*Pulses*			
	Gram	13,719 (4.38)	14,971 (7.80)	1,433,954 (8.00)
	Other rabi pulses	2,251 (0.72)	1,524 (0.70)	43,760 (0.24)
	Other Kharif pulses	11,587 (3.70)	12,890 (6.72)	2,952,406 (16.48)
	Tur/Arhar	5,632 (1.80)	1,854 (0.97)	19,115 (0.11)
	Sub-total	33,189 (10.60)	31,239 (16.29)	4,449,229 (24.84)
C	*Oil seeds*			
	Sesamum	634 (0.20)	1,453 (0.76)	512,766 (2.86)
	Rape seed and mustard	99 (0.03)	760 (0.40)	2,441,254 (13.63)
	Groundnut	205 (0.07)	41 (0.00)	418,110 (2.33)
	Sub-total	938 (0.30)	2,254 (1.18)	3,372,130 (18.82)
D	*Commercial crops*			
	Sugarcane	146 (0.05)	64 (0.00)	6415 (0.04)
	Cotton	11,954 (3.82)	1,033 (0.54)	96,717 (0.54)
	Sub-total	12,100 (3.87)	1,097 (0.97)	103,132 (0.58)
E	Total cropped area	31,363 (100.00)	191,814 (100.00)	17,914,166 (100.00)

Source Statistical abstract, 2012. Directorate of economics and statistics Rajasthan, Jaipur (figures in parentheses denotes percentage to total cropped area)

The share of livestock population in Banswara and Dungarpur districts in the total livestock population of Rajasthan state is 2.42 and 1.89%, respectively. The percentage of cattle population was the highest (42.89%) in Banswara district while goat population was found highest (38.24%) in Dungarpur while it is 36.17% in Banswara. The buffalo population has also shown significant presence (above 20%) in livestock population of Dungarpur and Banswara districts. Total state poultry birds population was 80.24 lakhs of which 2.22 and 3.35% were found in Banswara

Table 3 Livestock population in the study area (2007) (number)

S. No.	Livestock	Banswara district	Dungarpur district	Rajasthan state
A	Animal husbandry			
	Goat	504,758 (36.17)	416,729 (38.24)	21,665,939 (37.53)
	Sheep	7,207 (0.52)	62,652 (5.75)	9,079,702 (15.73)
	Buffalo	282,438 (20.24)	232,133 (21.30)	12,976,095 (22.48)
	Cattle	598,453 (42.89)	375,023 (34.42)	13,324,462 (23.08)
	Pig	125 (0.2)	38 (0.00)	237,674 (0.04)
	Camel	558 (0.04)	1,672 (0.15)	325,713 (0.56)
	Donkey	1,713 (0.12)	1,214 (0.11)	81,468 (0.14)
	Horse	165 (0.01)	138 (0.01)	37,776 (0.07)
	Total livestock	1,395,418 (100.00)	1,089,600 (100.00)	57,732,204 (100.00)
B	Poultry	268,707 (19.26)	177,807 (16.32)	8,024,424 (13.90)

Source Statistical abstract, 2012. Directorate of economics and statistics Rajasthan, Jaipur (figures in the parentheses are percentage to the column total in livestock and row total in poultry)

and Dungarpur, respectively. Thus, on the basis of population of milking animals, farming system of crop production with dairy animals predominantly existed to maximize the farm income in the tribal area of southern Rajasthan.

Farming Systems

There are number of farming systems that existed in the study area. Farming system is a combination of crops, vegetables, orchards, dairy enterprise, and poultry to maximize the farm income. In this study, irrespective of the rain-fed and irrigated condition as well as location of the selected districts, four farming systems were observed.

$$\text{FS-I} : \text{Crops} + \text{Vegetable}(\text{C} + \text{V})$$
$$\text{FS-II} : \text{Crops} + \text{Dairy}(\text{C} + \text{D})$$
$$\text{FS-III} : \text{Crops} + \text{Dairy} + \text{Goat}(\text{C} + \text{D} + \text{G})$$
$$\text{FS-IV} : \text{Crops} + \text{Goat} + \text{Poultry} + \text{Orchard}(\text{C} + \text{G} + \text{Po} + \text{O})$$

The farming systems that existed in Dungarpur and Banswara districts in rain-fed and irrigated conditions were studied separately. A condition in which the crops depend on the rainfall and rainfall is below 500 mm per annum as well as the irrigation is not assured was considered as rain-fed condition. The irrigated condition was one where assured irrigation facilities for growing the crops exist and crops do not depend on rainfall. The discussion on various aspects of farming systems in rain-fed and irrigated condition is on the average per farm.

Farming Systems in Dungarpur District

The farming system existing in rain-fed and irrigated conditions was studied and discussed under the heads of number of farmers, cropping intensity, cropping pattern, and non-crop enterprises in different farming systems. The Sagwadra tehsil was selected as rainfed and Aspur was selected as irrigated tehsil.

Rain-Fed Condition

Farming system in rain-fed area of Dungarpur district is presented in Table 4. It is evident from the table that in this area, crops were grown only in kharif season and dairy, goat and poultry are also taken up by the farmers along with crops and vegetables. It is evident from Table 4, that out of 30 farmers, the maximum number of farmers (40%) adopted FS-II (Crop + Dairy) followed by FS-I (Crop + Vegetable) 23.33%, FS-III (Crop + Dairy + Goat) and only 13.33% farmers followed FS-IV (Crop + Goat + Poultry).

The cropping intensity ranged from 94.34 to 98.00%. The highest cropping intensity was observed in FS-I whereas lowest cropping intensity was found in FS-IV. There was a very little variation in the cropping intensity among four farming systems because of rain-fed condition.

Maize and soybean were prominent crops grown in all the four farming systems, black gram in three farming systems while green gram and vegetables had sown their presence only in one farming system. In FS-I, the average total cropped area was 1.14 ha of which kharif crops were put under 1.04 ha (91.2%) and only 0.1 ha was kept under vegetables. In FS-II, the crops were same as grown in FS-I. Out of the gross cropped area of 1.17 ha, soybean was grown in the maximum area (0.52 ha) followed by maize (0.45 ha) and black gram (0.20 ha). Similarly in FS-III with the gross cropped area was 0.60 ha, the crops grown were maize (0.30 ha), soybean (0.20 ha), and black gram (0.10 ha), while in FS-IV gross cropped area was 0.50 ha and black gram crop which found a first place in all three farming systems was replaced by green gram in this farming system and other two crops, i.e., maize and soybean remained the same. The area under crops in all the four systems was 3.41 ha in which FS-II contributed the maximum area (34.31%) followed by FS-I (33.43%). Thus, it can be concluded that maize and soybean were the important crops in all the farming systems and vegetables got its place only in FS-I.

Non-crop enterprises such as dairy cattle, goat, and poultry were found only in three farming systems, i.e., FS-II, FS-III, and FS-IV and in FS-I non-crop enterprises have not occupied any place. Thus, it can be concluded that non-crop enterprises has shown their existence only in three systems. Dairy enterprises (FS-II and FS-III) and goat enterprises (FS-III and FS-IV) were taken up in two farming

Table 4 Existing farming systems in rain-fed condition of Dungarpur district (per farm)

Farming system	No. of farmers	Gross cropped area (ha)	Cultivated area (ha)	Cropping intensity (%)	Crops		Vegetable		Dairy cattle (No)	Goat (No)	Poultry (No)
					Name	Area (ha)	Name	Area (ha)			
FS-I (C+V)	7 (23.33)	1.14 (33.43)	1.16	98.28	Maize	0.38	Okra	0.05	–	–	–
					Soybean	0.34					
					Black gram	0.32	Bottle gourd	0.05			
FS-II (C+D)	12 (40.00)	1.17 (34.31)	1.20	97.50	Maize	0.45	–	–	2.08	–	–
					Soybean	0.52					
					Black gram	0.20					
FS-III (C+D+G)	7 (23.33)	0.60 (17.60)	0.62	96.77	Maize	0.30	–	–	2.40	8.50	–
					Soybean	0.20					
					Black gram	0.10					
FS-IV (C+G+Po)	4 (13.33)	0.50 (14.66)	0.53	94.34	Maize	0.20	–	–	–	7.04	52.85
					Greengram	0.10					
					Soybean	0.10					
					Fallow land	0.10					
Total	30 (100)	3.41 (100)	3.51	96.72	–	3.31	–	0.10	4.48	15.54	52.85

C Crop, *D* dairy, *Po* poultry, *G* goat (figures in parentheses are percentage of column total)

Table 5 Existing farming systems in irrigated condition of Dungarpur district (per farm)

Farming system	No. of farmers	Gross cropped area (ha)	Cultivated area (ha)	Cropping intensity (%)	Crops		Vegetable		Dairy cattle (No)	Goat (No)	Poultry (No)
					Name	Area (ha)	Name	Area (ha)			
FS-I (C+V)	6 (20.23)	2.24 (26.23)	1.16	193.10	Maize	0.45	Tomato	0.22	–	–	–
					Cotton	0.15	Okra	0.30			
					Wheat	0.65	Bo.gourd	017			
					Mustard	0.29	Bi.gourd	0.01			
FS-II (C+D)	14 (46.67)	2.30 (26.93)	1.20	191.67	Maize	0.80			3.50	–	–
					Soybean	0.20					
					Cotton	0.20					
					Wheat	0.95					
					Barley	0.05					
					Mustard	0.10					
FS-III (C+D +G)	3 (10.00)	2.15 (25.18)	1.10	195.45	Maize	0.65			2.60	8.50	–
					Soybean	0.35					
					Black gram	0.25					
					Wheat	0.70					
					Mustard	0.20					
FS-IV (C+G +Po)	4 (23.33)	1.85 (21.66)	0.98	188.77	Maize	0.50			–	7.04	52.85
					Greengram	0.20					
					Cotton	0.25					
					Soybean	0.20					
					Wheat	0.55					
					Mustard	0.15					
Total	30 (100)	8.54 (100)	4.44	192.25		7.84		0.70	6.10	120	37.55

C Crop, *D* dairy, *Po* poultry, *G* goat (figures in parentheses are percentage of column total)

systems while poultry birds existed only in one farming system (FS-IV). Ram Rao et al. (2005) also identified crop + livestock integrated farming system in rain-fed area of Chhattisgarh in Central India. Similar results were also reported by Nayak (2003).

The various farming systems prevalent in irrigated area of Dungarpur district is presented in Table 5. Results showed that maximum number (46.67%) of farmers adopted FS-II (C+D) followed by FS-IV (23.23), FS-I (20%), and the lowest in FS-III (10.00%).

Data revealed that cropping intensity varied from 188.77 to 195.45%. The highest cropping intensity was observed in FS-III (195.45%) followed by FS-I (193.10%), FS-II (191.67%), and the lowest in FS-IV (188.77%). Under irrigated condition, the cropping intensity in all the farming systems was more than 100 because farmers take crops and vegetables in both the seasons due to availability of irrigation facilities.

The total cropped area in FS-I, FS-II, FS-III, and FS-IV were 2.24, 2.30, 2.15, and 1.85 ha, respectively. Thus, in all the farming systems, there was a good scope to cultivate the crops in both the seasons because of assured irrigation facility. It is observed from Table 5, that non-crop enterprises were found only in three systems, i.e., FS-II, FS-III, and FS-IV. In FS-II, the average number of dairy cattle was 3.50 where as it was 2.60 in FS-III. The average number of goats was 16.00 in FS-III whereas in FS-IV in addition to 11.50 average number of goats, 120 fruit plants in the orchard were also maintained. Thus, it can be concluded that non-crop enterprises showed their presence in three farming systems. Dairy cattle were found in two systems (FS-II and FS-III), goat was also found in two systems (i.e., FS-III and FS-IV), while orchard was found only in one farming system, i.e., FS-IV. Similar findings were found by Baishya et al. (2007), both in irrigated and rain-fed farming system in Karupt and Borpeta districts of Assam. The main components of the farming system in these districts were crops, dairy, goat, and piggery.

Farming Systems in Banswara District

Farming systems in Banswara district were studied in both, i.e., rain-fed and irrigated condition. The selected rain-fed tehsil was Kushalgarh, Gaddi, and Banswara.

Rain-Fed Condition

In rain-fed condition farmers of Banswara were taking only crops and onion nursery in kharif season while in rabi season land were kept fallow due to unavailability of irrigation facilities. Data revealed that maximum number of farmers (43.33%) falls in FS-II (C+D) and minimum number of farmers (13.33%) adopted FS-I (C+ON) as

Table 6 Existing farming systems in rain-fed condition of Banswara district (per farm)

Farming system	No. of farmer	Gross cropped area (ha)	Cultivated area (ha)	C.I. (%)	Crops		Onion nursery		Dairy cattle (No)	Goat (No)	Poultry (No)
					Name	Area (ha)	Name	Area (ha)			
FS-I (C+V)	4 (13.33)	0.58 (25.00)	0.62	93.55	Maize	0.25	Onion Nursery	0.08	–	–	–
					Soybean	0.10					
					Paddy	0.15					
FS-II (C+D)	13 (43.33)	0.85 (36.63)	0.90	94.44	Maize	0.45	–	–	3.06	–	–
					Soybean	0.10					
					Paddy	0.15					
FS-III (C+D +G)	7 (23.33)	0.45 (19.39)	0.51	88.23	Maize	0.20	–	–	3.00	17.00	–
					Soybean	0.15					
					Black gram	0.10					
FS-IV (C +Po)	6 (20.00)	0.44 (18.96)	0.48	91.66	Maize	0.24	–	–	–	–	55.75
					Paddy	0.20					
Total	30 (100.0)	2.32 (100.0)	2.51	91.97	–	2.24	–	0.08	6.06	17.00	55.75

C Crop, *V* vegetable, *D* dairy, *Po* poultry, *G* goat (figures in parenthesis is percentage of column total) depicted in Table 6. One-fourth farmers adopted FS-IV and 23.33% farmers adopted FS-III in the rain-fed area of Banswara district.

The cropping intensity in rain-fed condition varied from 88.23 to 94.44%. The highest cropping intensity was observed in FS-II (94.44%) followed by FS-I (93.55%) and FS-IV (91.66%). The lowest cropping intensity was seen in FS-III (88.23%). In rain-fed area, cropping intensity was low due to unavailability of irrigation facilities and farmers were forced to take crops only in one season. Data revealed that the four farming systems were adopted in this area. In FS-I apart from the crops, onion nursery was also raised. Crops viz. maize, soybean, and paddy were grown in 86.20% of the total cropped area and in the remaining area onion nursery was raised. In FS-II crops grown were maize (0.60 ha), soybean (0.10 ha) and paddy (0.15 ha) while farmers of FS-III cultivated maize in 0.20 ha followed by soybean (0.15 ha) and black gram (0.10 ha). The farmers of FS-IV produces maize (0.24 ha) and paddy (0.20 ha). Thus, it can be concluded that maize is main crop grown in all the farming systems as its by-product is used for dairy animals for diversification followed by paddy, soybean, and black gram.

Table 6 revealed that non-crop enterprises were taken up in three farming systems, i.e., FS-II, FS-III, and FS-IV. In FS-II, farmers reared 3.06 dairy cattle while in FS-III average number of cattle and goats were 3.00 and 17.00, respectively. Farmers who belonged to FS-IV adopted only poultry (55.75 average number of birds). Thus it can be concluded that in rain-fed farming systems non-crop enterprises were dairy, goat, and poultry. Similar finding were reported by Baishya et al. (2007), Sharma (2007).

Irrigated Condition

Table 7 shows farming four systems adopted in irrigated condition of Banswara district. Maximum number of farmers adopted FS-II (33.33%) followed by FS-III and FS-I (26.67% each) and the lowest in FS-IV (13.33%). Cropping intensity in irrigated condition of Banswara district varied from 181.81 to 196.36%. In FS-II maize (0.45 ha), paddy (0.35 ha), wheat (0.51 ha), and rapeseed and mustard (0.50 ha) were the major crops raised while in FS-III maize (0.35 ha), wheat (0.45 ha), and rapeseed and mustard (0.25 ha) were the major crops raised by farmers. FS-IV was the only system in which 50 orchard plants were raised besides growing the major crops like maize (0.20 ha), paddy (0.20 ha), and wheat (0.35 ha). Thus, it can be concluded from the above discussion that maize, paddy, wheat, and rapeseed and mustard were the major crops of this area. Besides crops in one farming system (FS-I) vegetables were also grown and orchard plants and poultry birds were also the part of another farming system (FS-IV). In other two farming systems, i.e., FS-II and FS-III besides crops, dairy and goats were also important components of farming system.

Table 7 Existing farming systems in irrigated condition of Banswara district (per farm)

Farming system	No. of farmers	Gross cropped area (ha)	Cultivated area (ha)	Cropping intensity (%)	Crops		Vegetable		Dairy cattle (No)	Orchard (No)	Goat (No)	Poultry (No)
					Name	Area (ha)	Name	Area (ha)				
FS-I (C+V)	8 (26.67)	1.60 (26.85)	0.86	186.04			Pea	0.06	–	–		–
					Maize	0.30	Okra	0.18				
					Paddy	0.25	Brinjal	005				
					Black gram	0.05	Tomato	0.27				
					Wheat	0.40	Green Coriander	0.04				
FS-II (C+D)	10 (33.33)	2.16 (36.24)	1.10	196.36	Maize	0.45			3.50	–		–
					Soybean	0.25						
					Paddy	0.35						
					Black gram	0.10						
					Wheat	0.51						
					Mustard	0.50						
FS-III (C+D +G)	8 (26.67)	1.20 (20.13)	0.63	190.47	Maize	0.35			2.60	–		11.55
					Black gram	0.15						
					Wheat	0.45						
					Mustard	0.25						
FS-IV (C+G +Po)	4 (13.13)	1.00 (16.78)	0.55	181.81	Maize				–	50.0		52.85
					Black gram	0.20						
					Paddy	0.10						
					Wheat	0.20 0.35						
					Barley	0.05						
					Mustard	0.10						
Total	30 (100)	8.54 (100)	3.14	188.67		5.36		0.60	6.10	50.0	11.55	46.0

C Crop, *D* dairy, *Po* poultry, *G* goat (figures in parentheses are percentage of column total)

Table 7 shows that three farming systems adopted non-crop enterprises. In FS-II the average number of dairy cattle maintained on farms was 3.50 in addition to crops. The FS-III possesses both goat and dairy cattle.

Conclusion

Out of 343 lakh hectares geographical area of Rajasthan state, Dungarpur district contributed 3.85 lakh hectares whereas Banswara district contributed 4.54 lakh hectares. The percentage of forest area to the total geographical area in both the districts is more than double compared to forest area at the state level. It is 16.00% in Dungarpur district and 20.12% in Banswara district compared to 8.02% at the state level. It is evident from the table that total cropped area in the state is 17,914,166 ha of which 55.76% (9,989,675 ha) kept under cereals and small millets, 24.84% (4,449,229 ha) under pulses, 18.82% (3,372,130 ha) under oilseeds crops and 0.58% (103,132 ha) under commercial crops. It is evident from the table that of the total 577 lakh animals in the state, the population of goat is the highest (37.53%) followed by cattle (23.08%), buffalo (22.48%), and sheep (15.73%). In the present study, irrespective of the rain-fed and irrigated condition as well as location of the selected districts, four farming systems were observed.

$$\text{FS-I : Crops} + \text{Vegetable}(C + V)$$
$$\text{FS-II : Crops} + \text{Dairy}(C + D)$$
$$\text{FS-III : Crops} + \text{Dairy} + \text{Goat}(C + D + G)$$
$$\text{FS-IV : Crops} + \text{Goat} + \text{Poultry} + \text{Orchard}(C + G + Po + O)$$

The maximum number of farmers adopted FS-II (C+D), i.e., 46.67% followed by FS-IV (C+D+G+O), i.e., 23.23%, FS-I (C+V), i.e., 20% and the lowest in FS-III (C+D+G), i.e., 10.00%.

References

International Rice Research Institute (Undated) World rice statistics. Department of Agricultural Economics, IRRI, Los Banos, Philippines

Nanda SS, Patro GK, Alim MA, Mahananda N (1999) Efficiency of rice based crop sequences under coastal eco-system. Indian J Agron 44(3):452–455

Poddar RS, Vijay Kumar HS, Shankramurty HG, Venketesh Murthy TN (1995) Impact of credit on cropping pattern and cropping intensity on farms in un-irrigated and irrigated villages in Bijapur district of Karnataka: application of linear programming technique. Agric Econ Res Rev 8(1)

Sen A, Sharma SN (2002) Zero tillage rice-wheat cropping system increases wheat production. Indian Farming, September, 2002, pp 3–5

Singh GB (1998) Natural resource management for sustainable agriculture in 21st century. Indian Farm 48(5):7

Energy Assessment of Rice Under Conventional and Drip Irrigation Systems

K. V. Ramana Rao, Suchi Gangwar, Arpna Bajpai, Lavesh Chourasia and Kumar Soni

Abstract Conventionally rice being irrigated under ponding conditions either by storing rain water or by pumping water or both by utilizing lot of electric/diesel energy especially when pumping is carried out. This paper examines the energy consumption of rice crop with different crop geometry by varying plant-to-plant and row-to-row spacing (25 × 25 cm) and irrigation practices by adopting drip irrigation in changed geometry. The data on energy required in cultivation of rice crop both in conventional practice were collected from primary data (number of laborers required, number of hours of operation of pumping system, number of weeding operations, etc.) and from secondary sources data (such as conversion of input data such as human labor, seed energy, fertilizer energy, etc.). The crop was cultivated during kharif season, by adopting drip irrigation as and when required. The total energy used in various farm operations during rice production was 95,117 MJ/ha in conventional practice, 78,678 MJ/ha in System of Rice Intensification practice (SRI) practices, and 54,877 MJ/ha in System of Rice Intensification practice (SRI) with drip practice. Average annual grain yield of rice in conventional practice were found as 3140, 4050 kg/ha in SRI and 7060 kg/ha SRI with drip. Calculated total energy outputs were 106,898 MJ/ha for conventional, 115,885 MJ/ha for System of Rice Intensification practice (SRI) and 165,155 MJ/ha for System of Rice Intensification practice (SRI) with drip irrigation. The highest average energy productivity was 0.128 kg/MJ for rice under SRI with drip, indicating more production per unit of energy consumed in rice production.

Keywords Energy · Rice · Input · Output · Direct and indirect

K. V. R. Rao (✉) · S. Gangwar · A. Bajpai · L. Chourasia · K. Soni
ICAR-Central Institute of Agricultural Engineering, Bhopal 462038, India
e-mail: kvramanarao1970@gmail.com

V. P. Singh et al. (eds.), *Water Resources Management*, Water Science and Technology Library 78, https://doi.org/10.1007/978-981-10-5711-3_2

Introduction

Energy is one of the most valuable inputs in production agriculture. In crop cultivation practices it is invested in various forms such as mechanical (farm machines, human labor), chemical fertilizer (pesticides, herbicides), electrical, etc. Efficient use of these energies helps to achieve increased production and productivity and contributes to the profitability and competitiveness of agriculture sustainability in rural living (Singh 2002). Energy use in agriculture has been increasing in response to increasing population, limited supply of arable land, and a desire for higher standards of living. Timely availability of the energy and its effective and efficient use are prerequisites for improved agricultural production, as the crop yields and food supplies are directly linked to energy. In the developed countries, increase in the crop yields was mainly due to increase in the commercial energy inputs in addition to improved crop varieties.

The increased use of inputs such as fertilizer, irrigation water, diesel, plant protection chemicals, electricity, etc., demand more energy in the form of human, animal, and machinery. The commercial energy used in agriculture increased nearly sixfold with growth rate of 11.8% between 1980–81 and 2000, but the share of agriculture in total energy consumption in the country increased 2.3–5.2% during the same period. In agriculture sector of India, the energy use pattern for unit production of crops has varied under different agroclimatic zones. The use of energy in crop production depends on availability of energy sources in particular region and also on the capacity of the farmers. There is a need to carry out energy analysis of crop production system (practices) and to establish optimum energy input at different levels of productivity. The appropriate use of energy input to crop production could originate from several types of conservation practices. The reduction, elimination, or combination at machinery operation will reduce energy (fuel) input and also may reduce the uses of labour and time.

Efficient use of input energy resources in agriculture not only saves fossil resources but also provides financial savings (Singh et al. 2004). Therefore, energy saving has been a crucial issue for sustainable development in agricultural systems. Development of energy-efficient agricultural systems with low input energy compared to the output of food can reduce the greenhouse gas emissions from agricultural production systems. Improvements in the efficiency of resource use in agriculture require not only spatial and temporal use of current resources but also the development of tightly defined and broadly applicable indices (Topp et al. 2007). Therefore, this study was taken up to estimate the energy requirement of conventional cultivation of Rice, SRI cultivation, SRI + Drip cultivation of Rice.

Materials and Methods

This work was carried out at ICAR—Central Institute of Agricultural Engineering, Bhopal. Soils of the experimental site are classified as heavy clay soils with clay content varying between 49.7 and 53.7% and with the field capacity ranging from 28.5 to 31%; which was situated at North of Bhopal at 77° 24′ 10″E, 23° 18′ 35″N at an elevation of 495 m above mean sea level. During winter, ambient temperature varies between 10 and 25 °C and in summer between 25 and 44 °C, with annual rainfall in the region is about 1080 mm, 90% of rainfall is received during two months (July–August) of a year. Date of sowing and rainfall input in rice production is totally not considered. Do you mean to say seasonal rainfall has no governance on production of paddy? In nursery raising also you applied drip? How much of total irrigation was given and how much was contributed by rain? Because it is highly variable, its implications should be discussed. Write three practices that were compared to make this article understandable. Give rainfall patterns and the dates of irrigation by flooding or drip irrigation. What was water regime created in different systems of paddy under the experiments? Two major parts of this study, operations and energy sources are examined below.

Operations

Energy consumption in rice production, such as tillage machinery, fertilizer broadcasters, sprayers, irrigation, transportation, and harvesting were determined in systems. The number and duration of operations, seed rate, pesticides, fertilizers and amount of human labor were investigated. From the literature review, equivalent energy inputs were determined for all input and output parameters for rice energy Sources.

Human: Human labor is used for almost every task on the farm, from driving machinery, maintenance, repair, irrigation, spraying, and fertilizer to management. Currently, the energy output for a labor is about 1.96 MJ/h (Khan et al. 2009) and which is the most expensive form of energy in field operations.

Fuel: Diesel fuel is the main source of fuel in agricultural machinery as well as motor pumps and water pumps. Fuel consumption was determined before and after any operation by filling the tractor fuel tank and recording the difference after the operation was completed. The energy output for analysis was determined (Khan et al. 2009) from fuel consumption per operation of one hectare of land times the fuel equivalent energy per liter as shown in equation.

$$\begin{aligned}\text{Energy}(\text{input})/\text{hectare} = {} & \text{Operation fuel consumption}\,(\text{L}/\text{ha}) \\ & \times \text{Fuel energy}\,(\text{MJ}/\text{L})\end{aligned}$$

Table 1 Management practices for rice

Practices/operations	Rice
Name of varieties	Pusa Sugandha-5
Land preparation tractor used	Cultivator, rotavator, land leveler and peddler
Land preparation period	May–June
Planting period	June
Fertilization period	June–September
Average number of fertilization	4
Period of weed control by manually	July–September
Average number of weeding	3
Irrigation period	June–September
Spraying period	July–August
Average number of sprayings	2
Harvesting time	October

Fertilizer: The use of inorganic fertilizers is the fastest growing form of energy consumption in agricultural production. Nitrogen fertilizer is the most important fertilizer in world agriculture, both in the amount of plant nutrient used and in energy requirements. The energy input for one kilogram of nitrogen, phosphate, and potassium are 66.14, 12.44, and 11.15 MJ/kg, respectively (Rafiee et al. 2010).

Agrichemical: Different methods of pest control, i.e., chemical and mechanical are usually applied to control or eliminate insects, weeds, and diseases. Chemical method mostly used for the control insects, diseases, and weeds.

Seed: Clean and proper seeds were provided in packages from seed produce companies and private institutes. Different factors, such as planting system, variety of seed and germination rate influence on amount of seed for rice.

Different agronomical practices for rice carried out in our research field showed in Table 1.

Method of Energy Calculation

Evaluation of Manual Energy Input

Manual energy (Em) expended as an input was determined in MJ by using (Khan et al. 2009) the following formula:

$$\mathrm{Em} = 1.96 \times \mathrm{Nm} \times \mathrm{Tm}$$

where,

Nm Number of labour spent on a farm activity
Tm Useful time spent by a labour on a farm activity, h

Table 2 Energy conversion factors used in agricultural production

S. No.	Power source	Equivalent energy (MJ)	Reference
A. *Inputs*			
1.	Human labor (h)	1.96	(Rafiee et al. 2010)
2.	Machinery (h)	62.7	(Canakci et al. 2005)
3.	Diesel fuel (L)	56.31	(Canakci et al. 2005)
4.	Chemicals (kg)	120	(Erdal et al. 2007)
5.	Fertilizer (kg)		
	(a) Nitrogen	66.14	(Rafiee et al. 2010)
	(b) Phosphate (P_2O_5)	12.44	
	(c) Potassium (K_2O)	11.15	
	(d) Zinc (Zn)	8.40	
	(e) Farmyard manure	0.30	
6.	Water for irrigation (m^3)	1.02	(Rafiee et al. 2010)
7.	Seed (kg)	3.60	(Beheshti Tabar et al. 2010)
8.	Electricity (kWh)	11.93	(Mobtaker et al. 2010)
B. *Output* (kg)			
	(a) Rice grain	15.70	(Ozkan et al. 2004)
	(b) Rice straw	12.50	

Table 2 shows the energy coefficients used in the calculations.

The total manual labour was recorded in each operation with working hours which was converted in man-hour. All other factors affecting manual energy were neglected.

Evaluation of Mechanical Energy Use

Mechanical energy input was evaluated by quantifying the amount of diesel fuel consumed during the tillage, sowing, threshing, winnowing, etc. The total time spent was also recorded. Diesel consumption in pump was also recorded during irrigation. Hence, for every farm operation, the diesel fuel energy input was determined in MJ (Khan et al. 2009) by:

$$E_f = 56.31 \times D,$$

where,

56.31 unit energy value of diesel, MJ L^{-1}
D amount of diesel consumed, L

Data Recorded for Energy Determination

Energy demand in agriculture can be divided into direct and indirect energies (Alam et al. 2005). The energetic efficiency of the agricultural system has been evaluated by the energy ratio between output and input. The inputs are human labor, machinery, diesel oil, fertilizer, pesticides, seed, electricity and irrigation and output yield values of rice crop has been used to estimate the energy ratio. The different field operations performed for completion of each activity in the experiment was measured in terms of time taken for human/machinery, fuel consumption, and expressed as energy input in megajoules (MJ) using corresponding constants (Lal et al. 2003; Binning et al. 1983; Alam 1986) as detailed in Table 2. The human labour energy equivalent was calculated taking seven working hours per day using figure of 1.96 MJ/man-hour for different operations. The sources of mechanical energy used on the selected farms included tractors and diesel oil. The mechanical energy was computed on the basis of fuel consumption (l/ha) in different operations. Therefore, the energy consumed was calculated using conversion factors (1 L diesel = 56.31 MJ) and expressed in MJ/ha. Indirect energy is the energy embodied in seeds, fertilizers, manure, chemicals, machinery while direct energy covers human labor, diesel, electricity, and irrigation.

Energy Indices

Based on the energy equivalents of the inputs and outputs (Table 2), the energy ratio (energy use efficiency), energy productivity and the specific energy were calculated (Demircan et al. 2006).

$$\text{Energy use efficiency} = \frac{\text{Total Energy Output (MJ/ha)}}{\text{Total Energy Input (MJ/ha)}}$$

$$\text{Energy productivity (kg/MJ)} = \frac{\text{Grain Output (kg/ha)}}{\text{Total Energy Input (MJ/ha)}}$$

$$\text{Specific energy (MJ/kg)} = \frac{\text{Total Energy Input (MJ/ha)}}{\text{Grain Output (kg/ha)}}$$

$$\text{Net energy (MJ/ha)} = \text{Energy Output (MJ/ha)} - \text{Energy Input (MJ/ha)s}$$

Results and Discussions

The energy input through human, seed, fertilizer, tractor, etc., were included in the study for analysis. Human labors were involved in land preparation, sowing, weeding, and fertilizer, pesticides application, harvesting, and threshing. The agronomic practices during the cultivation practices of rice along with the periods relevant were shown in Table 1. The operational energy input data of mechanized and traditional cropping of various operations for rice was collected from CIAE, Bhopal (Annual Report 2004–05). Collected data were analyzed for studying different energy use patterns and results were depicted in Table 3. The total input energy used in various farm operations during rice production were 95,117 MJ/ha for conventional practice, 78,678 MJ/ha for System of Rice Intensification practices and 54,877 MJ/ha for System of Rice Intensification with drip practice.

The result revealed that 445.0 h of human power and 40.0 h of machine power were required per hectare for rice production in conventional practice, where as in System of Rice Intensification practices with and without drip irrigation of rice requires slightly higher human power and slightly lower machine power (Table 3). The use of diesel fuel for operating tractors, combine harvester, water pumping systems, and transportation were calculated as (140 L) in conventional practice of rice which was more as compared to SRI and SRI with drip practices. The distribution of total fertilizers energy input was estimated as 73.72% nitrogen, 7.80% phosphorus, 2.75% potassium, 0.06% zinc and farmyard manure was 15.67% for rice. Similar studies have also reported that diesel fuel and fertilizer were the most intensive energy inputs (Erdal et al. 2007; Mobtaker et al. 2010). The total chemical (fungicide, insecticide, and weedicide) energy input was estimated as 372 MJ/ha for rice under all the practices. The total energy input of irrigation, electricity, and seeds were more in conventional practices of rice as compared to other practices (Fig. 1).

Average annual grain yield of rice obtained as 3140 kg/ha in conventional practices, 4050 kg/ha in System of Rice Intensification practice and 7060 kg/ha in System of Rice Intensification practice with drip and total energy output was calculated as 106,898 MJ/ha for conventional, 115,885 MJ/ha for SRI and 165,155 MJ/ha for SRI with drip. The energy input and output, yield, energy use efficiency, specific energy, energy productivity, and net energy of rice production were presented in Table 4. Energy use efficiency (energy ratio) was calculated as 1.12 for conventional practice, 1.47 for SRI and 3.00 for SRI with drip. Energy output to input ratio and specific energy are integrative indices indicating the potential environmental impacts associated with the production of crops (Khan et al. 2009) and these parameters can be used to determine the optimum intensity of land and crop management from an environmental point of view. Energy ratio in some agricultural crop productions were reported as 1.5 for sesame, 2.8 for wheat, 3.8 for maize, 4.8 for cotton in this study. The highest average energy productivity of farms was 0.128 kg/MJ for rice under SRI with drip. That means 0.128 grain

Table 3 Amount of inputs and outputs in rice production

S. No.	Quantity (inputs and outputs)	Conventional practices		SRI		SRI with drip	
		Quantity per unit area (ha)	Total energy equivalent (MJ/ha)	Quantity per unit area (ha)	Total energy equivalent (MJ/ha)	Quantity per unit area (ha)	Total energy equivalent (MJ/ha)
A. *Inputs*							
1	Human labour (h)	445	872.2	565	1,058.4	541	1,060.36
2	Machinery (h) are both not same?	40	2,508.0	32	2,006.4	32	2,006.4
3	Diesel fuel (L)	140	7,883.4	112	6,306.72	112	6,306.72
4	Chemical fertilizers (kg)						
	(a) Nitrogen	325.5	21,528.57	325.5	21,528.57	325.5	21,528.57
	(b) Phosphate	183.2	2279.00	183.2	2279.00	183.2	2279.00
	(c) Potassium	72.0	802.8	72.0	802.8	72.0	802.8
	(d) Zinc	2.30	19.32	2.30	19.32	2.30	19.32
5	Farmyard manure (kg)	15,245.46	4573.6	15245.46	4573.6	15,245.6	4573.6
6	Chemicals (kg)	3.10	372.0	3.10	372.0	3.10	372.0
7	Water for irrigation (m^3)	17,000	17,340	12,500	12,750	5000	5100
8	Electricity (kWh) are not the same?	3060	36,505.8	2250	26,842.5	900	10,737
9	Seeds (kg)	120	432.0	25	90.0	25	90.0
	Total energy input (MJ)		95,116.69		78,678.31		54,875.77
B. *Outputs*							
1	Grain (kg)	3140	49,298.0	4050	63,585.0	7060	110,842.0
2	Straw (kg)	4608	57,600.0	4184	52,300.0	4345	54,312.5
	Total energy output (MJ)		106,898.0		115,885.0		165,154.5

output was obtained per unit energy. The highest specific energy was obtained 30.29 in conventional practice as compared to other practices but the highest net energy was obtained 110,278.73 for rice in SRI with drip practices.

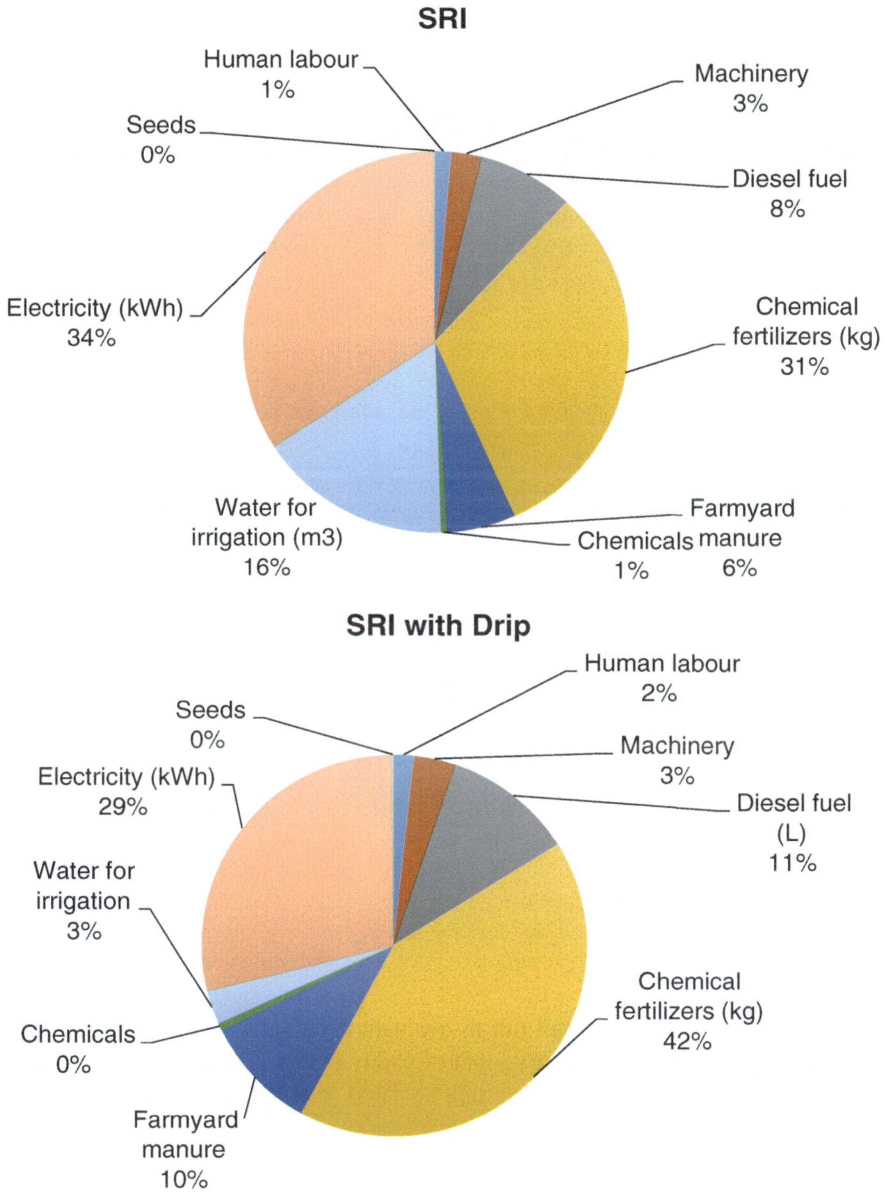

Fig. 1 Total energy equivalent (MJ/ha) of rice

The distribution of total energy input in direct and indirect forms is presented in Table 5. The total energy input could be classified as direct energy (65.81, 59.68, and 42.28%) and indirect energy (34.18, 40.32 and 57.72%) for rice. The result of

Table 4 Energy input–output ratio in rice

Items	Conventional practices	SRI	SRI with drip
Energy input (MJ/ha)	95,116.69	78,678.31	54,875.77
Energy output (MJ/ha)	106,898.0	115,885.0	165,154.5
Grain yield (kg/ha)	3140	4050	7060
Energy use efficiency	1.12	1.47	3.00
Specific energy (MJ/kg)	30.29	19.42	7.77
Energy productivity (kg/MJ)	0.033	0.051	0.128
Net energy (MJ/ha)	11,781.31	37,206.69	110,278.73

Table 5 Total energy input in the form of direct and indirect for rice production (MJ/ha)

Form of energy (MJ/ha)	Conventional practices	%[a]	SRI practices	%[a]	SRI with drip	%[a]
Direct energy[b]	62,601.4	65.81	46,957.62	59.68	23,204.08	42.28
Indirect energy[c]	32,515.29	34.18	31,720.69	40.32	31,671.69	57.72
Total energy input	95,116.69	100	78,678.31	100	54,875.77	100

SRI System of rice intensification (SRI)
[a]Indicates percentage of total energy input
[b]Includes human labor, diesel, electricity, irrigation
[c]Includes seeds, fertilizers, manure, chemicals, machinery

this research clearly showed that the rice s production is mainly depends on electricity, irrigation, and chemical fertilizers.

Conclusions

Energy analysis has been carried out to evaluate the energy requirements in rice. The results revealed that the energy input of paddy production for the conventional practice was the highest. The energy input related to water, electricity, and fertilizers contributed the biggest share of the total energy inputs in production systems. The results indicated that the net energy output for conventional rice production systems was the lowest, indicating inefficient utilization of energy in conventional system of rice cultivation. The study concludes that the SRI with drip irrigation can be adopted for achieving more crop productivity from unit energy (0.128 kg/MJ) incurred on its cultivation.

References

Alam A (1986) Energy requirement of food production and utilization in the rural sector, Satus paper presented at second international symposium on food energy nexus. New Delhi Feb, pp 12–14
Alam MS, Alam MR, Islam KK (2005) Energy flow in agricultural: Bangladesh. Am J Environ Sci 1(3):213–220
Beheshti Tabar I, Keyhani A, Rafiee S (2010) Energy balance in Iran's agronomy. Renew Sustain Energy Rev 14:849–855
Binning AS, Pathak BS, Panesar V (1983) The energy audit of crop production system research report, School of energy studies for agriculture, Panjab Agricultural University, Ludhiana, Panjab (India)
Canakci M, Topakci M, Akinci I, Ozmerzi A (2005) Energy use pattern of some field crops and vegetable production: case study for Antalya region, Turkey. Energy Convers Manag 46:655–666
Demircam V, Ekinci K, Keener HM, Akbolat D, Ekinci C (2006) Energy and economic analysis of sweet cherry production in Turkey: a case study from Isparta province. Energy Convers Manage 47:1761–1769
Erdal G, Esengün K, Erdal H, Gündüz O (2007) Energy use and economical analysis of sugar beet production in Tokat province of Turkey. Energy 32 (1):35–41
Khan S, Khan MA, Hanjra MA, Mu J (2009) Pathways to reduce the environmental footprints of water and energy inputs in food production. Food Policy 34:141–149
Lal B, Rajput DS, Tamhankar MB, Agarwal I, Sharma MS (2003) Energy use and output assessment of food-forage production systems. J Agron Crop Sci 189:57–62
Mobtaker HG, Keyhani A, Mohammadi A, Rafiee S, Akram A (2010) Sensitivity analysis of energy inputs for barley production in Hamedan province of Iran. Agr Ecosyst Environ 137:367–372
Ozkan B, Akcaoz H, Fert C (2004) Energy input-output analysis in Turkish agriculture. Renew Energy 29:39–51
Rafiee S, Mousavi Avval SH, Mohammadi A (2010) Modeling and sensitivity analysis of energy inputs for apple production in Iran. Energy 35:3301–3306
Singh G (2002) Energy conservation through efficient mechanized farming. Agric Eng Today 24(2)
Singh G, Singh S, Singh J (2004) Optimization of energy inputs for wheat crop in Punjab. Energy Convers Manag 45:453–465
Topp CFE, Stockdale EA, Watson CA, Rees RM (2007) Estimating resource use efficiencies in organic agriculture: a review of budgeting approaches used. J Sci Food Agric 87:2782–2790

Response of Okra Under Microsprinkler and Surface Methods of Irrigation

S. P. Gupta

Abstract Investigations were carried out to study the effect of microsprinkler on water economy, growth and yield of okra at H.R.C., Patharchatta, G.B. Pant University of Agriculture and Technology, Pantnagar during 2001–2002. The crop was grown as an intercrop in between the rows of litchi. Irrigation requirement was met based on depletion in available soil moisture and ET of the crop for surface and microsprinkler irrigation, respectively. Results indicated that higher grain yield of okra 109.85 q/ha was recorded for microsprinkler irrigation as against 86.60 q/ha in surface irrigation methods of irrigation. Significant improvements in biometric growth and yield parameters were recorded for microsprinkler irrigation as compared to surface irrigation. Significant saving of water (38%) was achieved for microsprinkler irrigation resulting in higher water use efficiency of 15.60 kg/ha-mm as against 10.53 kg/ha-mm in case of surface irrigation. The gross BC ratio of 1.81, 1.55, and 1.90 were achieved for surface irrigation, microsprinkler irrigation considering with system cost and microsprinkler irrigation considering without system cost, respectively. The net profit achieved per cm application of irrigation water used was highest in case of microsprinkler irrigation considering without system cost, respectively. The net profit achieved per cm application of irrigation water used was highest in case of microsprinkler irrigation considering without system cost (Rs. 1.30) followed by microsprinkler irrigation with system cost (Rs. 1.07) and lowest in case of surface irrigation (Rs. 0.50).

Introduction

Land and water are the two most important resources in India for agricultural development and economic advancement of the country. With the low per capita availability of land and water compared to other countries of the world, there is an

S. P. Gupta (✉)
Department of Irrigation & Drainage Engineering, College of Agricultural Engineering, Dr. Rajendra Prasad Central Agricultural University, Pusa, Samastipur 848125, Bihar, India
e-mail: spg.spgupta@gmail.com

V. P. Singh et al. (eds.), *Water Resources Management*, Water Science and Technology Library 78, https://doi.org/10.1007/978-981-10-5711-3_3

increasing scarcity of these two resources in India. It is well recognized that adequate and timely supply of irrigation water is a crucial factor for the agricultural productivity. It has been estimated that the food requirement was about 220 MT in 2005 AD and it would increase to about 400 MT by the year 2050 AD. The present average production is about 2.2 t/ha under irrigated conditions and about 0.75 t/ha in non-irrigated areas. This can be easily increased to 3.5–4.0 and 1.5–2.0 t/ha for irrigated and non-irrigated lands, respectively (Sivanappan 1998).

India is the second largest producer of vegetables in the world, next to China. Vegetable production has touched new heights in recent past, producing 95 MT. However, as the country's population is increasing at 1.8%, our vegetable requirement up to 2010 will be around 135 MT (The Hindu, Survey of Indian Agriculture 2002).

Materials and Methods

Field experiments were conducted at Horticulture Research Centre, Patharchatta, G. B. Pant University of Agril. & Tech, Pantnagar. The research center falls in the tarai belt adjoining the foothills of Himalayas with an altitude of 243.8 m above mean sea level. The experimental site consists of patharchatta sandy loam soil with an average bulk density, particle density, and field capacity of 1.46, 2.63 g/cm^3, and 18.67% by weight basis, respectively. Studies were conducted in an area of 2640 m^2, which was divided into two parts for the study of microsprinkler and surface irrigation methods. Microsprinklers having nozzle size of 1.5 mm, design discharge of 120 lph and wetted diameter of 10 m at an operating pressure of 2 kg/cm^2 were used for the field investigations. Microsprinklers were placed at a regular grid interval of 4 m × 4 m. Okra (*Abelmoschus esculentus*) cultivar Prabhani Kranti was sown in between litchi rows as an intercrop on February 28, 2001 at a spacing of 45 cm × 20 cm and the final harvesting of matured pods was done on April 28; 2001. The experiment details are summarized in Table 1.

Soil samples were taken before and after each irrigation and its moisture status was monitored using gravimetric method. Moisture content before irrigation was range from 14.12 to 14.89% on weight basis whereas moisture content after irrigation ranges from 22.72 to 23.31%. Various growth parameters such as plant height number of branches and dry matter were recorded periodically at an interval of 15 days starting for 30 days after sowing. The harvesting of tender okra was done seven times at 3 days interval starting for 50 days after sowing.

In surface irrigation, check basins were made and measured quantity of water using Parshall flume was applied for irrigation. Irrigation frequency was based on 50% depletion of total available soil moisture in the root zone. The depth of irrigation water applied ranged from 5.5 to 6.7 cm. Whereas scheduling of

Table 1 Experimental layout for microsprinkler and surface methods of irrigation for okra crop

S. No.	Particulars	Method of irrigation	
		Microsprinkler	Surface
1	Size of plot (m × m)	48 × 24	40 × 24
2	Spacing of plants (m)	0.45 × 0.20	0.45 × 0.20
3	No. of rows in each plot	53	53
4	No. of plants in each row	240	200
5	No. of plants in each plot	12,720	10,600
6	No. of laterals in each plot	7	–
7	Spacing between laterals (m)	4	–
8	Spacing between microsprinkler (m)	4	–
9	No. of microsprinklers in corresponding plot	77	–
10	Basin size (m × m)		4 × 4

Table 2 Effect of microsprinkler and surface method of irrigation on plant height of okra

Crop duration (days)	Mean plant height (cm)		Per cent increase over surface irrigation	Standard error of mean		*t*
	Microsprinkler irrigation	Surface irrigation		Microsprinkler irrigation	Surface irrigation	
45	19.95	18.25	15.65	1.367	2.176	0.64*
60	25.30	22.50	12.44	2.468	2.478	11.22
75	34.50	30.80	12.00	3.038	2.800	10.09
90	62.60	58.80	6.46	4.564	4.459	13.08
105	64.60	61.70	4.70	3.824	3.393	3.93
120	63.10	58.90	7.13	3.900	3.725	11.70

* Non Significant
Significant at 5% level of probability

irrigation for microsprinkler was based on the estimated average ET rate of crop, which is calculated using $ET_t = E_p\ K_p\ K_c$ where ET_t is average evapo-transpiration rate of crop, mm/day, E_p is pan evaporation, mm/day, K_p is pan coefficient, a factor and K_c is crop coefficient, a factor.

Results and Discussions

The various biometric observations, grain yield, growth and yield parameters, water economy, water use efficiency and benefit—cost ratio of okra and related results of the study of microsprinkler and surface irrigation methods were analyzed and summarized in table form (Tables 2, 3, 4, 5, 6, 7, 8, 9, 10 and 11).

Table 3 Response of microsprinkler and surface irrigation methods on number of leaves of okra

Crop duration (days)	Mean no. of leaves/plant		Per cent increase over surface irrigation	Standard error of mean		*t*
	Microsprinkler irrigation	Surface irrigation		Microsprinkler irrigation	Surface irrigation	
45	10.40	8.00	30.00	0.686	0.615	10.85
60	16.90	13.60	24.26	1.792	1.477	7.80
75	20.10	16.90	18.93	3.009	2.483	4.82
90	28.80	23.90	20.50	2.909	2.660	1.08
105	33.20	28.80	15.28	1.873	1.937	10.31
120	29.80	24.90	19.68	1.672	1.595	21.00

Table 4 Response of microsprinkler and surface irrigation methods on stem diameter of okra

Crop duration (days)	Mean stem diameter (mm)		Per cent increase over surface irrigation	Standard error of mean		*t*
	Microsprinkler irrigation	Surface irrigation		Microsprinkler irrigation	Surface irrigation	
45	9.20	7.20	27.78	0.593	0.628	2.30
60	11.30	8.70	29.89	0.746	0.817	9.75
75	14.40	11.40	26.32	0.884	0.657	14.23
90	16.90	13.50	25.19	0.971	0.968	9.16
105	19.30	15.40	25.32	0.857	0.921	12.40
120	16.90	12.40	36.29	0.849	0.872	3.48

Table 5 Response of microsprinkler and surface irrigation methods on pod length of okra

Crop duration (days)	Mean pod length (cm)		Per cent increase over surface irrigation	Standard error of mean		*t*
	Microsprinkler irrigation	Surface irrigation		Microsprinkler irrigation	Surface irrigation	
45	7.75	6.10	27.05	0.876	0.809	9.85
60	11.66	8.40	38.81	0.936	0.733	8.90
75	13.65	10.70	27.57	1.241	0.967	6.15
90	14.57	11.80	23.47	1.210	0.987	9.96
105	16.80	13.40	25.37	1.062	1.024	12.75
120	15.50	12.60	23.02	1.088	0.921	10.47

Table 6 Response of microsprinkler and surface irrigation methods on pod weight per plant of okra

Crop duration (days)	Mean pod/plant (gm)		Per cent increase over surface irrigation	Standard error of mean		*t*
	Microsprinkler irrigation	Surface irrigation		Microsprinkler irrigation	Surface irrigation	
60	40.10	37.40	7.22	0.875	0.872	7.36
70	84.10	78.60	7.00	2.063	1.204	5.37
80	123.80	11.60	10.93	2.342	1.447	4.12
90	230.80	210.70	9.54	4.404	3.581	2.84
100	203.60	183.10	11.20	2.468	2.527	6.54
110	198.20	173.50	14.24	2.520	2.891	9.35

Table 7 Effect of methods of irrigation on grain weight of okra

Irrigation methods	Grain weight/pod of okra (gm)										Standard error of mean	*t*
	Observation plots											
	1	2	3	4	5	6	7	8	9	10		
Microsprinkler	3.5	3.6	3.4	3.3	3.2	3.6	3.2	3.1	3.5	3.7	0.64	8.91
Surface	3.2	3.1	3.0	3.1	3.0	3.2	3.0	2.9	3.1	3.3	0.38	

Table 8 Effect of methods of irrigation on grain yield of okra

Irrigation methods	Grain yield/plant of okra (gm)										Standard error of mean	*t*
	Observation plots											
	1	2	3	4	5	6	7	8	9	10		
Microsprinkler	99	97	108	99	95	97	97	105	106	94	1.54	5.74
Surface	94	93	92	91	90	92	93	95	97	90	0.70	

Table 9 Growth and yield parameters of okra

S. No.	Particulars	Unit	Method of irrigation		Per cent increase over surface irrigation
			Microsprinkler	Surface	
1	Plant height	cm	45.01	41.69	7.96
2	Branches/plant	No.	2.15	1.32	62.88
3	Leaves/plant	No.	23.20	19.35	19.90
4	Stem diameter	mm	14.67	11.43	28.35
5	Internodes/plant	No.	3.17	2.92	8.56
6	Length of internodes	cm	7.38	6.78	8.85
7	Pod length	cm	13.32	10.5	26.86
8	Pod thickness	mm	17.17	14.12	21.60
9	Weight/edible pod	gm	30.10	28.15	6.93
10	Matured pod/plant	No.	4.58	3.30	38.89
11	Pod weight/plant	gm	146.77	132.48	10.78

Table 10 Water applied, water saved, and water use efficiency of okra

S. No.	Particulars	Methods of microsprinkler	Irrigation surface	Present saving or increase over SIM
1	Water applied through irrigation (mm)	191.4	310.0	38.3
2	Effective rainfall (mm)	512.54	512.54	–
3	Total water applied (mm)	703.94	822.5	14.42
4	Yield of okra (q/ha)	109.85	86.6	26.85
5	Water use efficiency (kg/ha mm)	15.6	10.5	48.57

SIM Surface irrigation method

Table 11 Benefit–cost ratio of okra

S. No.	Crop: okra (February–July 2001)			
			Irrigation system	
			Microsprinkler	
	Cost economics (Rs. ,000/ha)	Surface	Without system cost	With system cost
1	Fixed cost (Rs.)	–	–	37.90
2	Seasonal Fixed cost (Rs.)	–	–	3.927
3	Cost of cultivation (Rs.)	19.100	19.100	19.100
4	Seasonal total cost (Rs.)	19.100	19.100	23.416
5	Irrigation water applied (cm)	31.0	19.14	19.14
6	Yield of produce (q/ha)	86.6	109.9	109.9
7	Selling price (Rs./q)	0.40	0.40	0.40
8	Gross cost of production (Rs.)	19.100	23.200	28.416
9	Gross income (Rs.)	36.64	43.96	43.96
10	Gross benefit–cost ratio	1.81	1.90	1.55
11	Net profit/cm of water applied (Rs.)	0.50	1.30	1.07

Summary and Conclusions

From the above results the following conclusions were drawn:

1. The average grain yield of okra was found 109.85 and 86.60 q/ha in microsprinkler and surface methods of irrigation, respectively thus showing a net increase of 26.85% in case of microsprinkler irrigation in comparison to surface irrigation method.
2. Significant improvements in biometric growth and yield parameters were recorded for microsprinkler irrigation as compared to surface irrigation. The higher test weight and harvest indices were also recorded for microsprinkler irrigation as compared to surface irrigation.

3. Significant saving of water (38.3%) was achieved for microsprinkler irrigation resulting in higher water use efficiency of 15.60 kg/ha-mm as against 10.50 kg/ha-mm in case of surface irrigation.
4. The net profit achieved per cm application of irrigation water used was highest in case of microsprinkler irrigation without considering system cost (Rs. 1.30) followed by microsprinkler irrigation with considering system cost (Rs. 1.07) and lowest in case of surface irrigation (Rs. 0.50).

This study was undertaken to assess the comparative performance of microsprinkler and surface irrigation methods on okra. The best performance of the system was obtained under microsprinkler irrigation method.

Acknowledgements The work was carried out with the financial assistance provided by National Committee on the use of Plastics in Agriculture. Govt. of India, New Delhi and other facilities extended by the G.B. Pant University of Agriculture and Technology. Pantnagar, India is duly acknowledged.

References

Sivanappan RK (1998) Key note address. In: Proceedings of workshop on micro irrigation and sprinkler irrigation systems, New Delhi, pp ix–xiv

The Hindu (2002) Survey of Indian agriculture. Published by National Press, Kasturi Buildings, Chennai, 229 p

Development of Universal Ultimate Total Green Chemistry and Eco-Agriculture for Sustainable Productivity

R. C. Yadav

Abstract In agriculture, most of the chemical reactions emit several green house gases (GHGs), viz., carbon dioxide (CO_2), methane (CH_4), Nitrous oxide (N_2O), and Chloroflouro carbons (CFCs). The detritus consumers cause environmental nuisances of bad odors, bitter stings, bites, itching, fever, dysentery, malaria, dengue, etc. In this study, new eradicating routes were carved to reduce the emission of GHGs. Innovative application of scientific facts of nitrogen cycle, carbon cycle, sulfur cycle, P, K, etc., hydrology, chemistry, biology, environmental sciences and environmental engineering were embedded for creating bed configuration of raised beds and furrows that conduct all essential functions to produce primary and secondary natural resources to enhance productivity. It is named as smart, alive and enthusiastic nature agriculture that fosters total ultimate green chemistry. Subsequently, the components of the smart nature agriculture have been innovatively devised to produce scientific basis of technology of crop and cropping patterns, manures and fertilizers, weeds and crop residues management, bed configuration, precision planting, ultimate green irrigation, zero weeding, and toil free eco-agriculture, harvesting, postharvesting storage of crops and cultivation of land for nitrogen management. The enhancement in productive capacity (REY 572 q/ha unimaginable yield level) and reductions of GHGs were sufficiently substantiated and several practices reformed based on the scientific bases. This smart nature agriculture practice proves to be a Sun technology, which is universally applicable for all soil, climate, ecosystems, crops, cropping practices for irrigated and rainfed agriculture, in contrast with so far acquired bright spots. Thus, this study gave total solution for enhancing productivity in the coming global warming scenario and climate change and showed new way of getting free from the worry of grim global food situation in present and posterity. It warrants development of seeding/planting combine, and equally important industrial support, as has been done for the development of combine harvester.

R. C. Yadav (✉)
CSWCRTI, Research Centre, Agra 282006, Uttar Pradesh, India
e-mail: ramcyadav@rediffmail.com

V. P. Singh et al. (eds.), *Water Resources Management*, Water Science and Technology Library 78, https://doi.org/10.1007/978-981-10-5711-3_4

Keywords Crops and cropping pattern · GHGs · Ecosystem · Eco-agriculture Environment · Input managements · Mechanization · Raised beds and furrows

Introduction

Food is the basic physiological necessity of all human. While developing countries are striving hard to produce food to feed their nations, the developed countries are attempting to accomplish physical quality of living (PQLI) (Thomas and Evans 2011). The large global populations in the countries will increase to a new large number by the year 2050. The increase will change the sequential orders of thickly populated countries in year 2003 to a different order of population of sequences. India will be the first and Ethiopia will be the tenth most thickly populated country of the world by 2050 (Thomas and Evans 2011).

Plants require 14 essential elements (Guftafson 1939/2010). The process of uptake is universal. World agriculture has been suffering from misdeeds of realization of faults in agricultural technologies in vogue. Water and environment interactions have created emission of greenhouse gases (GHGs), which have been now accepted as factors causing global warming and climate changes. So far how anthropogenic sources have been emitting the GHGS, have received scientific attention. However, how the water and environment interactions have bearing on the emission of the GHGs need to be understood. The chemical reactions which do not emit any of the aforesaid GHGs are called the green chemistry. Some innovative developments are developed for eradication excessive concentration of CO_2, (Yadav 2010, 2012a, b, c) methane (CH_4) (Yadav 2013a) In the recent years, dominance of the GHG nitrous oxide (N_2O) has received formidable challenge warranted in GHG, viz., nitrous oxide (N_2O), which emanates from agriculture and causes formation of acid rains, viz., nitric oxide and depletion of ozone layers (Wuebble 2009; Yadav 2014b). In the interactions, many elements, viz., N, P, K, O, C, and water are governed by scientific facts of environmental engineering and environmental sciences, hydrology, and plant sciences. The chemistry involves water and environment interactions. Suitable paths need to be carved by creating conditions for development of green chemistry in agriculture. The objective of the present study was to take stock of situation and create development of total green chemistry. The study will enable scientific basis for innovative technology for enhancing sustainable productivity, environment conservation, reduction of energy use in agriculture and creating societal benefits helpful in carrying out public governance.

Source of nutrient, dose of application, temporal and spatial placement, which bring highest productivity are the nutrient management. Element in Table 1 get involved in different chemical reactions under water and environment interaction, which need to be essentially of green chemistry. In doing so, the importance of the elements has changed. For example, sulfur and zinc are applied but without consideration of the chemical reactions, it brings deleterious effect on productivity of crop (Yadav and Ambekar 2011).

Table 1 Elements needed for plant growth

S. No.	Element	Symbol	Atomic weight	Common valance	Equivalent weight
1	Nitrogen	N	14	3^-	–
2	Phosphorus	P	31	5^+	6.0
3	Potassium	K	39.1	1^+	39.1
4	Calcium	Ca	40.1	2^+	20.0
5	Magnesium	Mg	24.3	2^+	12.2
6	Iron	Fe	55.8	2^+	27.9
7	Manganese	Mn	54.9	2^+	27.5
8	Boron	B	10.8	3^+	3.6
9	Sulfur	S	32.1	2^-	16.0
10	Zinc	Zn	65.4	2^+	32.7
11	Copper	Cu	63.5	2^+	31.8
12	Hydrogen	H	1.0	1^+	1.0
13	Oxygen	O	16.0	2^-	8.0
14	Carbon	C	12.0	4^-	–

Equivalent weight (combining weight) is equal to atomic weight divided by valance (Hammer and Hammer 2005)

Materials and Method

In order to deal with green chemistry, it is imperative to review the essential elements for plant growth (Table 1), (Gustafson 1939/2010). The plants absorb these 14 essential elements through the root hairs in a specific ionized form (Ehlers and Goss 2003). Detailed enumerations are made for green chemistry, universal fact of plant nutrition, removal/elimination of inimical products and development of gray or black chemistry. The irrotational mechanics, design specifications, and the technology module have been sufficiently dealt with by this author's earlier study (Yadav 2016). Detailed account of concept involved in the green chemistry universality of plant science, physics, chemistry, and knowledge of hydrology, soil science, manures and fertilizer, agricultural engineering, viz., water management and drainage engineering, selection of crops and cropping practices, bed configuration, precision seeding/planting, intercultural, weeding, ultimate irrigation, harvesting, and postharvesting operations (Yadav 2013a, b; Yadav and Chaudhary 2014). The present study is devoted to make further improvements in the third generation (3G) practices to make it a fourth generation (4G) technologies of innovations in agriculture. These process-based studies enable to devise new rationale for eco-agriculture, reduction in tillage, and required input in agriculture, weeding, and maintaining condition for nutrient buildup and reduction of the nitrous oxides. These practices have direct bearing on environment, nutrient buildup in the soil, eliminating nutrient degradation due to deficiency of plant nutrients. This development will be the enforcement of quantum mechanics in agriculture, i.e., creating a fixed path without any wilderness in agriculture (Yadav 2016). Thus, the study will provide basis of developments in agriculture, practices, benefits enhancement in productivity, quality, reduction in inputs, and reduction of wastages that undergo decomposition and cause the emission of GHGs. The development of eco-agriculture was made by seeing experimental results of other researches that led to perceiving line of new innovation. This aspect has been capitalized and various aspects were created by management of nitrogen cycle. The presentations in the result part will be self clarifying.

Results

The emission of CO_2, CH_4, N_2O, CFC, PFC, and SF_6, which are trace gases and cause green house effect and induce climate change. The chemical reactions are involved in the development of either gray or black chemistry. In the present research, all efforts were concentrated to eliminate condition that produces gray or black chemistry and accomplishes challenges (Wuebbles 2009) of green chemistry necessary for bringing sustainable production and environment protection.

The Green Chemistry

As indicated earlier, reference is to be made that those chemical reactions which do not emit any of the green house gas to the atmosphere, a necessary condition that does not promote global warming and climate changes, is classified as green chemistry. The green chemistry involved in all the 14 elements covering wide spectrum is described as follows. Earlier study of author (Yadav 2016) presented detailed description of cycles of N, P, K, S, C, O, and moisture. The basic scientific facts involved in development of the green chemistry are included in the presentation for the benefits of comprehension of the subject. The basic concept and further advancement over and above the previous study are presented here.

Table 1 presents the priority of essential elements and their facilitative elements in the order of the priority. Among them nitrogen N is of prime importance, which affects productivity, sustainability and nutrient buildup in the soil to eliminate land degradation due to lack of plant nutrients. The nitrogen cycle has to revolve in the production system to supplement RNh, ammonium, and the nitrate, which get absorbed by the plants. Figure 1 presents the nitrogen cycle, which depicts good and bad paths that create green and gray chemistry. There exist only two ways of getting the high concentration of nitrogen gas in atmosphere by thunder and rains and by microbial actions of nitrogen fixation. The out flow (loss) of the nitrogen from any site is by three ways, viz., nitrate leaching, surface runoff, and reverting back to atmosphere. The nitrate leaching pollutes ground water and runoff in the water bodies, which cause eutrophication and spoil quality of water in the water

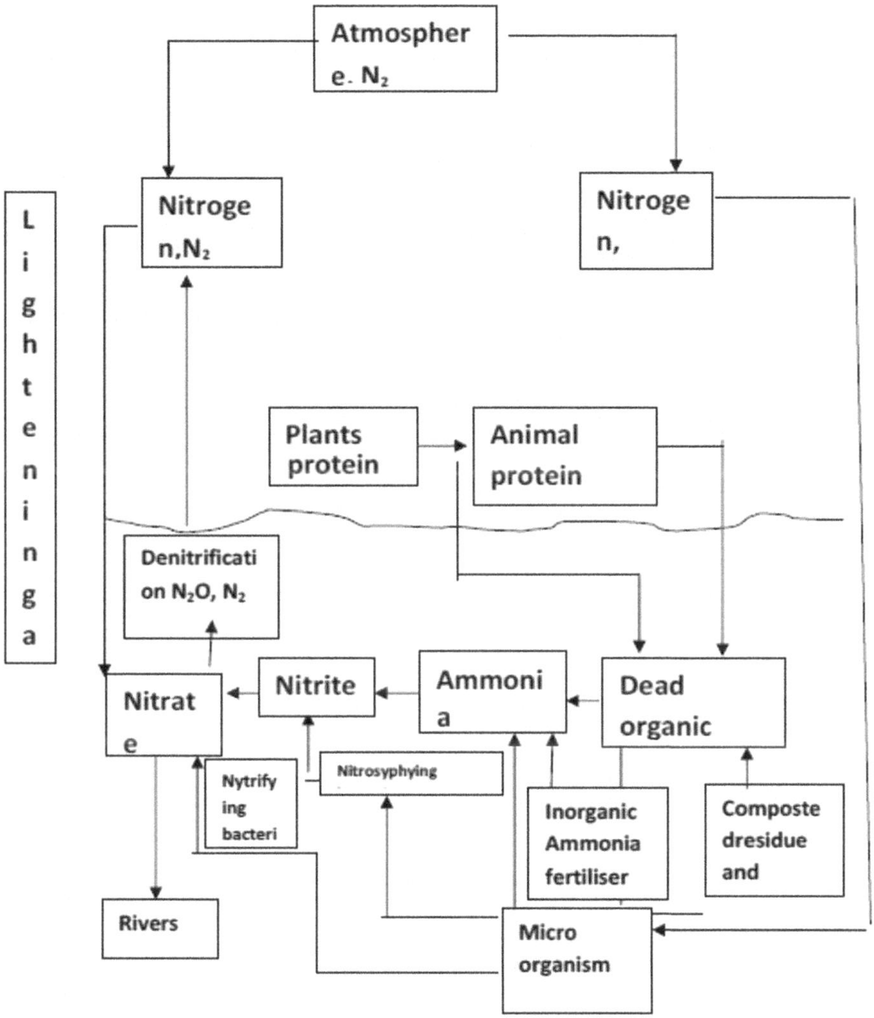

Fig. 1 Nitrogen cycle after De (2010) and updating

bodies. The gaseous loss of nitrogen causes environmental problems of acid rains and depletion of oxygen, a severe condition in many North European Countries. Deficiency of the N in the soil causes decrease in the plant productivity, for example, in the Gulf countries. So severe loss of N from soil creates misbalance situation of surplus and deficit in the nitrogen cycle under different set of conditions. That means, it requires management of nitrogen cycle to create productivity and protection of environment (Yadav 2014b).

Chemistry in Production System

Photosynthesis—a prominent green chemistry

$$CO_2 + H_2O^{hv} \rightarrow (CH_2O) + O_2 + \text{Energy} \tag{1}$$

Gray chemistry

$$CH_2O + H_2O \rightarrow CO_2 + 4H + \text{Energy} \tag{2}$$

Black chemistry

$$CO_2 + 8H\ + 8e \rightarrow CH_4 + 2H_2O \tag{3}$$

Nitrogen fixation

$$3(CH_2O) + 2N_2 + 3H_2O + 4H \rightarrow 3CO_2 + 4NH_4. \tag{4}$$

Salt Formation

Acid reacts with base

$$NaOH + HCL \rightarrow NaCL + H_2O \tag{5}$$

Base reacts with strong acid

$$Na_2CO_3 + 2HCL \rightarrow 2NaCL + H_2O + CO_2 \tag{6}$$

When reactive metal dissolves in acid

$$Zn + H_2SO_4 \rightarrow ZnSO_4 + H_2 \tag{7}$$

Salts of weak oxides

$$CH_3COO^- \ + \ H_2O \rightarrow CH_3COOH + OH^- \tag{8}$$

$$NH_4 + H_2O \rightarrow NH_3 + H_2O + H \tag{9}$$

Ithionic acid and Ammonides have no effect on pH.

Hydroxide ions regulate alkalinity and sodic acid regulate pH on pH meter.

Studies have shown that the presence of sulfur cycle enhances N usage efficiency. Hence, depiction is made (Fig. 2) for the aerobic (good) and anaerobic (bad) paths of sulfur cycle. The simple way of creating the functioning of the sculpture cycle (Yadav 2012a, b, c) to add aerobically decomposed compost made from the residues etc.

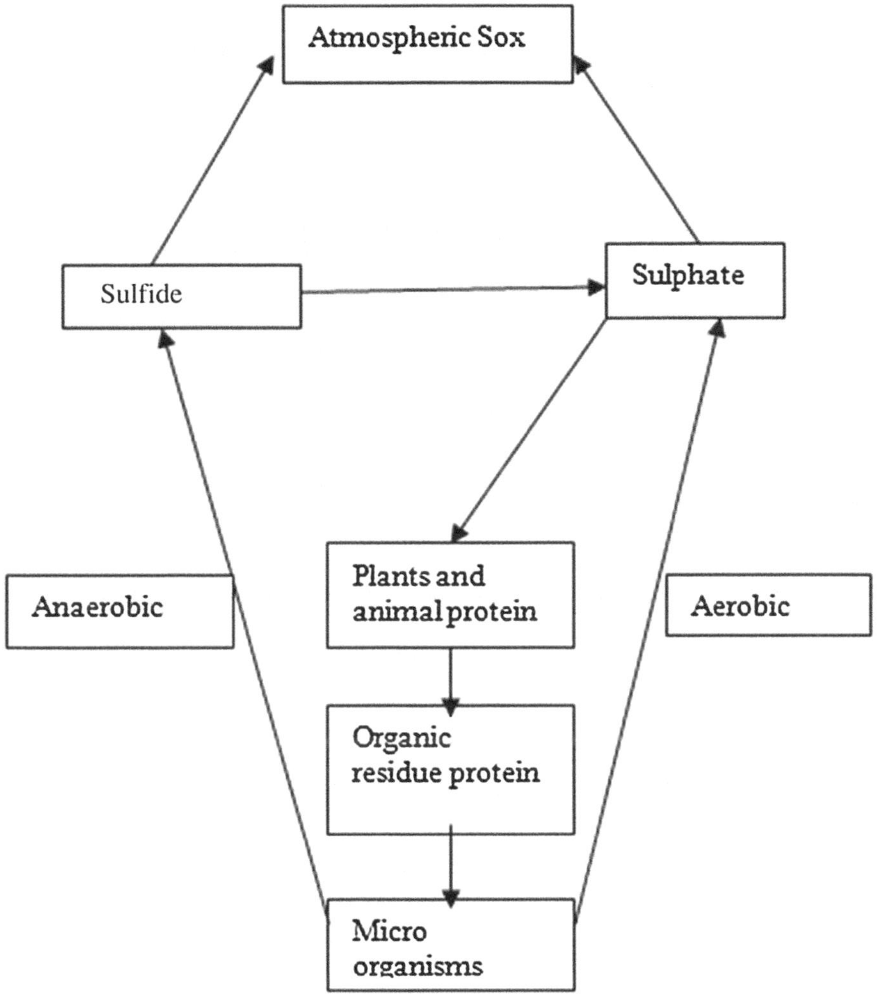

Fig. 2 Sulfur cycle after De (2010), Liu and Li (2008)

The new scientific innovative technology produces both primary and secondary natural resources. Table 2 presents the crisp of the development and benefits of the green chemistry.

It is evident that 12 beneficial aspects develop due to creation of the green chemistry, which needs research and developments. Further, the quantum mechanics results in setting of the line of actions and removes wilderness in agriculture (Yadav 2013a, b, 2016; Yadav and Chaudhary 2014). Hence, the research needs have been reduced to optimization of the variables.

Table 2 New technology induced primary and secondary natural resources, product, knowledge status, and engineering application

S. No.	Scientific fact	Useful	Harmful	Status of knowledge	Engineering technology
1	Nitrogen cycle	Nitrate	Nitrite	Known	Mixed cropping, inter cropping crop rotation, green manuring
2	Phosphorus cycle	Phosphate	Phosphide	Known	Band placement
3	Potash cycle	Potash	–	–	–
4	Sulfur cycle	Sulfate	Sulfide	Not well known	Conduct aerobic decomposition of residue
5	Absorption	Removal of inimical substances	Pollution, degradation	Not well perceived and applied	Not existing in agriculture, but it can be innovated
6	Adsorption	Removal of inimical substances	Pollution, degradation	Not well perceived and applied	Not existing in agriculture
7	Pollution control	The biomass is converted in aerobically decomposed compost	Biomass burning produces CO and SO_2 and NO	Unable to follow the right path	Setup for Aerobic decomposition and right type of mechanization
8	Plant nutrient conservation	Nitrogen content mineralization	Toxic salts and gases, nutrient deficiency	Not so precisely and of enmass awareness	Apply eco-agriculture
9	Energy use reduction	Save energy input for tillage, weeding and moisture conservation	Wastage of energy, economic loss, N loss, N_2O pollution	The system does not exist in agriculture practices	Develop machineries for the eco-seeding
10	Joint strength	Enhance and produce uniform germination	Crust strength restricts germination	Known, but way to exercise these facts not widely applied	Emergence gets enhanced by collective strength
11	Crop diversification	Crop diversification will enable availability of commodities	Discrepancy in food nutrition	Known, but there exists some short of helplessness	Food processing to prepare complete food
12	Fixed resources use, viz., land	Intensification and better use, enable N, P, K, S, C and O	Release of GHGs	Poor scenes, GHG emission and bad odors	Provide good drainage

Measures to Overcome the Toxic Heavy Metal and Toxic Gases Generated in the Production Process

It needs to be clarified that because agriculture domain is open ended, it will involve some part of gray or black chemistry to function. The examples have been dealt with in detail elsewhere by different researches (Dhillon and Dhillon 2010; Guilette et al. 1998; Jacks et al. 2005; Herbal et al. 2002; Heera et al. 2004; Jaglan and Qureshi 1996; Krishnamoorty 1976). Different methods of decomposition by composting produce the lowest content of heavy metal of Fe and Mn, but enhance in copper Cu and Zn. It means the organic farming utilizing organic manures cannot be 100% free from heavy toxic metals and toxic gases (Table 3). The organic manure primarily farm yard manure contains highest contents of Fe, Mn, Zn, and Cu. The aerobic composting (NADEP) enhances N, reduces Fe and Mn. But content of zinc and copper becomes more than that of vermin compost. In the vermin, compost N content gets enhanced, but there is contrast reduction of volume of cow dung compared to compost. Further, the vermin compost is totally dependent on cow dung as base material which is getting scarce due to mechanization of agriculture. Preparation of vermin compost requires intensive care against the least care for compost and some care for NADEP. It may be cautioned that craze for organic farming, which largely is based on compost, will emit the substantial volume of methane having 21 time warming against that merely 1 from carbon dioxide. Fascination for organic food will adversely affect environment. The yield is also going to get adversely affected that will add worry for creating food provision for remaining part of public after the organic food fascination driven people. The toxic metals, the toxic gases, and chlorinated toxic organics can be reduced by inactivation through application of activated charcoals as depicted by data in Table 4. The bio-inactivation/ remediation is workable for all cases of poor quality of soil, water, and environment. The racy nature technology has the provision of

Table 3 Gray and black chemistry involved in organic manure; elemental composition of organic manures, average values

Organic manures	GHG emission chemistry	Macronutrients			Some selected heavy metals			
		N	P	K	Fe	Mn	Cu	Zn
		% wet weight basis			mg/kg dry weight basis			
Farm yard manure	CO_2 and CH_4, gray and black	0.54	0.31	0.51	440	155	10	78
Aerobically decomposed	CO_2 gray	0.93	0.52	1.15	215	96	25	56
Vermin compost	CO_2 gray	1.36	0.48	0.65	619	245	16	45

The chemical analysis data based on reference (Biswas et al. 2012)

Note There was no visualization of building of sulfur in the organic manures, supporting lack of visualization of working of sulfur cycle in present day agriculture

Table 4 Removal of some toxic chlorinated organics by activated charcoal treatment (De 2010)

Compound	Concentration, μg/100 g		
	Initial	After	Removal efficiency, %
Aldrin	48	<1.0	99^+
Dieldrin	19	0.05	99^+
Endrin	62	0.05	99^+
DDT	41	0.1	99^+
Arochlor 1942 (PCB)	45	<0.5	99^+

addition of activated bio-char ranging in 5–10% of organic manure application. Thus, challenge of creating new method for controlling the toxic heavy metals and toxic gases can be fulfilled by simple remedial measures of soil management amendment, viz., the activated charcoal. The utility of organic farming has already described in detail in another study (Yadav 2013b). Further this aspect will be taken up in the forth coming section of this study.

The ideal requirement is the best known and developed situation which conduct green chemistry should always remain in function in situation under all changing scenario of climate, both exogenic and endogenic, biological alteration or supplementation, etc. These aspects will be presented in detail in the following section of results and discussion.

The Racy Nature Agriculture Technology

The Technology

Considering application of the prescribed specifications an innovative technology named as racy (smart, alive and enthusiastic) nature agriculture was developed, which fulfills all aspirations of the green chemistry (De 2010; Yadav et al. 2008, Yadav 2014a, b, c, 2016) challenge for sustainable productivity. The basic physical appearance of the technology is displayed in Fig. 3a, b. It comprises raised bed and furrow which changes in the hydrologic response to provide well drained moist condition and sufficient aeration to keep the green chemistry (Yadav 2014c, 2015d, 2016) functioning all the time of the cropping season. In addition to earlier known practice of raised bed and furrows (Lauren et al. 2006; Kenneth 2006; Yadav 2016), the green chemistry (Yadav 2010, 2012a, b, 2016), not visualized before, is now endowed by the technology (Yadav 2010, 2012a, b, 2013a, b, 2014a, b, c, d, 2015a, b, c, d). The cropping patterns (Acharya et al. 2008; Yadav 2015e), precision planting (Yadav 2015f), intercultural and ecosystem (Yadav 2015c) for zero weeding (Yadav 2015f), harvesting, and postharvest practices (Yadav 2016) are

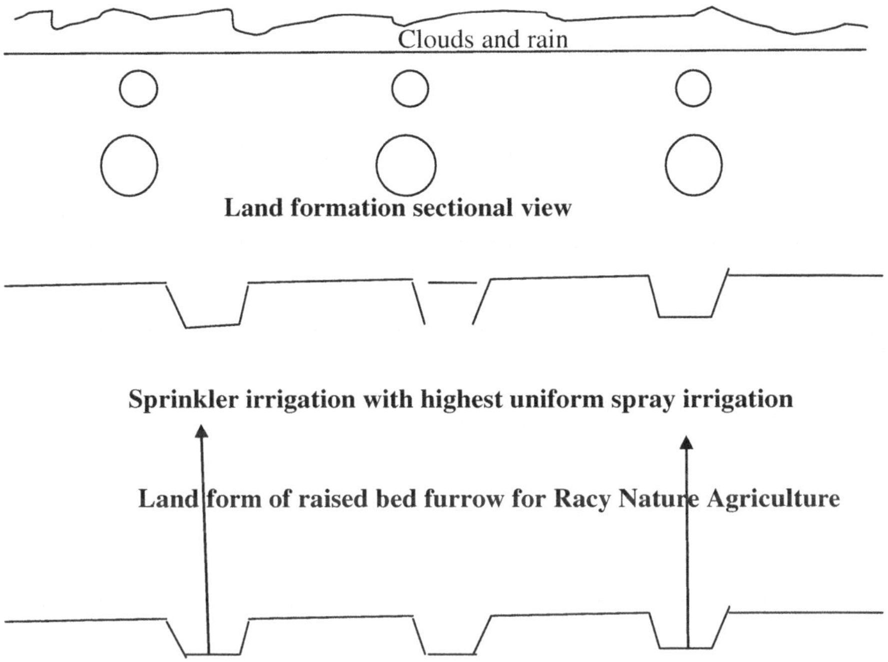

Fig. 3 **a** Land formation of raised bed and furrow for racy nature agriculture under rainfed situation. **b** Land form of raised bed and furrow and sprinkler irrigation for racy nature agriculture

innovatively devised. The technology is suitable for all crops, cropping patterns, both irrigated and rainfed situations. It is equally suitable for poly houses and controlled environmental condition.

Technology Module

The raised bed furrow land form supplements adequate oxygen diffusion in the root zone, increased moisture and nutrient reserve for plants under water logged as well as dry condition. Its local customization is to be researched upon.

The sprinkler spray application of irrigation water will increase oxygen content; it will supplement the raised bed enhanced storage of nutrients and moisture and sufficiently aerated, occasionally saturated and drain off the excess water to keep always convene aerobic decomposition of organic and cellulose. This will supplement plant nutrient by way of enabling sulfur cycle to function. This situation brings good water and air interaction.

The enforcement of green chemistry and contribution of different practices constituting the technology as well as the composite effects are summarized in Table 5. The individual component practices have capacity to enhance yield in range of 5–15%. The composite effects comes out to be almost bringing the double the present harvested yields.

Table 5 Enforcement of green chemistry, suitably designed practices and conservative assessment of yield enhancement by racy nature agriculture practices components and their justification

S. No.	Green chemistry	Racy nature agriculture practice component	Possible increase in yield, %	Basis and justification
1	Nitrogen cycle, sulfur cycle	Aerobically decomposed manure, application	15	Aerobically decomposed cellulose by the sulfur bacteria produce sulfate, directly taken by plants for building body tissues and promoting growth
2	Green Mechanization	Plowing	5	Plowing creates aeration that convenes aerobic decomposition
3	Supplementation of oxygen	Formation of raised bed furrow system	10	RBC increase soil depth to larger volume of moisture and air in the root zone. The additional moisture and aeration make plant growth under both the condition of water logging and drought. This situation permits crop diversification in the low lands where only paddy cultivation is possible
4	Adsorption and absorption	Application of activated charcoal for bio remediation	Improved quality	Hard metals such as iron, zinc, manganese, and copper etc. get reduced by the NADEP compost application. Further, these toxic metals, toxic gases, and chlorinated organics get absorbed by charcoal application in practice
5	N cycle management	Precision sowing	5	The precision sowing enable harness yields from entire space of the field under crop. Both raised bed and furrow can be sown/planted to extract utility high and low moisture and the oxygen supply
6	Plant science, nutrient building and nutrient use	Maintenance of optimum plant density	10	Plant density that will be produced by the crop variety and crop should be optimized by generation II (2G) research

(continued)

Table 5 (continued)

S. No.	Green chemistry	Racy nature agriculture practice component	Possible increase in yield, %	Basis and justification
7	Eco-balance and reduction n cost of cultivation	Establishing zero weeding pulse based ecology	15	Zero weeding will reduce the cost of cultivation, eliminate land degradation by plant nutrient deficiency, and sustainable agriculture
8	Irrigation involving green chemistry	Ultimate irrigation	10	Sprinkler irrigation freshens irrigation water by eliminating hydrogen sulfide and methane like harmful gases in water, thus create adequate oxygen supply and save irrigation water
9	Reduction of GHG emission by weeds	Weeding	–	No weeding is required. Effect of pulse based ecology is indicated under S. No. 7
10	Supplementation of oxygen	Inter-culture	10	The inter-culture again enhances aeration during the crop growing in the field. Raised bed and furrow creates sufficient supply of oxygen
11	Conservation and buildup of N in the soil profile	Subsequent cropping system	5	Subsequent cropping 11 utilizes land which emits GHGs contained in soil (at least 10% of total yearly GHGs emission). The subsequent cropping enables function of nutrient cycle under aerobic condition and supplement the soil to be harnessed by the paddy crop
12	Joint strength	Uniform stand establishment	–	Ensure optimum desired stand establishment under varying condition, a condition necessary for getting yields from all part of field under crop
	Composite enhancement additive index		85	Combined additive effect of all factors enumerated above
	Multiplicative index		2.247	Combined multiplicative effect of above all factors
	Average[a]		2.0485	(1.85 + 2.247)/2 = 2.0485

[a]The composite factor needs research study for exact value. As an acceptable middle path mean value is adopted in this study to deal with in the further developments

Validation of the Technology

Crops and Cropping Patterns

In the green chemistry based yearly cropping patterns, some nutrient mineral cycle do function. Thus, the cropping pattern should be designed in such a way that during crop growth period, concerned nutrient cycle should function. This fact is substantiated by presenting data in Table 6. Crop rotation, nitrogen cycle, nitrogen and sulfur cycle, and sulfur cycle followed by nitrogen cycle could be different options. In the treatment C4 cropping pattern rice-onion-cowpea, the sulfur cycle worked on abundant rice residues and produced sulfate, which was efficiently utilized by onion crop (Acharya et al. 2008). Following this, cowpea was sown that fixed nitrogen to be utilized by following crop of rice. In this treatment sale of cowpea pods also produced high economic return that enabled high rice equivalent yield. This fact is further substantiated by treatment C3-rice-cabbage-green gram, where green gram further added nitrogen that was utilized by the rice crop. Thus, it is evident that creation of functioning of nitrogen and sulfur cycle appears to be the most effective in maintaining crop yield during the crop year.

The example crops referred to in Table 6 were grown on the flat bed configurations and not on raised bed and furrow land formations. The raised bed and furrow land formations provide border effect, enhanced soil depth, more moisture storage, and functioning of green chemistry (Yadav 2013b, 2016), for longer duration than those with flat bed. Thus, multiple benefits will culminate in enhancing harvest index of crops. This fact is substantiated by values ascertained in last column of Table 6. Although there is enhancement in rice equivalent yield, the maximum yield comes for the treatment where nitrogen cycle functioned during the cropping season. The racy nature technology will result in better utilization of resources in agriculture, where resources constraints are emerging and environmental degradations occurring.

Yearly Yield Enhancement and Water Conservation

The concept of promoting functioning of nutrient element cycle in any cropping system of region can be ascertained by yield responses of crops in different sequences for the entire year. This fact is substantiated by taking crop yields of established cropping pattern and some alternative cropping pattern in resources constraint, viz., water in the state of Punjab. Due to rice-wheat cropping pattern, the ground water depth had been decreasing at an alarming rate that caused political concern for agriculture.

For the purpose of assessment yield advantage by the new technology under the present study, existing present harvested yield (1G) and the level of yield acquired in racy nature agriculture (2G) are presented in Table 7. The crop yields are likely

Table 6 Rice based cropping system attributes and productivity that supports functioning of different nutrient elements' cycle

Treatments	Involvement of nutrient element cycles	Yield of rice, Tonnes/ha	Rice equivalent yield, Tonnes/ha		System productivity, Tonnes/ha	Prod efficiency, kg REY/ha/d	Likely enhancement in yield due to eco-agriculture by factor 2.0485
Cropping sequence		Rainy season	Winter season	Summer season			
C1 (rice-potato-sesame)	Crop rotation	4.2	21.5	2.7	28.4	97.5	58.17
C2 (rice-rapeseed-groundnut)	Nitrogen cycle	4.3	4.9	7.6	16.8	52.8	24.4
C3 (rice-cabbage-green gram)	Crop rotation and nitrogen cycle	4.4	22.8	3.2	30.4	105.6	62.47
C4 (rice-onion-cowpea)	Sulfur cycle and nitrogen cycle	4.4	26.7	9.3	40.4	126.5	82.76
CD ($P = 0.05$)		0.069	4.132	3.342	7.608	20.82	15.58

Yield data extracted from (Acharya et al. 2008)

to increase based on the pre crop land use in the yearly yield assessment. This fact will further induce level of confidence in technology under the present study.

Data presented in Table 7 are the enhancement in crop yield, viz., base yield, increase in yield by racy nature agriculture (2G) and racy eco-green chemistry based agriculture (4G). The missing 3G technology was for Racy-SRI Combo agriculture. Using the yield data, yearly REY for different cropping patterns are drawn in Table 8.

The data presented in Table 8 substantiated that rice-wheat is the most remunerative and stable cropping pattern in Punjab. Rice is cultivated during rainy season when water availability is more; though distribution pattern is unpredictable. The wheat crop is grown during winter when water demand is low due to low evaporative demands. But, even under these situations wide spread withdrawal of ground water had caused alarming concern of fall in ground water table. Different methods of rice cultivations are being resorted to with a view to reducing water demand for rice. At the most alternate wetting and drawing practice of rice irrigation have been practiced. This author has developed an ultimate green irrigation practice (Yadav 2015a, d) which fulfills all necessary requirement of green chemistry and saves water. The new cropping pattern of maize-gram also produced REY equivalent to that of rice-wheat. The sprinkler irrigation is applied for uniform saturation of soil profile and during high water demand at crop stage of tillering, earhead emerging stage by over flooding the furrows. Thus, high yield can be harvested while keeping water consumption low (Yadav 2015a, b, c, d, e, f, g).

The water need for irrigation can be reduced by replacing wheat crop to mustard or gram. By change of crop of rainy season, viz., maize followed by gram will produce REY at par with that of rice-wheat (Table 7). There will be tremendous reduction in water demand for agriculture and enhancement in yearly yield. In the maize-gram cropping patterns, both the cycles of nitrogen and sulfur work

Table 7 Enhancement in crop yield under different generation of racy nature technology

Crops	Base yield, q/ha, (1G)	Accepted yield, q/ha (2G)	Racy eco-agriculture 2.0485 (4G)
Yield increase factors		**1.8985**	**2.0485**
Wheat	40	76	82
Rice	60	114	123
Maize	50	91	102
Pearl millet	22	43	45
Mustard	18	34	37
Soybean	26	50	53
Gram	24	46	49
Pigeon pea	20	38	41
Potato	200	380	410

[a]Table devised as per yields achieved in India

Table 8 Best identified yearly cropping sequences super imposed eco-agriculture and rice equivalent yields (REYs) for Punjab, India

Items	Racy nature agriculture (3G)		Total REY, q/ha	Racy eco-agriculture 2.045/1.8985+=1.972, (4G)		
Crops	Rainy season	Winter season		Increase		REY, q/ha
Cropping sequence rice-wheat						
Crops	Rice	Wheat		Rice	Wheat	
Yields q/ha	114	76		123	82	
REY	114	57	171	123	61	184
Cropping sequence maize-wheat						
Crops	Maize	Wheat		Maize	Wheat	
Yields, q/ha	91	76		102	82	
REY	57	57	114	85	61	146
Cropping sequence maize-mustard						
Crops	Maize	Mustard		Maize	Mustard	
Yields, q/ha	91	34		102	37	
REY	57	70	127	85	76	161
Cropping sequence soybean-wheat						
Crops	Soybean	Wheat		S	W	
Yield, q/ha	50	76		53	82	
REY	62	57	119	99	61	162
Cropping sequence maize-gram						
Crops	Maize	Gram		Maize	Gram	
Yields, q/ha	91	46		102	49	
REY	57	115	172	85	122	207

Price of commodity, Rs/q: Wheat 1200; Rice 1600; Maize 1000; Mustard 3300; Soybean 2000; Gram 4000

synergically. The racy nature agriculture will be highly successful crop management practice for Punjab with regard to saving in water.

In the earlier researches, scientist found encouraging responses of raised bed and furrow, but they could not make it sustainable as the concept of green chemistry was not endowed in it. This research has demonstrated ways and simple means to acquire it.

Figure 4 clearly shows that cropping pattern, viz., maize-gram creates the highest REY 207 q/ha, more than that of rice-wheat (184 q/ha). Water demand is drastically reduced with maize-gram crops. There is soil nutrient reserve and good soil health under the maize and nitrogen fixation by gram, thus, in addition to water

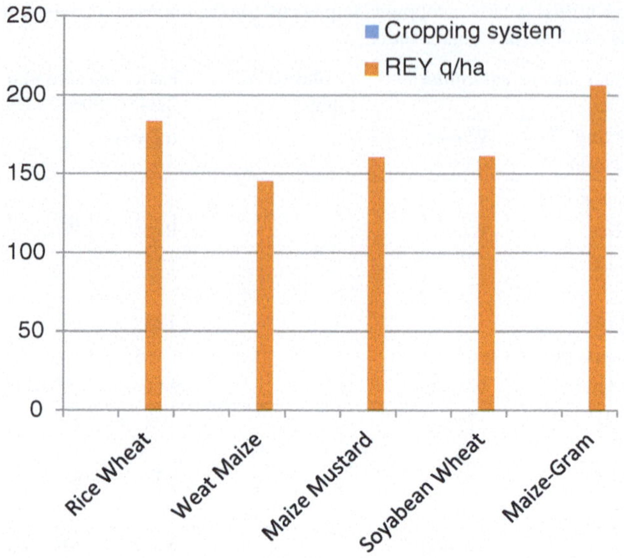

Fig. 4 Rice equivalent yield (REY) of different cropping patterns of racy eco-nature agriculture as applied for agriculture in Punjab, India

there is also saving in nutrient requirement of crops. The management of the input can also bring some advantages. In addition to increase in REY, the reduction in cost of cultivation will be considerable. The agricultural activity will get converted from wet foot to dry foot, hence it will reduce drudgery of working in agriculture. That means work efficiency will also improve. Nevertheless, there will be tremendous opportunity to conduct debate on the benefit of the cropping pattern maize-wheat in due course of time. Rice-wheat a first priority cropping pattern that demanded a change because of high demand for water can be easily replaced by the new cropping pattern.

Maize-Gram: since these crops were considered as sole crop cultivated in respective season, further, enhancement in REY can be made possible. Many other benefits of eco-agriculture will make the present technology still more effective for enhancing the yield more than what is revealed in Fig. 4.

The fact of sprinkler irrigation on wheat raised on flat bed is depicted by uniform crop under sprinkler irrigation (Fig. 5). The raised bed and furrow land form will equip the soil with more moisture, nutrient and perform green chemistry for longer duration than a flat bed. Further, furrow over irrigation at time of high water demand will become ultimate irrigation that will conserve water and produce high harvest index. Utility of raised bed and furrow, NADEPED composting, and ultimate green irrigation, Yadav (2013b, 2015d), which are devised and substantiated in the present study, fortify the claim of achieving high yield and conserving water and nutrients by application of innovative technology of racy nature eco-agriculture.

New Cropping Practice Based on Nitrogen Cycle Management

Different crop sowing planting combinations have been devised to create, eco for zero weeding, intra row cropping, inter cropping, and opportunity cropping to fix nitrogen and bring synergic effects in increasing yield. The practices of intra- and inter-row cropping have been amply experimented (Yadav et al. 2013) The ecosystem and eco-weed management and nutrient buildup by nitrogen cycle are new concept developed by the author in the present study (Yadav 2015c). The opportunity cropping is also new vision and practice, adopted for biomass production for dairy cattle fodder. The optimization of seed rate has been carried out and this practice is helpful in food production. The likely land equivalent ratios (LERs) are recorded in the Table 9 . Since these are subsequent additions in the same cropping systems they become additive to give high LER. That means acquisition of substantial vertical growth in agriculture by nitrogen and sulfur cycle management.

Let us consider the case of maize-gram cropping pattern, as it has proven the best cropping pattern with least water demand. Earlier REY has reached to 207 q/ha/ taking relook of LER for this cropping pattern in Table 9, there is clear-cut visibility of enhancing crop yield to another level (LER 4.0 and above) for both rainy and winter season. The base yield of maize was 50 q/ha which after eco-agriculture it becomes $50 \times 4 = 200$ q/ha. Likewise, the base yield of gram 24 q/ha with eco-agriculture will increase to $24 \times 4 = 96$ q/ha. With these corresponding yields, the rice equivalent for maize would be 125 q/ha and for gram will be 240 q/ha. Thus, the yearly cropping pattern REY will be 365 q/ha. When this eco-agriculture is overlaid on the racy nature land and water management practice, the total REY for this cropping pattern will be $207 + 365 = 572$ q/ha. Thus, such high yield of green chemistry based eco-agriculture will produce REY 572/ha. Earlier when high yield of rice was reported from Bihar, India about 240 q/ha, it was refuted by Chinese scientist. Here, the technology has been created which will produce yield not so far visualized by the researchers. This result is a breakthrough of the world food productivity 572 q/ha. In Table 9, many such cropping patterns are presented.

Improvement in Quality

The yield of cereal crops largely contains nitrogen proportional to the harvest index. The nitrogen content gets converted into essential amino acids (Horobin 2003, Einsminger et.al. (1994)), which later forms the protein in human body. The racy nature technology uses NADEP (aerobically decomposed compost) which produces sulfate, readily absorbed by plants and enhances the sulfur content of the yield (Yadav 2013b, 2015a, b, c, d, e, f, g, 2016). The sulfur contents form the energy

Table 9 New cropping practice based on nitrogen cycle management

S. No.	Main crops	Eco-agriculture, likely + LER	Intra row[a] likely LER	Opportunity cropping	Total LER
Rice-based nitrogen cycle managed crop					
1	Paddy	–	–	–	1.0
2	Paddy + Black gram	BG@50(0.15)	BG@50 (1.51)	–	1.66
3	Paddy + Black gram, 1:1 (25 × 10), 25 × 10)	GG@50(0.15)	BG@50 (1.51)	BG@100(1.54) + Paddy@50%(0.6)	3.80
4	Paddy + Sisbania (25 × 10), Sisbania 25%	GG/BB@50 (0.15)	BY@50 (1.51)	–	1.66
5	Maize (50 × 15), Paddy (25 × 10)	BG@50(0.15)	M + Paddy@50 (1.51)	Paddy@100(1.54) + BG@50(0.6)	3.80
Wheat-based cropping pattern					
6	Wheat (22.5 × 7) Regular sowing	–	–	–	1
	Wheat + Lentil (22.5 × 7) + (22.5 × 7)	Lentil@50(0.15)	Lentil@50 (1.51)	Lentil@100 (1.61) + Wheat @50(0.7)	3.97
7	M (45 × 115) + Wheat (22,5 × 10) (25 × 7 cm), 1:1 + GG	Lintel @50 (0.15)	Lintel@50 (1.51)	Paddy@100 (1.54) + Lintel @50(0.5)	3.70
Maize in rainy season					
8	Maize (60 × 20)	–	–	–	1
9	Maize (60 × 20), GBG (0 × 15) + FM	GG@50(0.15)	GG@50 (1.51)	GG@100(1.54) FM@50 (0.8)	4.00
Gram in winter					
10	Gram Regular (30 × 15)	Gram@50(0.16)	Gram@50 (1.51)	–	1.77
11	Gram + Wheat 1:1	Gram@50(0.16)	Gram@50 (1.75)	Gram@100 (1.75) + Wheat@50 (0.8)	4.46

compound, thus, quality of food is enhanced by the green chemistry endowment in the racy nature agriculture (Yadav 2013b, 2015a, b, c, d, 2016).

Further, in the arsenic dominant soil and water or in both, the sulfate becomes competing (FAO RAP 2006) element against arsenic uptake by plants (Chapagain and Hoekstra 2010; FAO RAP 2006). This is scientific interpretation and experimental substantiations are yet to be established. However, this opens an avenue for research for the scientific community. In the racy nature agriculture technology, use of activated charcoal brings bio-inactivation of heavy metals and toxic gases, thus, up take of the undesirable heavy metals and toxic gases are reduced. All these measures improve the quality of food products produced by the green chemistry involving racy nature agriculture.

The fact that raised bed and furrow create condition for the diversification is getting prominence again to produce diversified food and allow leguminous crop to grow, the main constituent in eco-agriculture. The opportunity of crop diversification allows conducting intra-and inter-row cropping in the wetland paddy production, which forms the aerobic rice, an important practice for reducing water demand for paddy production. The creating path for development of green chemistry eliminates huge problems of health hazards (Yadav 2015a). Thus, universal problems of developments of bad scenarios of health hazards (Bossio and Geheb 2008; Yadav 2015a) for which technologies of only white spots could be created that will get overshadowed by the Sun technology of racy nature eco-agriculture (Yadav 2013b, 2016; Yadav and Chaudhary 2014).

Reduction in GHG Releases

The racy nature agriculture comprises practices which largely operate under aerobic condition created by well drainage maintained in soil (Davis and Cornwell 1997, Yadav 2014a, c, 2015a). Oxygen content is enhanced that enhances biological respiration and aerobic reactions as described in Table 2 (Yadav 2013b, 2014c, 2015a). These facts have been substantiated in detail by this author's studies (Lauren et al. 1994; Yadav 2010, 2012c, 2014a, b, c, d, e, f, g, 2016). The substantiation was done by making contrast of production technologies against racy nature agriculture (yadav 2013b). Thus, racy nature green eco-zero weeding technology reduces emission of GHGs, enhances building of nutrients, reduces energy input in tillage and inter-culture, and postharvest cultivation. The earlier postharvest tillage practices were not on scientific basis and this study provides basis for what should be done to conserve fertility during the intervening period as well as during the cropping. This makes agriculture productive, toil free, energy conserving, and sustainable over the 2G (Yadav 2016) by this study technology 4G.

Discussion

Worsening situation of resources, viz., land, water, precipitation, and environmental condition makes the thinkers to worry of food sufficiency for increasing global population (Thomas and Evans 2011). Among several factors, the chemistry of reactions and question of sustainability became a major concern. In any production systems, some useful and wastage products develop (Yadav 2012a, b), which undergo different chemical reactions that produce detritus food chains to become source for the environmental nuisances. The good products get utilized in creating primary produces which are consumed by the primary consumers. The dynamics of increasing population and consumption in quantity and quality are becoming issue for sustainability of production system and protection of environment. The global warming and climate change are being realized in forms of extremes of occurrences of floods and droughts, rise and fall in extreme temperature and wind blows, on settings of monsoon seasons of years that affect crops and fruit production etc. These environmental factors disturb time of occurrence of seasons and retard productivity and processing of agriculture and horticultural crops. The renewable resource of water for agriculture in general is getting scarce and in general fail to follow its established season. Lately, challenges of green chemistry and sustainable productivity have been carved with great expectations of development of innovative measures/practices/ product to overcome the situation described aforesaid.

In the present study, these important aspects have been dealt with and substantial progresses made. The innovation largely rests on the management of the N, P, K, S, C, O, and ultimate green irrigation bound smart agriculture. This helps decide the nature of tillage and other operations involving tillage and input of energy in agriculture. In this regard, zero weeding eco-agriculture is entirely a new innovative concept whereas the ultimate green irrigation was already created in previous studies (Yadav 2015b, c, d, e, f, g, 2016) that becomes saviors of crops under the rainfed condition, where when rains continue for some days weeding cannot be carried out in wetland condition. That means weeds overtake crop and crop will fail or if matures, drastic reduction in yield will occur. This, innovative practice will prove as a boon for sticky black soils and cultivation of soybean.

The technology components were selected based on the results of earlier researches. The technology of racy nature agriculture comprises raised bed and furrow, selection of crops and cropping practices, manures, fertilizers, precision sowing/planting, ultimate green irrigation, cultivation and weeding, harvesting, and practices for postharvest cultivation. The land formation practice of raised bed and furrows were studied and applied by earlier researchers (Kenneth 2006; Lauren et al. 2006) where in major thrust was on crop diversification. However, the aspects of green/gray and black chemistry were not analyzed then. Those practices did not justify for adoption and it remained largely at the research institutions. In the present study, the green chemistry producing processes are endowed to produce high quantity and quality of foods. The technology validations have been substantiated. Its productivity and environmental protection capability were contrasted. Many innovative refinements, viz., ultimate irrigation practice (Yadav 2015b) zero methane rice production technology (Yadav 2013a), designer rice production (Yadav 2014b) technology, and zero weeding toil free eco-agriculture have been developed (Yadav 2015b, c, f). The racy nature agriculture technology is a Sun technology (Yadav 2013b; Yadav and Chaudhary 2014) having a fixed mode and universal applicability. It involves cyclic irrational mechanics (Ramamrutham 2008; Yadav 2016) of season variation and keeps revolving working of nitrogen, sculpture, water and oxygen for almost all times to create green chemistry. The technology also comprises measures to inactivate the toxic heavy metals and toxic gases so as not to allow to move in the food chain (Yadav 2012b, c, 2015a, f). Thus, quality and quantity of production eliminates health hazardous products. It is clear that theory and principles enable the development of universal technology, which now at the most needs customization at local condition (Fig. 6). By optimization of

Fig. 5 A view of sprinkler irrigated wheat crop in 2013 at Central Institute of Agricultural Engineering farm, Bhopal, India. Note this field was sown on flat bed and irrigated by sprinkler irrigation without the furrow irrigation. The raised bed and furrow planting and overflow furrow irrigation at the critical irrigation stages will enhance the harvest index of crop, which will make the ultimate green irrigation practice

necessary inputs of production systems and any shortfall can be made up by scientific management in agriculture. Thus, wilderness in science of agriculture has been set to become to follow fixed route on the basis of quantum mechanics (Yadav 2013b ; Yadav and Chaudhary 2014).

The application of the green chemistry, which promotes sustainability and brings stability as well as some automation has enabled specifications for racy nature eco-agriculture. As such, it appears to be formidable task and not adoptable by the resource poor farmers. All aspects of the technology of eco-agriculture are free from scale dependence. It requires band of selective mechanization and suitably designed seeding/planting combines. It was an incomplete development in world agriculture to develop harvesting combines without considering improving seeding/planting. Ideally, there should have been effort to develop first seeding/planting combine to produce agriculture that can be harvested. The harvester combines have come from the industrialized countries. Now the scientific specifications are drawn. For use of them development of seeding/planting combines is highly warranted. It will complement the reform version of oriental saying that sow good then harvest good.

This study has clarified several aspects and created number of innovations, combing all it has made agriculture in grip of manoeuveration. The development of ultimate green irrigation practice enables lessen down effect of drought in rainy season cropping. Raised bed stores more moisture in soil profile that reduces number of supplementary irrigation. Conversely, during the continuous rainfall, the furrows work as auto drainage channel and bed condition remain aerobic. This land form is a smart and it supports smart agriculture. That is why the racy nature eco-zero weeding agriculture is real smart agriculture. Arriving to this stage, this author has created several innovations, which otherwise do not exist in literature of agricultural science. Therefore, in support of the theme, and statements, authors innovative researches are included. Without these references, the questions of why, how, and for what could have come, which got eliminated by the author's innovative researches.

Detailed justification and descriptions of the technology have been devised and documented. The timeline of progressive developments is also indicated and substantiated (Yadav 2016). The technology is ready for universal application and optimization for its efficacy at global scale. Only one technology module will be able to produce food of excellent quality, which can be patented and geographical registry (GIr) identified. This will help the producers and consumers get right price and genuine products. This fact will fulfill the needs of industrialized country striving for physical quality of living (PQLI) and for developing country sustainable productivity of quantity and quality foods for present and posterity.

The racy nature agriculture is a technology at par with innovative technologies which affect daily life of people (Table 10). Since food is the first priority, it will be aspect of first attention in any given situation. The green chemistry technology is

Table 10 Technological changes[a] that have affected daily life through rapid pace and undeveloped areas

S. No.	Area of advancement	Specific area	Visible effects
1	Microelectronics for information processing	Computer	Computer graphics stopped TV goers accept as realistic
		Electronic mail	Displaced postal services
		World Wide Web	Displaced postal services
		Cable TV	Displaced TV antenna
2	Medical and Biotechnology	Biotechnology research	Development of antibiotics etc.
3	Magnetic resonance imaging and fiber optics technique	To image human anatomy	To make surgical repairs with only minimum amount of cutting of healthy tissues
4	Cloning of large animals and genetic modification of plants, microbes	Improvement in breeds etc.	Offer possibility of enhanced health and well being in future
5	Ancient babylonia evolved in scriptoria of medieval monasteries	Passed through the invention of printing movable type	Photography, lithography and computer desk top publishing
6	Land and water restoring eco-agriculture technology	Inconsistent, isolated and scanty bright spots existing advancements	Universally applicable forth generation (4G) Sun technology developed in the present study. This is a unique and equally strong innovation in agriculture for present and posterity

[a]Information extracted from [Yadav 2015c]

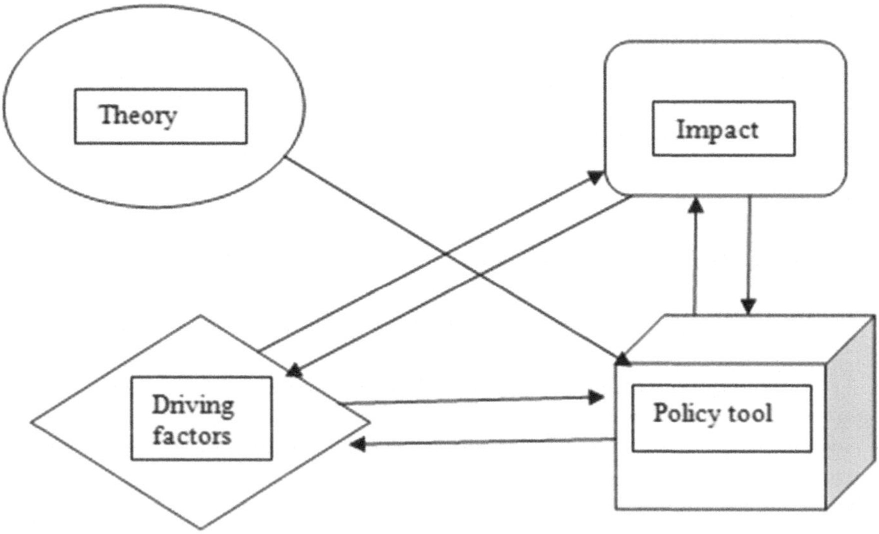

Fig. 6 Link network of theory, impact, driving factors, and policy tools

scientific principle based and policy issues can be framed. Further local discrepancies can be resolved by customization as depicted by Fig. 6 on interaction of technology, policy, and impact. Thus, the racy nature agriculture sets a new pattern of green chemistry and sustainable agriculture at global scenario. Many of the measures are non-monetary inputs and the technology is input sensitive so advantages of this research can be derived in proportion to component application under various resources.

Conclusion

The study has demonstratively substantiated development of total green chemistry eco-agriculture to enhance sustainable productivity, reduction of energy input, and protection of environment. This toil free eco-agriculture has the capability to create new yield plateau and sets new record of rice equivalent yield (REY) 572 q/ha and research direction to acquire it to make up any shortfall that occurred by inadequate scientific managements in agriculture. The new eco-agriculture by enforcement of nitrogen cycle and synergic interventions, largely involves non-monetary input in agriculture. It requires development of seeding/planting combines more urgent than harvester combine. Thus, this study creates new platform of agriculture science, from mere wildering science, to the quantum mechanics based fixed mode agriculture. This research is at par with any innovative development in science and engineering facilitating daily life of the people that shows way for food policy thinkers get out of worry on food situation.

Acknowledgements The author duly acknowledges reference and sources of scientific materials used to support the themes and statements in the study. It is certified that no institutional support was availed in conducting this research and preparing the manuscript.

References

Acharya DB, Mondal SS, Sasha M (2008) Production potential and profitability of different rice (*Oriza sativa*) based cropping systems under irrigated nutrient management condition in Indo—Gangetic plains of West Bengal. Indian J Agri Sci 78:569–572

Biswas H, Narayan D, Lakaria BL (2012) Effect of integrated nutrient management on soil properties and performance of aonla (Emblica officinalia Gaertn) based agri horti system in Bundel Khand region. Indian J Soil Conserv 40(2):141–146

Bossio D, Geheb K (2008) Conserving land protecting water: an introduction. In: Bossio D, Geheb K (ed) Conserving land protecting water comprehensive assessment in agriculture. Series CABI, in association with CGIAR challenge program on water and foods. International Water Management Institute Sri Lanka pp XI–XVIII

Chapagain AK, Hoekstra AY (2010) The green, blue and grey water foot prints of rice from both production and consumption perspective. Value of water. UNESCO Research report series No 40

Davis M, Cornwell DA (1997) Introduction to environmental engineering. McGra Hill Book series Inc. pp 33–36

De AK (2010) Environmental chemistry 7th edn. New Age International Publications, New Delhi, pp 7–13

Dhillon KS, Dhillon SK (2010) Selenium accumulation in sequentially grown wheat and rice as influenced by gypsum application in seleniferous soil. Plant Soil Nutr 227(1):243–248

Ehlers W, Goss M (2003) Water dynamics in plant production. CABI Publication, Walling Ford, U.K

Einsminger AH, Ensminger ME, Konlanola JE, Radson JRK (1994) Food and nutrition encyclopedia. CRC Press Inc., USA, p 1740

FAO RAP (2006) Arsenic contamination of irrigation water, soil and crops in Bangladesh: risk implications for sustainable agriculture and food safety in Asia. RAP Publication 2006/20 food and agriculture organization of the United Nations regional office for Asia and the pacific, Bangkok 2006

Guilette EA, Meza MM, Aqullar MG, Soto AD, Garcia, IE (1998) An anthropological approach to evaluation of preschool children exposed to pesticides in Mexico. Environ Health Prospective 106(6):347–358

Guftafson AE (1939/2010) Hand book of fertilizers, their sources, makes effect and use. 3rd, Agro Bios, Jodhpur, India p 15

Hammer M, Hammer M (2005) Water and Waste Water Technology. Printice Hall of India, New Delhi 110001

Heera CK, Partal K, Dhillon K (2004) Dietary selenium intake by men and women in high and low selenium area of Punjab. Public Health Nutr 7(1):39–43

Herbal MJ, Johnson TM, Tanji KK, Cao S, Bullen TD (2002) Selenium stable 180 take ratio in California. Agric Drainage Water Syst J Environ Qual 31:1146–1156

Horobin W (ed) (2003) Protein how it works? Science and technology, vol 13. Scavandish Corporation, USA, pp 1864–1866(2):141–146

Jacks G, Bhattacharya P, Chaudhury P, Singh VP (2005) Genesis and control of some Flouride in ground water in India. Appl Geochem 20(2):221–228

Jaglan MS, Qureshi MH (1996) Irrigation development and its consequences in arid region in India. Environ Manage 20:323–336

Krishnamoorty KAVR (1976) Further observations on the syndrome of endemic genuvalgum of south India. Indian J Med Res 6(2):284–291

Lauren JG, Pettygrove GS, Daxbury JM (1994) Methane emission associated with a green manuring amendment to flooded rice in California. Biogeochemistry 24:5–65

Lauren JG, Duxbury JM, Hossain MI, Shah G, Talukder ASMHM, Meisner CA (2006) Permanent raised bed cultivation improves nitrogen and water in rice wheat cropping system of South Asia. In: 18th World Congress of Soil Science, Phildelphia Pennsylvania USA 9–15 July 2006

Liu J, Li X (2008) Sulphur cycle in typical meadow *Calmagrostis ongustifolia* wetland ecosystem in the Sangang plain, north east China. Environ Sci (Miscellaneous paper) China 20(4): 470–475

Ramamrutham S (2008) Hydraulics, fluid mechanics and fluid machines. Dhanpat Rai Publishing Company, New Delhi, 110002 pp 254

Sayer KD (2006) Raised bed cultivation. Encycl Soil Sci 1–4

Thomas C, Evans T (2011) Poverty, development and hunger, In: Baylish J, Smith S, Oven P (eds) The globalisation of world politics: an introduction to world relations, Fifth Edition, Oxford University Press 471

Wuebbles DJ (2009) Nitrous oxide: no laughing matter. Science 326:56–57

Yadav RC (2010) A process for eradication of polluting gas, Filed through. Indian council of agricultural research. 12/11/2002 granted 25/2/2010

Yadav RC (2012a) Physically based geoengineering measures to counter global warming. Hydrology current research USA. 3(5):1000142

Yadav RC (2012b) Eradication of acid rain physically based process and product hydrology current research 20:100042

Yadav RC (2012c) Innovative application of scientific facts for nutrient recovery from waste water streams for sustainable agriculture and protection of environment. Hydrology: current research. (5):1000142

Yadav RC (2013a) Innovative applications for scientific facts for zero methane emission and enhancing productivity of paddy fields for sustainable global food security: a conceptual model. Front Environ Sci Eng, USA 2013

Yadav RC (2013b) Racy nature agriculture versus other like technologies: a technologies contrast. Am Eurasian J Agric Environ Sci 13(10):1412–1439. doi:10.5829/idosi.aejaes.2013.13.10.11253

Yadav RC (2014a) Innovative application of scientific facts for arresting GHG-N_2O and improvising lucrative ventures with enhanced land, water and nutrient use efficiency. J Energy Environ Photon 128:486–520

Yadav RC (2014b) Innovative application of scientific facts for production of designer quality rice. Intern J Adv Res Technol 3(10):150–189

Yadav RC (2014c) Drainage engineering: a savvier for sustainable resources use, protection of environment and professional development. J Civil Eng Photan 107:200–213

Yadav RC (2014d) Biological nitrogen harvesting from aquatic ecosystems—A new scientific vision. Fish Aquac J 5(3)

Yadav RC (2015a) Geology and geographical implications on food related health hazards and their remedial measures. In: 8th WSEA conference on geology and geography. Salerno University, Italy, 28–30 June 2015

Yadav RC (2015b) The ultimate green irrigation practice by innovative application of scientific facts. World J Agron Food Sci Technol 2(1):1–30

Yadav R (2015c) New ecology for eradicating GHG—nitrous oxide (N_2O). Open Int J Eco Syst 1 (1):1–2

Yadav RC (2015d) Biological phosphorus harvesting for multiple uses—A new scientific vision. J Aquac Mar Biol 2(2):00022. doi:10.15406/jamb.2015.02.00022

Yadav RC (2015e) Innovative application of scientific facts for devising maneuverable zero weeding and GHG emission ecology for enhancing productivity of cereals, oilseeds and pulses. J Agric, Photan, (Accepted)

Yadav RC (2015f) Toil free eco agriculture for bringing sufficiency of bread, dhal and onion for good public Governance. Danish J Agri Animal Sci

Yadav RC (2015g) Support of drainage for management of nitrogen cycle in Africa Iris Drainage System Engorge. USA 3:128

Yadav RC (2016) Innovative application of scientific facts and irrotational mechanics to develop green chemistry endowed racy nature sustainable agriculture. Int J Green Chem 59–81

Yadav RC, Ambekar VW (2011). Enhancing N, P, K use efficiency in paddy by application of Zinc and assuring production. I Alexir 2015

Yadav RC, Chaudhary MP (2014) Racy nature. A Sun technology towards quantum mechanics in agriculture. World J Agr Res Sci EP, U.K 2(5):223–227

Yadav RC, Yadav J (2013) Development of business out sourcing (BPO) in agriculture for sustainable productivity and environment protection. IJMSEM Taylor and Francis publishing Co. London. Published on line 12 Aug 2013

Yadav RC, Dube SK, Srivastav P (2008) Fatality in water well, the main source of water supplying the Brij Kshetra of north India in: National Seminar on house hold water supply water technology HSTC, Mathura, India

Landscape Gardening and Environmental Restoration: Concerns in Mughal Kashmir

Mumtaz Ahmad Numani

Abstract Landscape gardening ecology is definitely emerging as an ideal concept of art between science and the social sciences. It, today, has become an interdisciplinary significant subject of study. Any artist is, today, a specialist, with varied but favourite and specialised questions. Were the Mughals had been acquainted and conscious of this form of significant art? Or, had landscape gardening been the result of their taste for leisure and pleasure so far as presented and argued? Or, it were neither of these? Besides, what is now set forth as an ideal is an inclusive maintenance of 'biodiversity' that actually aims to highlight the concept of ecologically 'environmental sustainability'. Hence, a school that calls its approach 'Historical Ecology' is insisting to look upon the 'past' in order to insight the 'present' and guide the future. Thus, in-brief, of several, but two objectives may remain principal to define this paper. One, what was the Mughal perception of Kashmir ecology? Two, what was their contribution in landscape gardening and environmental restoration in Kashmir.

Keywords History · Ecology · Landscape gardening · Sustainability
Kashmir · Mughals

Landscape Garden Ecology

In pursuit of garden ecology, good sense prevails. Landscape gardening is definitely emerging as an ideal concept of study. It, today, has become an interdisciplinary significant subject of study between science and the social sciences. Any artist is today a specialist, with varied but favourite and specialised questions. Although, the word, 'landscape' has received multiple connotations over the passage of time, but, here in this paper, 'landscape' simply refers to the human–earth relationship. Thus, of several but a few select important questions immediately follow about. One, does

M. A. Numani (✉)
Centre of Advanced Study in History, A.M.U., Aligarh, India
e-mail: mumtaznumani@gmail.com

V. P. Singh et al. (eds.), *Water Resources Management*, Water Science and Technology Library 78, https://doi.org/10.1007/978-981-10-5711-3_5

landscape garden expresses personal and cultural values? Two, does landscape garden reflects deeper environmental concerns? Three, what role Mughal gardens do play in both?

Source and Methodology

Mughal rulers kept detailed memoirs and chronicles. They had a flourishing tradition of history writing (Thackston 1999). Kashmir became a part of the Mughal Empire under Emperor Akbar in the year 1586. Unlike Babur and Jahangir, although, Akbar personally did not take interest in writing for more than one reason, but he was best chronicled by his courtier Aalami Abu'l Fazl. The other two celebrated official historians (Muhammad bin Muhammad entitled Jalala-i-TabatabiZawariUrdistani and Muhammad Amin bin AbulQasimQazwini) under Shahjahan provide detailed descriptions of the land of Kashmir. They accompanied the imperial entourage to Kashmir and recorded their first-hand impressions. The accounts of Qazwini and Lahori contain descriptions of the ecology of the valley for which the pattern was already set by Emperor Jahangir. So far, the ecological concerns of Mughals are concerned, Jahangir is the most interesting character to be explored at length. And our best textual source on Jahangir is Jahangir himself in his ***Jahangirnama (Tuzuk-i-Jahangiri)***, an autobiography in which he reveals his multifaceted persona as a sovereign, naturalist-cum-ecologist, hunter, aesthete, patron of the arts and a collector (Jahangir 1624; Khan 1863–1864). The importance and complexity of this text begins only now to be fully understood by modern historians, has been earlier pointed out by C. Lefevre-Agrati, and subsequently highlighted by Koch (2009). Indeed, the interdisciplinary discourse between natural scientists and art historians is brought about by Jahangir himself to explain the advantages of a combined method, written and visual, in representing natural phenomena, and sees in it an improvement of his ancestor Babur's approach. Ebba Koch goes on to point out that: scientists have explored the ***Jahangirnama*** for its observations on biology, botany, geology, ornithology and zoology. And art historians have analysed how Jahangir directed his artists to turn his observations of natural phenomena into nature studies. If we consider Jahangir's methodology, we notice that, as a scientist, he has a selective approach; he investigates, observes, records, depicts, measures, enumerates and tests what he considers as noteworthy and outstanding (Koch 2009). All of which showcase that he was fond of 'scientific' experiments of his own devising. Jahangir had, for instance, the valley of Kashmir remeasured, to clarify the measurements of Abu'lFazl in the ***Akbarnama*** (Jahangir 1624; Thackston 1999; Koch 2009). Koch further adds that, if we compare Jahangir with Rudolf II, we note that Jahangir's scientific experiments were something of a one man show; he did not have a circle of scientists and scholars at his disposal. His personal involvement and his achievements thus deserve more admiration, and more importantly, Jahangir has come closer to what Frances Bacon saw in practiced science as a means to sharpen the faculties of a ruler to see through things and evaluate behaviour and situations, ‘to rule with a clear understanding of nature and mankind’ (Koch 2009).

In-short, here, of several but two closely different reasons identified: might engaged Mughal Emperors’ towards the landscape ecology of Kashmir. One, the landscape of Kashmir valley exhibits both: rich flora and fauna, which provides an ideal reason for its ecological concern. Two, the topogeographical outlook of the valley resembling too much with that of Central Asia, seen in the past, provided motivation to the Mughal emperors’ concern towards Kashmir.

Mughal Concerns

One insightful vision of Mughal Emperors’ and their nobles have been seen in their keen interest of laying flower gardens, maintenance of springs and waterfalls which got them closer to the holistic idea of preserving the treasure of Kashmir landscape ecology. Such a deep concern is visible in their choice of ‘selection’, ‘observation’ and ‘planning’. In brief, say for example, the landscape architectural design that the Mughals’ have applied in the maintenance of springs and gardens is significant for more than one, but two important reasons. One, it channelises the abundance of water with an appropriate movement and direction. Two, it prevents soil erosion. Both these ecocentric techniques thus serve an example of their attempt for ‘ecological landscape gardening and environmental protection’. Therefore, the Mughal Emperors’ did not look upon Kashmir just as a pleasure ground (the opposite of which so far has been argued widely), but their continuous intellectual engagement with its natural heritage suggests a deeper concern with what now constitute core ‘environmental issues’.

The relevance and significance of the theme continuing can further be substantiated in the words of James Wescoat. He views that, although climate was not a major topic, which is a significant point of negative evidence, some hydroclimatic incidents were recorded that led to infrastructure and policy adjustments. For example, the first Mughal ruler Babur complained bitterly about the climate, waters and culture of Hindustan immediately after the conquest in 1526 CE. To counter these deficiencies, he ordered the construction of waterworks, gardens and baths to make the capital city of Agra resemble the landscape of Kabul. Three decades later, Mughal documents began to present a favourable perspective on the climates of India (Wescoat 2013). Unlike other Subahs (provinces) of the Mughal empire, warm climate and the deficiency of water although was not an issue with Kashmir, despite that the descendants of Babur showed a deep ecological concern towards the Subah of Kashmir.

As said earlier, Kashmir became a part of the Mughal Empire under Emperor Akbar. But, proper emphasis on garden aesthetics returned with his son and successor Jahangir, who, like his great grandfather Babur, was keen naturalist, listing the flora and fauna of his travels which, in the words of Wescoat, were partly the basis for his garden projects. Also, like Babur, he kept a private journal that shed light on his personal reflections and aspirations (Wescoat 2011).

In the case of Kashmir, Jahangir improved the spring at Virnag (Fig. 1), the crystal clear source of the Jhelum River, in which he ornamented fish with pearl rings. Water flowing from the spring was directed down terrace gardens (Wescoat 2011). The following passage has come from Jahangir.

He writes: 'The source of the **Bahat** [Jhelum] is a spring in Kashmir called **Virnag**, since I [Jahangir] had heard that the depth of this spring was over a man's head, I told them to throw a rope with a stone tied to it into the spring. When it was measured, it turned out that it was no more than one and a half times the height of a man. After my accession I ordered the perimeter of the spring encased in stone, a garden made around it, straight waterways made, and porticos and chambers constructed around the spring. Thus, a place was created the likes of which few travellers can point to (Jahangir 1624; Khan 1863–1864; Thackston 1999). What can be written of the purity of the canal or of the greenery and the plants that sprout below the spring? Bitter herbs, aromatic herbs, various dark green and pale green herbs all grow together. One bush that was seen was as multi-coloured as a peacock's tail and shimmering like wavy water with isolated flowers blooming here and there. (…), it was ordered that plane trees should be planted on both sides of the canal' (Jahangir 1624; Khan 1863–1864; Thackston 1999).

Like Virnag, Jahangir also praises Achabal Spring (Fig. 2) which is located 58 km from the capital city Srinagar and 45 km from Virnag. For example, Jahangir says: 'The Achabal spring has even more water than the others, and it has a beautiful waterfall. Around it are fine plane trees and elegant poplars whose top branches have grown together. Delightful places to sit have also been provided. As far as the eye could see a splendid garden with Ja'fari flowers is in bloom. You would say it was a patch of paradise' (Jahangir 1624; Khan 1863–1864; Thackston 1999). Certainly, the Mughal Emperors' had a vision for the development of the natural heritage of Kashmir. But, unfortunately and sadly so, the trees (mostly fruit bearing) have largely disappeared and Achabal is a garden of open spaces now.

Jahangir also built the terraced garden that came to be known as Shalamar (Fig. 3), which stepped down from its hillside water source toward the level of Dal Lake, which it ultimately reached by a long perfectly straight canal. NurJahan, writes Wescoat, was said to have initiated the manufacture of the ***otto*** (essence) of

Fig. 1 Virnag Spring. Photograph by author

Fig. 2 Achabal Spring-cum-Garden

rose at Shalamar garden (Wescoat 2011). The following passage has thus come from Jahangir.

He records: 'In March 1620, these two or three days I [Jahangir] got into a boat and enjoyed touring and looking at the flowers in Phak and ***Shalamar*** (Phak, Jahangir informs, is the name of a pargana on the other side of Dal Lake). Shalamar is also adjacent to the lake and has a beautiful water channel that comes from the mountain and empties into Dal Lake. My son Khurram [later Shahjahan] ordered the waterfall to be stopped up, that to create a waterfall one might enjoy. And this spot is one of the scenic delights of Kashmir' (Jahangir 1624; Khan 1863–1864; Thackston 1999). Moreover, the Emperor called his garden ***Bagh-i-Farah Bakhsh*** and ***Bagh-i-FaizBakhsh***. The following passage from the pen of Qazwini (official historian of emperor Shahjahan) bears testimony.

He writes: 'In Kashmir, there are number of gardens the aesthetic beauty of which appeals to the heart and mind of every person. I [Qazwini] will explain the characteristics of few of them. The Farah-Baksh garden is beautiful and living. Its green plants immediately remind the Sidrah and Toubah. Its lively fountains are adding beauty to the water passage around. Its buildings imitate the symbol of real paradise. This garden [Farah-Baksh] has been constructed by Emperor Jahangir. Earlier it was popular by the name of Shalimar. Actually the foundation of Farah-Baksh was laid fourteen years ago when Emperor Jahangir was a prince. Its beauty consistently increases year after year. It was only when Emperor Jahangir came to see this paradise like garden; it was named Farah-Baksh. And hence it became familiar by this name. From the beginning to the end of this garden is a road of thirty yards (30 gaz) wide. During the days of prince Khurram (later Shahjahan) the workers of Emperor Jahangir had planted Chinar and Safaid (poplar) trees at a distance of two gaz between on the both sides of this road. Now when the emperor came to this garden again these plants have flourished well. A ten gaz wide water canal was also laid on by the name of "Shah-Nahr" which enters into the Farah-Baksh garden from the backside, and thus continues flowing further through the middle of the garden by passing a Villa (building). A thirty gaz houz (pool) built nearby of this Villa, has also got an extension of Chabutra (tower like structure) and a beautiful fountain in the middle of it. There are four other beautiful

Fig. 3 Shalamar garden. Photograph by author

fountains situating at the four corners of this houz. On the other side of the same Villa is a similar kind of houz from which three more water passages of thirty gaz wide run into Dal Lake. But Shah-Nahr is sixty two gaz wide. The like of which doesn't exist anywhere. There are chinar trees on both sides of the Shah-Nahr situating at equal distance from each other. On the backside of Farah-Baksh, Emperor Jahangir laid the foundation of one more garden known as Faiz-Baksh. And it was decided, a road would be constructed around the water canal through the midst of which Shah-Nahr also flows. Moreover, one houz (40 into 40 gaz) and one Villa (80 into 10 gaz) would also be constructed and then the periphery to be encased with stones from all sides' (Qazwani; Jafari 2009; Khan 1990).

In view of Francois Bernier, the most beautiful of all the gardens of Kashmir is the one belonging to the King, called Shalamar. In his journey to Kashmir, he therefore records: 'The entrance from the lake is through a spacious canal, bordered with green turf, and running between two rows of poplars. Its length is about five hundred paces, and it leads to a large summer-house placed in the middle of the garden. A second canal, still finer than the first, then conducts you to another summer-house, at the end of the garden. This canal is paved with large blocks of freestone, and its sloping sides are covered with the same. In the middle is a long row of fountains, fifteen paces asunder; besides which there are here and there large circular basins or reservoirs, out of which arise other fountains, formed into a variety of shapes and figures' (Bernier 1656–1668).

Asaf Khan, brother of NurJahan, built the nearby Nishat Bagh in Srinagar (Fig. 4), which had a single pavilion at the top and a magnificent set of terraces overlooking Dal Lake. Unlike the gardens of the plains, or those of Kashmir today, water supplies were abundant for these gardens, which, writes Wescoat, led to a shift away from narrow channels with subtle bubbling fountains and rippling cascades, and towards dramatic cascades and fountain displays. Wescoat further adds that, spatially, these Kashmiri gardens had extensive prospects with no visual boundaries, perhaps like the earliest Mughal gardens of montane Central Asia and

Fig. 4 Nishat garden. Photograph by author

Afghanistan (Wescoat 2011). The following passage from the pen of Qazwani records:

'There is one more garden constructed by Yaminad-Daula Asaf Khan on the south side of the Dal Lake known as Nishat Bagh. First, there was constructed a Villa (building) in this garden which was faced by the Dal Lake on one side and garden by the other side. The periphery around has got nine terraces one after the other and each terrace has got its own water fall (Aabshar). At the base of the mountain, one large Villa on the margins of Nishat was also constructed to which an extension of Chabutra (tower like structure), and nearby a large pool were set up. After Farah-Baksh, Nishat-Bagh is the most beautiful garden' (Qazwani).

Other Notable Gardens Established Around Dal Lake

Besides the existing world famous Mughal gardens of today, we are been told that there were hundreds of notable but lesser known gardens constructed either by the emperors themselves or by the princes/ princesses or nobles and other officials of the Mughal court in Kashmir. Out of these hundreds of notable gardens, many of them were laid out around the periphery of Dal Lake, which, however, physically have disappeared on ground with the passage of time unfortunately. The below cited long passage translated from the contemporary principal sources thus bears testimony for their existence during the Mughal period.

It says: 'Among the Emperor's gardens, there is one more garden known as "Behr-Ara Bagh". It is divided into two parts. One part of it situates on high slope and the other situates near Dal-which has four chinar trees situating in the midst. Two beautiful Halls passing through these chinar trees have been built. There is one more garden known as "Daulat Khana Aali" which is famous by the name of "Noor Afza". One can hardly find an example of its beauteous scenery, cleanliness and flourishing flowering plants anywhere else on the earth. The garden of "Aish Abad" also situates at the banks of Dal Lake. It too has got a good number of chinar trees in it. Similarly, "Baghi Illahi" which was laid out by Mirza Yousuf Khan Mustahdi at the corners of canal Lar has one large chinar tree in it. And the canal of Lar flows in the midst of this garden. A throne like wooden platform has also been raised in front of the roots of this chinar tree. Nearby a 10 into 10 gaz pool has been constructed—which has a beautiful wide water passage connected-through which water flows out. There are three more gardens which belong to the princess Jahaniyan Begum [Jahan Ara]. The first one garden is known as "Jahan Ara" which situates at the middle of Dal. It was laid out by Khawaja Sara Jawahar Khan-one of the eunuchs of the princess. Moreover, for the construction of buildings and the plantation of flowering trees, Jahan Ara appointed Firasat Khan—who performed his duty with great care. The second is known as "Noor Bagh" (also called Bagh-i-Noor Afshan) which situates at the banks of river Jhelum and was constructed under the sponsorship of Noor Jahan Begum. An accomplished beautiful Villa has been constructed in it. Also on the banks of the river stood beautiful maple trees-the appearance of which to the viewer is not less than the exact replica of heavenly garden. The third one is known as "Safa Bagh" which is constructed on the northern side of the city at Safa-poor-seventy Krohs away from the city. Also on the eastern side of the garden situates a three Kroh pond. Among the gardens constructed by princess and workers, one garden is known as "Bagh-i Shah Abad" (also called Karna). This Bagh was bestowed on Dara Shikoh. The structure and construction of which can hardly be described in words. It also situates at Dal. The second garden is known as "Bagh-i Moraad" which is associated with sultan Moraad Baksh. It too situates at Dal. One more garden is known as "Bagh-i Naseem". It was constructed by Azam Khan on the northern side of Dal Lake. One more garden is known as "Bagh-i Afzal Abad" which was constructed by Afzal Khan near to Bagh-i Naseem. Both these gardens possess different varieties of flowering plants in them' (Qazwani; Jafari 2009; Khan 1990). Besides these, Inayat Khan informs that his father Zafar Khan had constructed two notable gardens known as Bagh-i Zafar Abad and Bagh-i-Husn Abad. The Bagh-i Zafar Abad cost three lake rupees in its construction during the 12 years of Zafar Khan's government in Kashmir. It was laid out on the margin of the lake Zadibal-overlooking the environs of the Eid'gah (Khan 1990).

Of Water and Plants in the Mughal Gardens of Kashmir

Besides the other components, the known fact is that: water and flowering plants remain the most important ingredients of the ecosystem. These two remain the basis for sustaining the landscape of any territorial unit. Thus, two things are important to deal with. One, what was the principal source of water to the Mughal gardens of Kashmir? And how did the Mughals channelise it? Second, what kind of plant material did the Mughals plant in the gardens in Kashmir?

Significance of Water and the water Channels

In their pioneering research of Mughal gardens, garden historians have put it right that: Mughal gardens in Kashmir are the natural terraced garden type. Thus, unlike Mughal gardens of other places in India, in Kashmir, the principal source of water to the gardens was natural springs. This fact can also be corroborated by a brief description of Bernier, who in his visit to Kashmir writes: 'from the sides of all these mountains gush forth innumerable springs and streams of water, which are conducted by means of embanked earthen channels even to the top of the numerous hillocks in the Valley; thereby enabling the inhabitants to irrigate their fields of rice. (…), the numberless streams which issue from the mountains maintain the Valley and the hillocks in the most delightful venture. Thus the whole kingdom wears the appearance of a fertile and highly cultivated garden' (Bernier 1656–1668). But it has to be borne in mind, in Kashmir, the Mughals were not the earliest rulers who laid out canals to water their gardens (Khuihami 1885), but they were indeed the first to make better use of the old canals and to build up the new ones in order to sustain the world famous gardens constructed by them in Kashmir. For example, Hassan Khuihami, the well-known author of ***Tarikh-i-Hassan*** reports that before the coming of the Mughals many canals (like; ***Lar-Kol-Nhr, Dab-Kol-Nhr, Ara-Kol-Nhr,*** *and* ***Zikr-Kol-Nhr***) were laid out in Kashmir either for the use of irrigation or for the general use of public. For example, sultan ZainulAbidin dug the canals named ***Lar-i Nhr*** (which prospered the Safapur Village) and ***Shah Ju-i Nhr*** (the waters of which was used for the gardens and for the use of general public) (Khuihami 1885). Later on, under the sponsorship of the Mughal Emperors' the nobles repaired the ***Shah Ju-i Nhr*** and brought its waters to the gardens of ***Bagh-i Ilahi, Bagh-i Bahr Ara, Bagh-i Gulshan, Darshini Bagh and Bagh-i Inayat*** into the Zafarabad (Khuihami 1885). Moreover, a few more canals laid out are worth to mention here. For example, a canal was laid out by Emperor Jahangir from the upper hills of Harwan to water Shalamar garden, and Asaf Khan, the governor of Kashmir during the reign of Shahjahan, carried out it further to water Nishat garden (Khuihami 1885). Jahangir also sponsored thirty thousand rupees to Haydar Malik to reroute the stream from ***Lar*** (Sind) to ***Nur'afza garden*** (Jahangir 1624; Khan 1863–1864; Thackstan 1999). During the reign of Shahjahan, an official decree states that Asaf Khan was allowed to take a branch of the canal/stream named ***Shah Nhr*** to Nishat garden on the condition that it should not cause trouble to the peasants of Dachigam and adjacent villages by reducing their share of water for irrigation (Khuihami 1885).

Thus to connect the streams/canals with the gardens, it was given a considerable amount of attention by the Mughals and their nobles. If the garden historians further search (or research) for the 'water-architect theory' of the Mughal gardens of Kashmir in depth, they would indeed figure out that the Mughals were highly conscious about what we call now, 'landscape water architecture'.

Conclusion

In this age or before, one can hardly think of a natural system that has not been considerably altered, for better or worse, by human culture. The contemporary reports so far documented are replete with references in this matter. And Mughal Emperors like other human fellows do not provide a place of exception in this regard. They also have brought some immense changes in the landscape of Kashmir. But these changes (especially in case of landscape gardening) are visible for better than worse. However, an irony of the fact is that, mediaeval Indian scholarship so far has understood, interpreted and appreciated the Mughal landscape gardening concerns in Kashmir as only their pursuit of 'leisure' and 'pleasure', which, however, completely negates the richness of their intellectual wisdom, and therefore, say in modern terms, underplays their character of attempting 'ecological landscape gardening and sustainable environmental development in Kashmir'. In other words, the Mughal Emperors' did not look upon Kashmir just as a pleasure ground, as has been so far exaggerated, rather their persistent ecological engagement with its landscape suggests a somewhat deeper concern with what now constitute core 'environmental issues'. These gardens, if one has to count for, aesthetically, functionally, symbolically and ecologically, reflect deeper 'environmental concerns' in our age of 'environmental crisis'. Therefore, one must not fail to recognise the continuity and proximity of Mughal landscape gardening attributes in this age of 'environmental crisis' everywhere.

References

Alvi MA, Rahman A (1968) Jahangir-The Naturalist, National Institute of Sciences of India, New Delhi, pp. 106–108, 100–102

Bernier F (1656–1668) Constable A (trans) (1983) Travels in the Mogul empire. Oriental Books Reprint, Delhi, pp 399–400, 413, 296–297

Crown S, Haywood S (London, 1972) The gardens of Mughal India. Reprint, Srinagar 2006, pp 191–192

Jafari, Sayid MY (2009) (ed) Shahjahan Nama compiled by Mohammad bin Mohammad entitled Jalala-i- Tabatabai Zawari Urdistani, Al-Hoda International Publishers and Distributors, Office of Cultural Counsellor, New Delhi, pp. 277–279, 280–282, 303–304

Jahangir NM (1624), Khan SA (ed) (1863–64) *Tuzuk-i-Jahangiri.* Sir Syed Academy Reprint in 2007, Aligarh Muslim University, Aligarh, pp 297–298, 299, 313, 302–303, 347–48, 300–307

Khan I (ed), Begley WE, Desai SZA (1990) *Shah JahanNama.* Oxford University Press, New Delhi, pp 124–125, 123, 125, 126

Khuihami PH (1885), Shah SH (ed) (Srinagar, 1954) *Tarikh-i-Hassan*, Shams-Uddin (Urdu trans) (Srinagar, 2003) *Shams-ut-Tawarikh*, vol 1, pp 296–99, 142–157, 147–49, 149–50, 146–47, 302–307

Khuihami, Pir Ghulam Hassan, Tarikh-i-Hassan, (Srinagar, 1954), (Ed.), Sahibzada Hassan Shah, Vol. 1. The Research and Publication Department, Srinagar, pp. 138–148, 146–148, 142–143, 291–297

Koch E (2009) Jahangir as Frances Bacon's ideal of the king as an observer and investigator of nature. J Royal Asiatic Soc London (Third Series) 19(3):298, 297, 331, 328

Numani MA (2015) Emperor Jahangir's method of observation and approaches to investigation of Kashmir ecology: an appraisal of his 'deep sense of sensitivity' towards nature. J Ecol Nat Environ 7(3):76–79

Qazwini, Muhammad Amin bin Abul Qasim, Padshahnama/Shahjahan Nama, transcript of Reza Library Rampur available at Centre of Advanced Study in History Seminar Library, Aligarh Muslim University, Aligarh, ff. 602–604,605–606

Thackston WM (trans and Ed.) (1999) The Jahangirnama: memoire of Jahangir, emperor of India. Oxford University Press, New York, pp XXI, 331, 332, 346, 335, 345, 381, 333–341

Wescoat Jr. JL (2011) The changing cultural space of mughal gardens. Blackwell Publishing Ltd., Hoboken, pp 201, 212, 215

Wescoat Jr. JL (2013) Water, climate, and the limits of human wisdom: historical-geographic analogies between early mughal and modern south Asia in the professional geographer. Routledge, p 2

Eco-Treatment Zone in Open Drain

Sayali Joshi and Pallavi Patil

Abstract In India, of the 233 Class-I cities situated in 14 major river basins, only 24% are having proper sewerage systems and Class-II cities do not have any sewerage system for collecting domestic wastewater. Just collection of the sewage is not enough; further treatment facility to purify the sewage is also a necessity. So, all these city wastewaters are naturally taken to the nearby rivers and lakes by nallas and odhas (streams—natural drains). These natural drains in the cities are serving as sewerage lines. The studies published by CPCB in 2013 reveal the pathetic condition and inadequacy of pollution treatment infrastructure in India. Then Planning Commission has suggested developing "treatment zones" in city drains. It is a very innovative approach acknowledged by Government of India which needs to be strengthened further by very scientific implementations by scientists who have mastered the art of eco-treatment of polluted waters. From the case studies of Udaipur's ecological restoration of Ahar River and Allahabad's ecological treatment Rasoolabad Stream Complex, it can be said that such types of ecotechnological in situ treatment—Green Bridge—system deliver more than expectations in the form of extended social and ecological capitals. The capital and operational costs of ecological treatment processes are comparatively less than conventional engineering approaches.

Keywords Eco-treatment · Ecotechnology · Nonpoint sources
Green bridge

S. Joshi (✉) · P. Patil
Shrishti Eco-Research Institute (SERI), B-106, Devgiri, Opp. P. L. Deshpande Garden, Near Ganesh Mala, Sinhagad Road, Pune 411030, Maharashtra, India
e-mail: sayali@seriecotech.com

P. Patil
e-mail: pallavi.patil@seriecotech.com

V. P. Singh et al. (eds.), *Water Resources Management*, Water Science and Technology Library 78, https://doi.org/10.1007/978-981-10-5711-3_6

Introduction

In the developing cities, it is estimated that more than 90% of sewage is discharged directly into rivers, lakes, and coastal waters without treatment of any kind (www.nyop.unep.org; www.unep.or.jp). Over 2 billion people are expected to live in metropolises, mega-cities of developing countries leading to aggravation of problems of river and lake pollution. In most of the developing countries, governments are fighting to eliminate faecal contamination from rivers and lakes which are sources of drinking water. In India, cities produce nearly 40,000 million litres of sewage every day and barely 20% of it is treated[1].

Central Pollution Control Board (CPCB) carried out study on status of municipal wastewater generation and treatment capacity in metropolitan cities, Class-I cities and Class-II towns of India and published a document. As per the CPCB report, there are 498 Class-I Cities (including Metropolitan cities) having population more than 1 Lac as per 2001 census. Sewage generated in Class-I cities is estimated about 35558.12 MLD total sewage treatment capacity of Class-I cities is 11553.68 MLD. Out of 11553.68 MLD sewage treatment capacities in Class-I Cities, only 8040 MLD exists in 35 metropolitan cities, i.e., 69%. The capacity of sewage treatment in remaining 463 Class-I cities is only 31%. While total sewage generation in Class-II towns is 2696.70 MLD, out of which total sewage treatment capacity in Class-II towns is 233.7 MLD which is just 8% of the total sewage generation. Actual sewage treatment, due to inadequacy of the sewage collection system, shall be low compare to capacity (CPCB 2013).

The 150-year-old conventional aerobic and anaerobic treatment systems are yet to be accepted worldwide as they are cost intensive and complicated to maintain. Investment of crores of rupees in the Ganga Action Plan could not give the results due to unavailability of electricity to run the modern facilities of state-of-the-art treatment systems, technologies, and inadequate sewer collection and conveyance facility. Same was the case with Yamuna Action Plan and Dal Lake Pollution Control Plan.

Centralization of sewage treatment facility has many engineering and management difficulties such as:

1. Disorganized construction of buildings, townships, and roads leading to backlog of sewage treatment facilities
2. Already constructed areas do not have effective sewage collection, conveyance, and treatment systems

Due to lack of proper conveyance/infrastructure facility to collect all sewage generated from already developed/constructed area or remote places in the city, sewage is directly coming into the surface fresh water bodies like lakes and rivers through channelized and unchannelized drains. Consequently all city's storm water or natural drains are converted into the sewage carrying drains. If we are able to treat and reuse this water for nonconsumptive purposes, our fresh water demand will be reduced and we will be able to save our fresh water resources from pollution which will lead to sustainable development of our cities.

Ex Planning Commission, Government of India's **Report of the Working Group on Urban and Industrial Water Supply and Sanitation for the Twelfth Five-Year-Plan (2012–2017)** spells out explicitly the guidelines and line of action to reduce the pollution of rivers and lakes due to inadequate sewage and effluent treatment facilities in urban and industrial sectors. The studies published by prime pollution authority of India—Central Pollution Control Board (CBPC) in 2013 reveal the pathetic condition and inadequacy of pollution treatment infrastructure in India. May be because of that, Ex-Planning Commission have suggested to develop "treatment zones" in city drains. It is a very innovative approach acknowledged by Government of India which needs to be strengthened further by very scientific implementations by scientists who have mastered the art of eco-treatment of polluted waters. The original text of Ex-Planning Commission's report is cited here.

Make Drains Treatment Zones

Sanitary engineers-turned-pollution managers have a one-size fits all solution—first build underground sewerage network (however long it takes), then connect households to the system (even if there is resistance or delays) and then once the pipeline has been officially inaugurated, it will transport official waste to the treatment plant (built earlier but not working because of lack of sewage). This will be done and pollution will be controlled.

So, the question is how the waste—generated in households and conveyed through open drains and then into the river can be cleaned? The drains exist—lead to stench, disease and unlivable conditions. Instead of waiting for the end-day when the drain will be transformed into the storm water carrier it was meant to be and the sewage will disappear mysteriously into underground chambers, new solutions can be found. The drain, open and unhygienic, can be used as a treatment zone. The sewage can be treated in the open drain, intercepted in the open drain and then conveyed for after-treatment to the already built sewage plant. This is not to say that this open-air treatment will clean sewage and turn it into drinking water. But it will certainly reduce pollution and also turn the drain, from a stinky and dirty sewer to a planted waterway, which will be part of the city's landscape.

Again, this is not a tried or easy solution. But experiments to clean stretches of drains, using bioremediation technologies have been conducted, with success. The challenge is now to up-scale this approach and to integrate it into the pollution plans of the country. It is also a challenge to compute the costs of this emerging technology and to develop indicators for its performance so that projects do not become new scams, this time in the name of pollution. The bottom line is that the city has to invest in sewage management, but it has to invest to do things differently. (Reference—Report of the Working Group on Urban and Industrial Water Supply and Sanitation for the Twelfth Five-Year-Plan (2012–2017) November 2011).

Environomics—environmental economics of sewage treatment with a changing urban scenario and pressure of clean technologies due to climate change, one is looking for the better option in sewage treatment technologies which are economic, sustainable, and eco-friendly. Based on the experience of huge spending in Ganga Action Plan, National River Conservation Directorate (NRCD) has strongly recommended using energy less methods to treat sewage. The ecotechnologies are much cheaper than energy intensive conventional mechanistic sewage technologies. It is normally observed that conventional systems cannot deliver as desired, if not properly maintained with designed electric supply and skilled man power (SERInews 2009).

As for the existing treatment system, the treatment process can be selected as per the output requirements but even with the most sophisticated treatment facility, a recycle and reuse of the treated water is must, as it will not only reduce the fresh water demand but also will reduce the sewage generation. The idea of target specific and tailor made solutions are important for prevention of pollution in local water bodies. Also awareness among people for the proper maintenance of public water bodies is of at most importance and social and cultural differences should not come in its way.

Ecotechnologies

Core of ecotechnological treatment system is based on ecosystem approach. Conventional systems generally exploit elements of nature—a few groups of microbes supported by infrastructure, energy, and chemicals for degradation of waste matters while ecotechnological treatment systems rely on ecological interactions and biological improvement in natural web of life.

Natural streams, rivers and lakes have their own in-built purification system which is comprised of biotic–abiotic factors such as winds, natural slopes, stones for biological growth, and complex food web help in the purification process. This food web is nothing but utilization of one's waste by another as its own food. Nature has her own living machinery of detritivorous microbes and other living species to consume wastes. These principles have been harnessed in the treatment of polluted streams using STAC (Saprobic Trophic Absorption and Cycling) system comprised of grafting of ecotechnological horizontal eco-filtration—Green Bridges.

Detritus food chain in the nature has the capacity to assimilate sewage constituents and transfers them into ecological cycles of nutrients. There is need to bring paradigm shift in design concepts from calculable concentration models and performance criteria to ecosystem approach of using detritus-induced complex food chain and nutrient cycles. Use of ecological processes in treating and assimilating nutrients from the sewage reduces capital as well as operational costs substantially. Conventional treatment systems reduce carbonaceous BOD only but remaining COD and non-carbonaceous BOD then lead to permanent undesirable changes in ecosystem of receiving water bodies. This can be evaded by use of ecosystem approach and ecological engineering to treat the sewage to convert into ecologically corrected water (SERI News 2012).

There is two types of actions involves in the natural system, one is consumption of pollutants as nutrient source in detritus food chain while other is to use of wastes generated from this process are useful for green plants growth.

The scheme involved application of ecological engineering to remove organic and inorganic pollutants from the water and to utilize them as nutrient in the ecological cycles. Green Bridge is developed using filtration power of cellulose/fibrous material with stones. All the floatable and suspended solids are trapped in this biological bridge and the turbidity of flowing water is reduced. Green plants on the bridges increase the DO level in water, which in turn facilitates the growth of aerobic organisms, which degrade organic pollutants.

Vegetation being significant ecological elements of any landscape, biomass, and diversity play key role in ecosystem dynamics and global cycles. Vegetation is a biological sink for atmospheric carbon (CO_2), as 50% of their standing biomass is carbon only.

Applications of ecological engineering principles, environmental chemistry, microbiology, interactions of organisms and succession of biological communities are very useful to consume organic and inorganic pollutants from the wastewaters and bioconvert them into nontoxic form, finally transferring the elements in the ecological cycles. These eco-transformations, eco-conversions and degradation or bio-utilization of pollutants—nutrients are the part of ecological cycles—biogeochemical cycles. In the ecotechnology, attempt has been made to apply natural flora and fauna in well-designed manner to develop technologies like Green Bridge, Green Lake Eco-Systems, Green channel, Biox (biological oxidation) and Stream Eco-Systems.

Advantages

- Availability of pollution-free water for nonconsumptive use
- Clean water for agriculture reducing the accumulation of toxic metals into crops and grains, thereby improving the production efficiency, quality and price
- Increased biodiversity
- Improvement in groundwater quality over a period of time
- Control of nuisance insects and odour
- Improvement in healthy environmental conditions for the population in the adjacent areas
- No failure of system due to breakdowns and nonavailability of electricity
- Site for the ecological tourism and education

Though ecotechnology is comparatively new option, polluter has got a very cost-effective technique to control the pollution and convert it into resources. Ecotechnology harnesses bio-powers to assimilate anthropogenic wastes into ecological cycles without putting demand for man-made electricity. Conventional waste management systems need a lot of electricity which in turn does not become a candidate for carbon credits. But, ecotechnological treatment units having multiple uses like carbon sink, reduction in use of electricity and minimizing release of methane like GHGs are more useful in getting carbon credits at international levels. These techniques are more useful for the developing countries which cannot afford the cost of sophisticated mechanized auto-control techniques to manage the waste.

While talking about the various process parameters for studying performance of the treatment system, the significance of dissolved oxygen (DO) as indicator of quality of treated water. Maintaining the proper level of DO in the local public water body not only signifies the quality of water, but also plays a significant role in sustaining the entire food chain depending on the water body, which will also improve the overall aquatic ecosystem.

Assessment criteria will involve ecological–ecosystem quality indices, health indices and socio-economic parameters. These indices shall be involved in order to cost futuristic trends for reforming action plans, resetting targets and revisiting policy principles.

Case Study I

Ecological Restoration of Rasoolabad Stream Complex on the Bbanks of Ganga River, Allahabad (Patil et al. 2012)

Rasoolabad is a cremation Ghat also known as Chandrashekhar Aazad Ghat which is situated on banks of holy River Ganga in Allahabad UP. Rasoolabad stream complex was group of 3–5 slender lined–unlined streams. It passes through settlements (Jhondwal, Rasoolabad and Mehdori Colony) near Rasoolabad Ghat and carries the raw sewage and domestic effluents from the residential complexes and drain into the River Ganga.

Ganga Seva Abhiyan is complete peaceful and non-violence movement under the devotional guidance of Swami Avimukteshwaranand Saraswati ji (disciple and representative of Sri Shankaracharya Swami Swaroopanand Saraswati ji Maharaj). The main object of this Rasoolabad Stream Complex Project was to prevent the drainage of industrial waste and human pollutants flowing into holy Ganga River along its long flow line. With clear intention of making Ganga River free of pollution, Ganga Seva Abhiyan decided to take up some demonstration projects on five selected nallah using ecotechnology treatment system, so they invited Shrishti Eco-Research Institute, Pune to design and implement the restoration of 5 polluted streams in Rasoolabad Area. It was implemented by Green Infrastructure as per the design given by Shrishti Eco-Research Institute.

Each stream traversing through dense population joined Ganga River, emptying in it a huge quantum of pollution. Each stream/drain became brownish/black coloured and emitting foul odour because of city wastewaters and massive solid waste disposal. At all sites, it is observed that due to lack of oxygen and absence of biodiversity of phytoplankton and zooplankton, anaerobic degradation makes water unsuitable for any type of use. Overall organic load in the Rasoolabad stream complex was about 3 tonnes per day (50–60% is biodegradable by conventional methods), and load of suspended solids is about 2.16 tonnes per day.

Ecological Treatment System Installed Under Rasoolabad Stream Complex

- On three line drains—green rolls in different structures were installed to sustain the high velocity of flow and wetland for the streams having flow in the range of 0.5–0.8 MLD
- Zero electricity, no skilled maintenance and in-built eco-equilibrium of bio-degradation and bio-absorption processes using green rolls, and wetland systems
- Combination of plants and bacteria for eco-remediation of Rasoolabad stream complex

Eco-restoration project at Rasoolabad stream complex shows improved water quality of nallahs. Reduction in TSS and Colour improves physical appearance along with Increased DO. Biodiversity status along nallah premises also gets improved. Social engineering helps to create public awareness regarding environment issue and bound locals to project. This is the unique eco-restoration project that shows the effective consortium of saints, technology and localities with mandatory permissions from government. Important point to be noted about eco-restoration by ecotechnology is that it is very low maintenance and low capital cost system that improves water quality by 70–80% without electricity and which may act as filler that was ignored in Ganga action plan mainly the unavailability of electricity and improper waste management systems.

Some Important Highlight of the Project

- Social engineering of local residents and cooperation from women to reduce the plastic waste in drains
- Support from local administration to provide bins to collect everyday's solid waste
- Control of foul odour and black colour in the drains
- Reduction (50–90%) in pollution draining into Ganga River by five waste streams (Graphical presentation given below)

The quality of water of drain is clearly visible in photographs of before and after installation of ecological treatment system in below figures:

Before

After

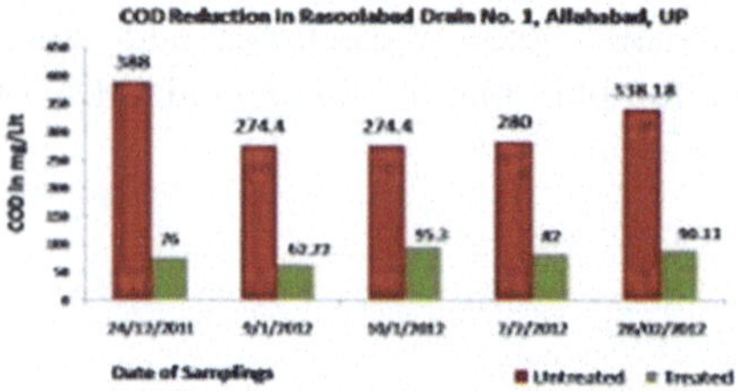
COD Reduction in Rasoolabad Drain No. 1, Allahabad, UP
COD in mg/Lit
388
76
274.4
274.4
95.3
280
338.18
90.11
24/12/2011
Date of Samplings
Untreated
Treated

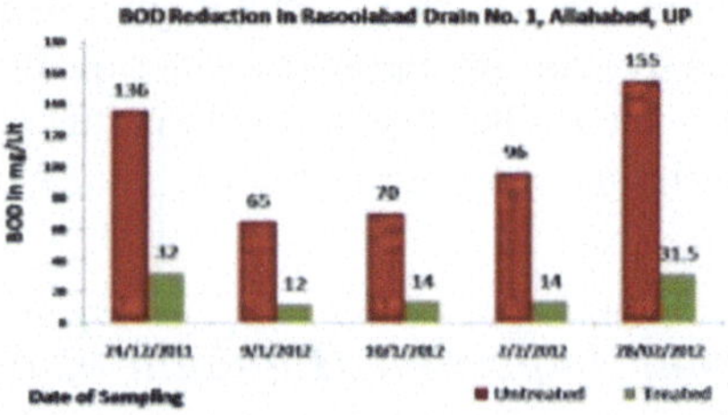
BOD Reduction in Rasoolabad Drain No. 1, Allahabad, UP
BOD in mg/Lit
136
32
65
12
70
14
96
14
155
31.5
Date of Sampling
Untreated
Treated

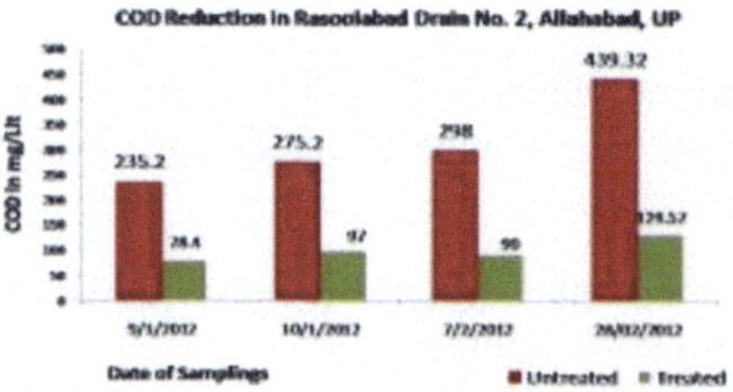
COD Reduction in Rasoolabad Drain No. 2, Allahabad, UP
COD in mg/Lit
235.2
275.2
298
439.32
Date of Samplings
Untreated
Treated

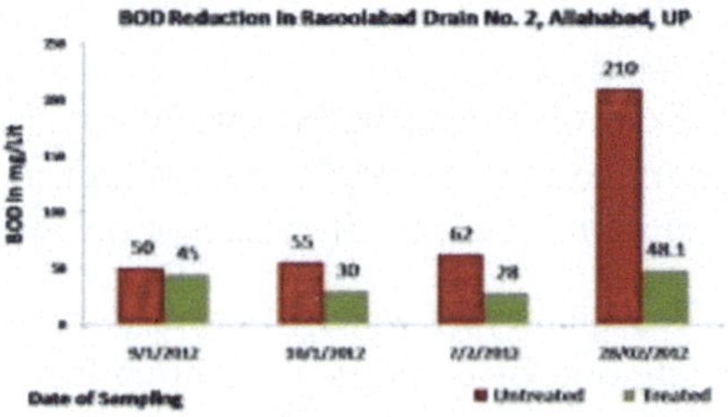
BOD Reduction in Rasoolabad Drain No. 2, Allahabad, UP
BOD in mg/Lit
50
45
55
30
62
28
210
48.1
Date of Sampling
Untreated
Treated

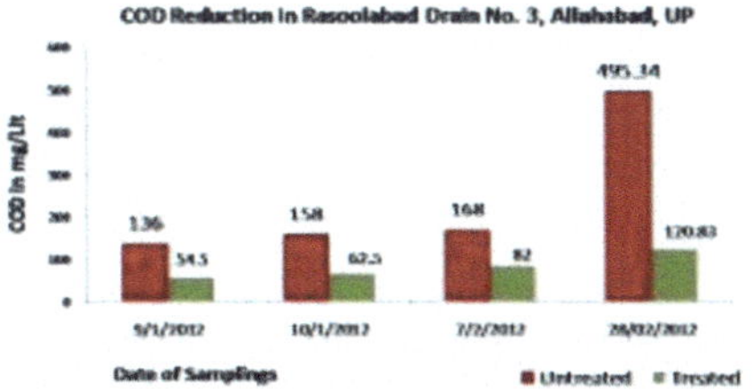

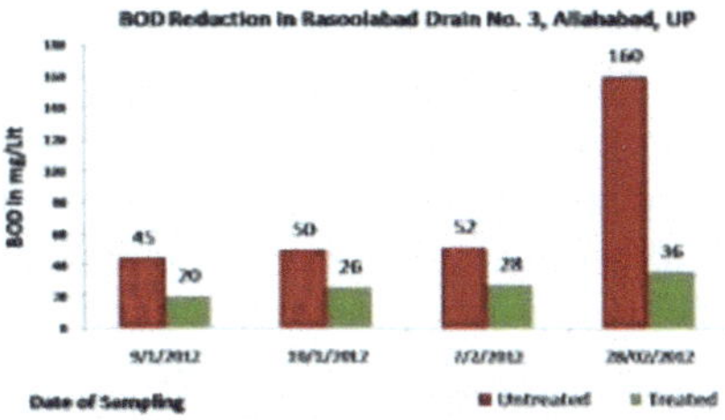

Case Study II

Ecological Restoration of Udaipur's Ahar River (Kodarkar and Joshi 2010)

Ahar River is flowing through Udaipur city, Rajasthan (I) and its non-monsoon flow is 150 MLD (Million Litres per Day). Ahar River was highly polluted due to enormous discharge of city's untreated wastewater and industrial wastewater into the river. It affected the ecological health of river and downstream water reservoir—Udaisagar Lake—a source of livelihood of villagers. The river water was not suitable for any activities also lost its biodiversity and developed bad odour due to anaerobic degradation. There was no suitable water quality for survival of any aquatic organisms as well as river was infested with water hyacinth in certain stretches leading to elimination of other resident species of the river like turtles, water snakes, fishes and freshwater micro-invertebrates (2) (Kodarkar and Joshi 2010).

A Udaipur based nongovernment organization—Jheel Sanrakshan Samiti (JSS) initiated eco-restoration project by creating social awareness of the masses to support this river restoration programme with technological support and guidance by experts of Shrishti Eco-Research Institute (SERI) of Pune—a research organization having experience of 16 years in the ecotechnological systems for point and nonpoint sources of pollution. This project was funded by Udaipur Chamber of Commerce and Industries (UCCI).

The treatment site was selected before the confluence with receiving water body, i.e., Udaisagar Lake to instal ecotechnological based treatment units for water purification and revival of ecological health of river. Ecological restoration activities started with removal of aquatic weed water hyacinth mingled with non-biodegradable plastic material at a selected stretch, 10 km downstream of Udaipur. One screen made up of MS with anti-corrosive painting shall be installed upstream of Kanpur Pulia. The treatment scheme is comprised of six horizontal in situ eco-filtration system—Green Bridges (Two near Kanpur Pulia and four before Sukha Naka Bridge) were developed in the course of river in a distance of about 1.6 km. Green Bridges were seeded with mixed bacterial cultures helpful in treating organic and inorganic wastes and local green plants were grown to support the activity of microorganisms symbiotically.

The self-purification capacity of the Ahar River was increased by improved level of dissolved oxygen up to 8 ppm in the river previously which was zero. Dissolved oxygen content also increased multifold triggering growth of aerobic organisms. This resulted in exponential increase in phytoplanktons and zooplanktons which attracted the subsequent trophic levels including fish as toxicity of wastes neared to zero. It is a bio-indication of reduction in pollution levels in Ahar River. The entire river stretch got its life again with return of turtles, snakes and increased number of bird species in and around the river. Change in Ahar River after installation Green Bridge shown in following pictures:

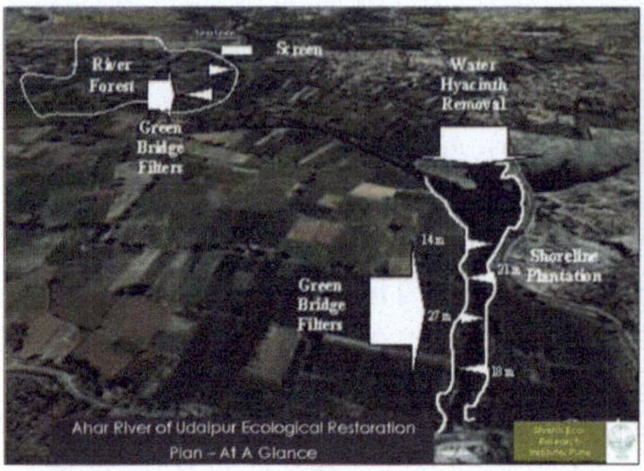

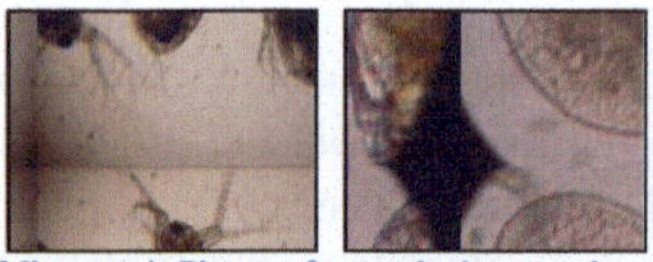

Microscopic Picture of zoo planktons and phytoplanktons

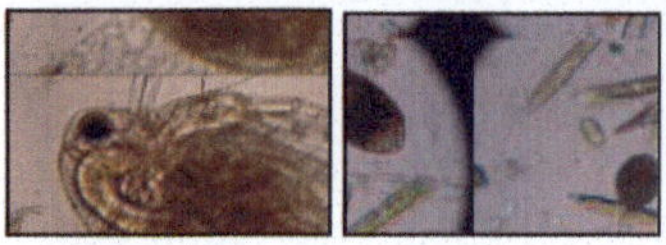

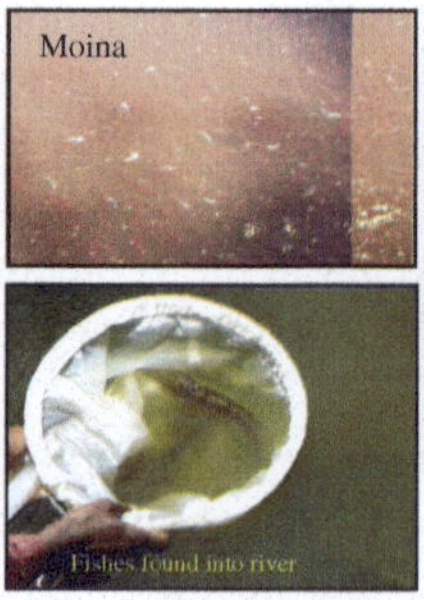

On-site testing of water carried out by CPCB Team, They noted that D.O Level was 8 ppm of treated water while before Green Bridge installation it was zero.

Foam reduction before and after implementation of Green Bridge system

Analysis of Ahar River

Dissolved Oxygen of sample water at various depths and locations
Sample taken between 1.40 pm and 4.30 pm

S. No.	Sampling station	Left bank			Middle			Right bank		
		0.1 m	0.25 m	Bottom	0.1 m	0.25 m	Bottom	0.1 m	0.25 m	Bottom
1	Before screen	7.6	7.1	6.3	6.4	5.2	NA	7.3	6.2	NA
2	After screen	10.6	10.2	9.8	10.3	10.2	9.8	10.1	9.7	7.4
3	After green bridge 1	10.8	9.7	NA	11.2	10.9	NA	7.9	7.6	NA
4	After green bridge 2	10.7	10.1	NA	7.9	6.7	NA	8.6	8.4	NA
5	After green bridge 3	11.2	10.9	10.6	7.8	7.6	6.7	10.2	10.1	NA
6	After green bridge 4	10.8	10.4	9.4	10.3	10.2	8.9	10.5	10.4	NA

Transparency of River at Various Locations

S. No.	Sampling station	Unit	Transparency
1	Before screen	cm	12.6
2	After screen	cm	14.9
3	After green bridge 1	cm	15.4
4	After green bridge 2	cm	17.6
5	After green bridge 3	cm	21.5
6	After green bridge 4	cm	24.2

Important points to be noted about eco-restoration of any water bodies by ecotechnological treatment systems are, a. These are very low maintenance systems and require less capital cost, b. It improves water quality by 70–80% without electricity which may act as filler that was ignored in Ganga Action Plan mainly the unavailability of electricity and improper waste management systems.

Conclusion

1. The role of eco-treatment is supplementary, complimentary for the existing or upcoming treatment facility; as it will enhance the efficiency as well as it will make the entire treatment process more eco-friendly by reducing land and energy footprint.
2. Acknowledgement and encouragement of eco-treatment for city drains being zero energy footprint, negligible space footprint, eco-friendly process.

We require cost-effective and sustainable solutions like ecotechnology that not only corrects water ecologically but also saves energy and improves biodiversity and help water body to regain its self-purification capacity along with benefits for people attached with water body. Ecological treatment can give economical and sustainable treatment option to make wastewater reusable. This ecologically corrected water will be more beneficial for agriculture, landscaping, etc.

References

Central Pollution Control Board (2013) Performance evaluation of sewage treatment plants in India funded under NRCD

Kodarkar M, Joshi S (2010) ILBM impact story—ecological restoration of highly polluted stretch of Ahar river, Udaipur and ecological improvement of Udaipsagar lake, Rajasthan, India. Presented in final review meeting and international symposium of a project entitled "Intengrated Lake Basin Management (ILBM), Basin Governance, Challenges and Prospects", Nov. 2–7, 2010. International Lake Environment Committee (ILEC) Foundation, Headquarters, Kusatsu, Japan

Patil P, Agre P, Pramod S, Joshi S (2012) Water Quality assessment at Rasoolabad stream complex eco-restoration project, Allahabad, UP. India paper presented in International conference Anthropogenic Impact on Environment and Conservation Strategy held at Ranchi, Jharkhand

SERInews vol 3 No. 7, 2009 Mar

SERI news vol. 6 No. 8, 2012 Apr

www.nyop.unep.org; www.unep.or.jp

Part II
Water Resources Planning and Management

Water Crisis: Issues and Challenges in Punjab

Ravishankar Kumar, Upma Vaid and Sunil Mittal

Abstract Punjab, an agricultural state of India, is facing a severe water crisis due to lesser annual rainfall than normal (700 mm) since 1998. Further, Punjab is not getting adequate amount of river water due to political reasons like Indus treaty, damming and diversion of river water, water conflict with Haryana, Rajasthan, and central government. However, the irrigation water demand (4.45 m ham) is significantly more than total irrigation water availability (3.04 m ham). Hence, in most parts of the Punjab state, groundwater is being overexploited for irrigational purpose. Apart from this water scarcity or depletion problem, water quality is also being deteriorated and not suitable for drinking purpose. Basic groundwater parameters such as salinity, electrical conductivity (EC), chloride (Cl^-), and nitrate (NO_3^-) have surpassed the maximum permissible limit in most of the parts of this state. Even toxic heavy metals [like selenium, uranium, arsenic, and lead] and pesticides have also been reported in groundwater samples of several regions of Punjab. Intake of this heavy metals and pesticides contaminated water is affecting the health of native people. The condition of groundwater depletion and quality deterioration is most severe in Malwa region of Punjab. The poor water quality and presence of toxic heavy metal may be linked with the prevailing health issues in this region. Government is taking several initiatives regarding this issue and passed the Punjab Preservation of Sub-Soil Water Act (2009). Government is also providing subsidy to individual farmer to lay down underground pipeline, drip and sprinklers systems for irrigation. Additionally, government is promoting and appreciating preventive measures like watershed management and rainwater harvesting.

Keywords Groundwater · Punjab · Heavy metals · Malwa region

R. Kumar · U. Vaid · S. Mittal (✉)
Centre for Environmental Science and Technology, Central University of Punjab, Mansa Road, Bathinda 151001, Punjab, India
e-mail: sunil.cevs@gmail.com

V. P. Singh et al. (eds.), *Water Resources Management*, Water Science and Technology Library 78, https://doi.org/10.1007/978-981-10-5711-3_7

Introduction

Water is one of the fundamental resources and indispensable element of life on the earth. Out of total available water on earth, 97.2% water is saline and only 2.8% is fresh water. Even out of this 2.8% fresh water, 2.1% is found in the form of ice/glaciers, while only 0.37% is available in the form groundwater, and 0.02% as surface water. In surface water, 87% is in lakes, 11% is in swamps, and only remaining 2% is in rivers (Fig. 1).

In the last few decades, overpopulation, industrialization and vast agricultural practices are making situation more critical and it is becoming difficult to meet the water demand for drinking and other purposes. Worldwide, around one billion people are devoid of safe drinking water. Half of the world's wetlands have been lost since 1900. It can be predicted on the basis of increasing water demand and declining freshwater availability that future conflicts over water will may result into water wars.

India has 16% of the world's population but only 4% of the world's renewable water sources. Hence, it is a big challenge to meet the demand of safe drinking water of such a huge population with inadequate water resources. Therefore, in our country, unsafe water and lack of basic sanitation is causing several types of diseases and killing many people every year. Even almost 30,000 children under 5 years of age die every week due to consumption of unsafe water and unhygienic living conditions.

Punjab is called as breadbasket of nation. It is a cocktail of heavy agricultural practices and industrialization activities. Now, this Indian state is coming under threats of deficit of water recourses. The water table is falling rapidly and it is a serious matter of concern not only from state point of view also from national. Additionally, the water quality of the state is also deteriorating rapidly. The interesting similitude is found between the states of Punjab and California (USA) (The Tribune 2015). Both are world's leading food producing states; providing annually 30–50% of rice and 40–75% of wheat to the central food pool. But currently, both are facing identical problems related to the water required to irrigate cropland.

The state of emergency has been declared consecutively fourth year in California regarding water scarcity. Punjab is also facing water crisis situation, but not like California. However, if current condition continues then Punjab will also attain similar condition as California. The NASA (National Aeronautics and Space

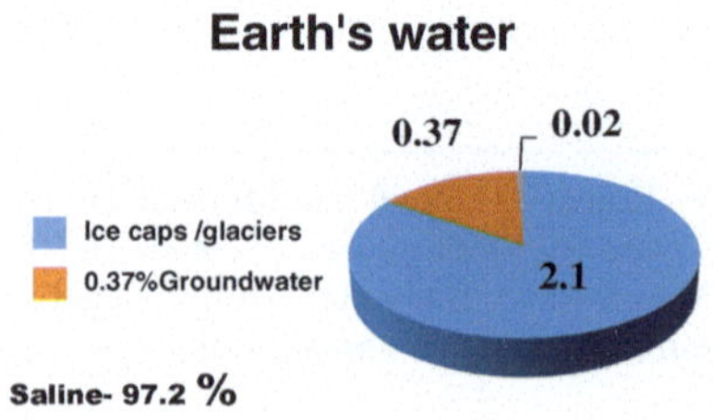

Fig. 1 Distribution of water on earth

Administration) have reported that beneath the earth surface level, water is disappearing in the Northern part of India (The Tribune 2015), especially in the states like Punjab and Haryana and if the same continues then there are chances to collapse agriculture system of that region. The aim of this paper is to assess the current situation of water crisis in Punjab and their challenges. So, the water crisis issues and their challenges of Punjab state are described in detail in this current assessment. Further, preventive measures adopted against the current situation and the initiatives taken by the Punjab government and other agencies have also been discussed.

Sources of Water and Their Status in Punjab

Groundwater

Agricultural activities of Punjab have heavy requirement of water for irrigation purpose. Rivers, groundwater, canal water and rainfall are the major sources of water in Punjab. Canal and tube wells are the main sources of irrigation. The dominancy of rice and wheat cropping system has overexploited groundwater resources, resulting in rapid decline of groundwater table of the entire state (approximately). Tube wells are the major source of over exploitation of underground water reserves. The number of tube wells in the state has been increased from 1.92 Lakhs in 1970 to 13.8 Lakh in 2011 (ENVISb 2015). Approximately, one tube well is there for every 2 ha in this state. The irrigation water demand (4.45 m ham) is significantly more than total irrigation water availability (3.04 m ham) (Table 1). The estimated net available groundwater is around 20.35 Billion Cubic Meter (BCM), which is less than annual demand 34.66 BCM (Fig. 2). The annual deficit of groundwater is determined as 14.31 BCM (CGWB 2016; ENVISb 2015; SOE Punjab 2007).

Eighty percent of the total geographical area of the state (110 blocks) is categorized as overexploited, 3% area (5 blocks) as critical and semi-critical, whereas only 17% area (23 Blocks) is under safe category for groundwater development (CGWB 2009) (Fig. 3). During 2008–2012, the water table all across the state has receded at average annual rate of 0.70 m (Fig. 4) and range of water table decline varies from 0.10 to 4.0 m (Jain 2013). Sangrur and Patiala are the most affected districts in concern of water table. However, the water table is rising in some southwestern parts of the state. The reason behind that in those regions,

Table 1 Status of water in Punjab (*Source* CGWB 2009 ; ENVISb 2015; ENVIS Newsletter 2014–2015; SOE Punjab 2007)

Irrigation water demand	4.45 m ham
Surface water availability	1.43 m ham
Annual replenishable recharge	1.61 m ham
Total irrigation water availability	3.04 m ham
Irrigation water deficit	1.41 m ham

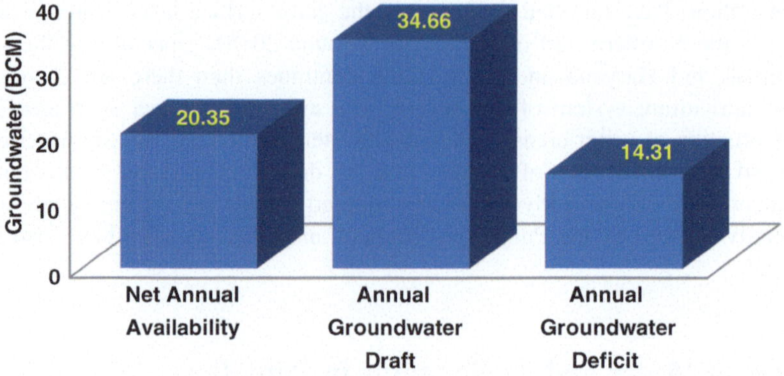

Fig. 2 Status of ground water resources of Punjab. *Source* Central Groundwater Board, Chandigarh (2009)

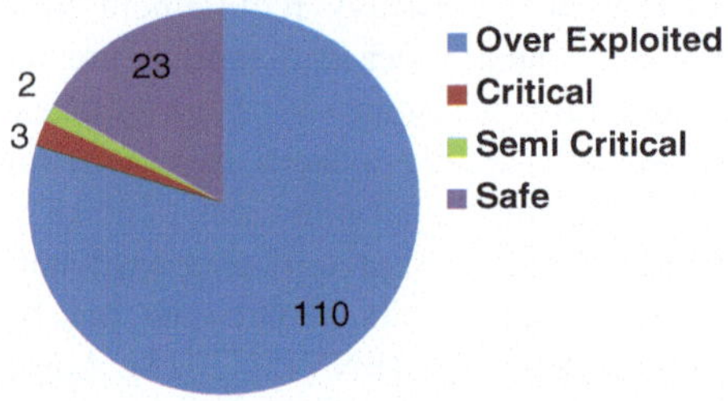

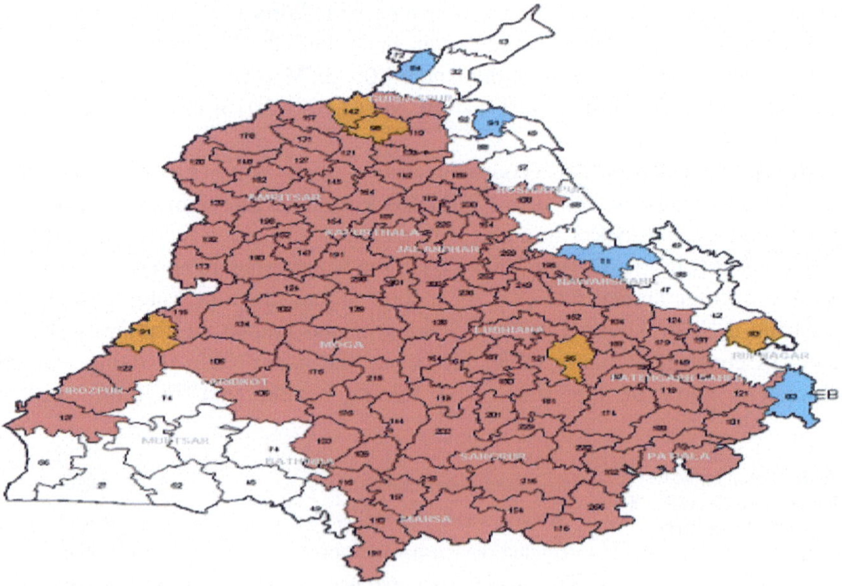

Fig. 3 Categorization of blocks in Punjab based on ground water development. *Source* Central Ground Water Board (2009)

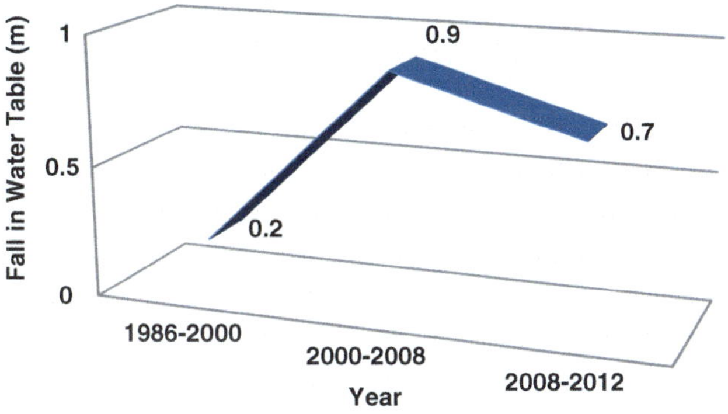

Fig. 4 Decline in water table (in meters) in Punjab. *Source* Jain (2013)

Table 2 Deficit monsoon rainfall year Punjab (2005–2014)

Monsoon year	Actual (mm)	Normal (mm)	Deficit %
2005	445.1	501.8	−11.3
2006	436.5	501.8	−13.0
2007	340.4	501.8	−32.2
2008	603.7	501.8	+20.3
2009	323.6	501.8	−35.5
2010	458.2	496.4	−7.5
2011	459.2	496.4	−7.5
2012	266.0	496.3	−46.4
2013	477.9	491.5	−2.8
2014	243.9	491.5	−50.4

Source Indian Metrological Department, Ministry of Earth Sciences, Session Report (2014)

groundwater extraction is limited for irrigation purposes due to brackish and saline quality of water.

Rainfall

The status of rainfall is less than normal (700 mm) since 1998. The state faced a severe drought due to low rainfall in this region. After 2005, the state is facing continuous low rainfall problems (Indian Metrological Department 2014). Except 2008, the rainfall was always below average in the state since 2004 (Table 2).

Table 3 Canal networks of Punjab (*Source* Department of Irrigation, Punjab)

S. No.	Name of canal system	Length of main canal (in km)
1	Sirhind Canal	59.44
2	Bist Doab Canal	43.00
3	Upper Bari Doab Canal	42.35
4	Sirhind Feeder	136.53
5	Eastern Canal	8.02
6	Bhakra Main line	161.36
7	Shahnehar Canal	24.23

Rivers and Canal

Beas, Chenab, Jhelum, Ravi, and Sutlej are the main rivers of Punjab state. All rivers are tributaries of river Indus. These five rivers are divided between India and Pakistan. However out of five, three rivers Sutluj, Beas, and Ravi flow through the Indian state of Punjab. Area of Punjab between rivers of Beas and Sutluj is called Doaba, which includes Jalandhar, Hoshiarpur, and Nawanshahr. The area of Majha lies between Beas and Chenab and also at both the sides of Ravi. The Majha part is called heart of Punjab and includes Amritsar, Gurdaspur, Faridkot, and Ferozepur cities. Malwa region is located in southern Punjab at range of Sutluj River and the major cities include Ludhiana, Patiala, Sangrur, Malerkotla, Shahabad, and Abohar. Although, Punjab has one of the largest canal systems of the country, but this state is not getting adequate amount of river water due to political reasons, i.e., Indus treaty, damming and diversion of river water, water conflict with Haryana, Rajasthan, and central government. Almost 30.88 lakh hectare of the cultivated area comes under canal networks. The estimation of river water is approximately 14.22 MAF, distributed into 7th main canal system of this state (Department of Irrigation, Punjab) (Table 3).

Further, an inadequate amount of rainfall is the major reason for the deficiency of required amount of water in the rivers as well as canals. The insufficient amount of canal water for irrigation purpose leads to overexploitation of groundwater.

Water Quality and Health Issues

Apart from depletion of water, quality of groundwater is also not suitable for drinking purposes. Rapid increases in population, urbanization, industrialization, and extensive agricultural practices have deteriorated water quality of this region. The intense agricultural activities of Punjab help to achieve national food security. But, large quantity of fertilizers and pesticides being used for better and enhanced crop yield, are also contaminating both groundwater as well as surface water bodies (Aulakh et al. 2009). Pesticides adversely affect human health and linked to certain

Table 4 Status of physicochemical parameters of districts of Punjab (*Source* ENVISa 2015a)

Contaminants	Districts affected (in part)
Salinity (EC > 3000 μS/cm at 25 °C)	Firozepur, Faridkot, Bhathinda, Mansa, Muktsar, Sangrur
Fluoride (>1.5 mg/l)	Amritsar, Bhatinda, Faridkot, Patiala, Fatehgarh Sahib, Firozepur, Sangrur, Gurdaspur, Mansa, Moga, Muktsar,
Chloride (>1000 mg/l)	Firozepur, Muktsar
Iron (>1.0 mg/l)	Bhathinda, Faridkot, Fatehgarh Sahib, Firozepur, Gurdaspur, Mansa, Hoshiarpur, Rupnagar, Sangrur
Nitrate (>45 mg/l)	Bhathinda, Faridkot, Fatehgarh Sahib, Firozepur, Gurdaspur, Patiala, Hoshiarpur, Jalandhar, Kapurthala, Ludhiana, Mansa, Moga, Muktsar, NawanShaher, Rupnagar, Sangrur

Table 5 Health effect of polluted water in Punjab (*Source* Bajwa et al. 2015; CGWB 2012 & 2016; Hundal et al. 2007, 2009; Hundal and Khurana 2013; Kochhar et al. 2007; Kumar 2005; Sharma 2012; Singh et al. 2003; Singh et al. 2013; Swarup et al. 1992)

Metals	Health effects	Reported areas of Punjab
Mercury	Abdominal pain, headache, diarrhea, haemolysis, chest problems	Presence in Budhanala water sample
Lead	Anemia, vomiting, loss of appetite, convulsions, damage of brain, liver and kidney	Highly reported in Bathinda and Ropar area
Arsenic	Lung cancer, disturbed peripheral circulation, metal problems, liver cirrhosis, hyper kurtosis, gastrointestinal tract ulcers, kidney damage	Bathinda and Mansa
Cadmium	Growth retardation, diarrhea, bone deformality, kidney damage, testicular atrophy, anemia, injury of central nervous system and liver, hypertension	In Canal water of Aboher and Bathinda area
Selenium	Damage of liver, kidney and spleen, fever, nervousness, vomiting, blood pressure, blindness, and even death	Ferojpur, Bathinda, Mukatsar District
Chromium (hexavalent)	Diseases in central nervous system, cancer, diarrhea, nephritis gastrointestinal ulceration	Scarce amount detection in some places of Malwa region
Uranium	Cancer, Infertility, diseases in central nervous system, mental abnormalities	Mostly in Villages of Bathinda, Firojpur, Mansa, and Sangrur District

types of abnormalities (Mittal et al. 2014). The enhanced level of basic physicochemical characteristics is detected in several districts of Punjab (Table 4). Contaminated water causes serious health hazards and people of this region are affected by various water-borne diseases (Table 5).

As per reports of Central Groundwater Water Board, water quality of rivers is being degraded at several locations. Ghaggar River flows during monsoon rains and the water quality of this river is generally observed worst at all locations. After that, deterioration in water quality of Sutluj and Beas rivers by discharge of industrial effluents and municipal waste has also been reported. Water-borne diseases have been reported in Moga, Jalandhar, and Barnala locations due to the presence of microorganisms in water (CGWB, Ministry of Water Resources Report 2014).

The groundwater pollution has crossed all the limits established by various national and international agencies. Several reports have already described the presence of high content of nitrate, fluoride, total dissolve solids, electric conductivity, chloride, and sulfate in the groundwater of this region. The presence of toxic heavy metals like arsenic (As), uranium (U), lead (Pb), cadmium (Cd) and selenium (Se) has also been reported in groundwater (Bajwa et al. 2015; CGWB 2007, 2011; Hundal et al. 2007, 2009; Hundal and Khurana 2013; Sharma 2012; Singh et al. 2013).

Heavy metals like As and Pb are classified as the carcinogenic chemical by WHO. Further, in case of U, reports suggested that its radiological risks may be linked with cancer. Due to the extensive deterioration of groundwater, the Malwa region of Punjab is facing serious health issues. The presence of U and As like carcinogenic chemicals of this region indicates their probable link with health issues (Bajwa et al. 2015; Blaurock-Busch et al. 2014; Sharma et al. 2013; Singh et al. 2013). The presence of U concentration in the hair samples of breast cancer patients was reported in this region (Blaurock-Busch et al. 2014). Not only anthropogenic activities, the natural geomorphologic structure of this region is also responsible for deterioration of water quality (Kochhar et al. 2007; Bajwa et al. 2015).

The water quality of canal system is considered better and more suitable for human consumption than groundwater of this region. So, Punjab government has decided to give more emphasis on use of canal water as portable water. Moga and Barnala districts have been chosen under 300-crore pilot project, to set up conventional water treatment plant for supplying potable water to residents of the villages affected by toxic heavy metals problems like arsenic and uranium.

Government Initiatives

The Central Ground Water Board, Punjab Irrigation Department and Agriculture Department have been monitoring the rapidly declining water table as well as water quality for the past many years. Further, various other central and state departments as well as R&D institutions are actively engaged in water quality assessment and monitoring. In 2012, the Central Ground Water Authority has notified 45 blocks in this state for restricting and banning the construction of new structures for extraction of groundwater for any use other than drinking. The government of Punjab took several initiatives regarding this issue and passed the Punjab Preservation of

Sub-Soil Water Act (2009) for saving water resources. According to this act, nursery of paddy should not be sown by the farmers before the 10th day of May and transplantation should not be done before the 16th day of June. The government of Punjab has taken initiatives to provide subsidy to the individual farmer to lay down underground pipelines, drip, and sprinkler systems. Additionally, government is promoting and appreciating preventive measures like watershed management and rainwater harvesting. Department of soil and water conservation of Punjab has successfully established Sahoran Kandi watershed (village Sahoran) and Fatehpur Kangar sub-watershed (Mohilpur block) in Hoshiarpur district under National Watershed Development Project for rainfed areas of integrated development. In other parts of Punjab like Kandi region, watershed development process is progress. The drainage project has been successfully established in village Mehatpur Oladini in Nawanshahar district and water table is rising after implementation. Four rain-water harvesting dams are built and the resultant of that 180 ha area is benefited (http://dswcpunjab.gov.in/contents/successstories.htm). In most of the drought-affected parts of Patiala, Sangrur, and Ropar, groundwater recharging projects are started by central groundwater board. Especially in Patiala district, a massive project for groundwater recharge is in progress.

Future Prospects

The preventive measures taken by the concerned authorities to meet the challanges regarding water related issues are insufficient. Now, time demands to take effective preventive measures massively to regulate the water usage and to introduce the water-saving measures like sprinkling system, rainwater harvesting, and underground piping system for irrigation on a large scale. The canal system requires to be revamped. The regulation or check on the farm and industrial sector on pumping out water should be imposed because almost more than 3 crore gallons of water is pumped out by industries in Ludhiana alone. The state government should constitute Water Regulatory Authority and also promote less water-consuming crops such as maize and sunflower. Role of public investment in water sector should be examined. The penalty or punisment should be strictly imposed on the person or industries, involved in contamination and wastage of water. The political reasons should be removed and the implementation of government policies should be done effectively with zero tolerance.

Conclusions

Rivers, canals, rainfall, and groundwater are main sources of irrigation in agricultural state of Punjab. The insufficient amount of rainfall is responsible for inadequate amount of water in rivers and canals. As a result, groundwater is being overexploited. This overexploitation of groundwater resources is a serious matter of concern. Almost all the drinking water parameter values exceed the acceptable limits in case of groundwater. The higher concentration of U, As and Pb like carcinogenic chemicals in groundwater may pose health risk to the local population of this region. Though efforts are being made by the concerned authorities to combat these issues and challenges, but there is an urgent need to take effective measures massively to regulate usage of water. Water-saving measures such as drip irrigation, sprinkling system, rainwater harvesting and underground piping system should be preferred for irrigation on a large scale. The canal system requires to be revamped. The state government should constitute Water Regulatory Authority and also promote less water-consuming crops such as maize and sunflower. Role of public investment in water should be examined. Hence, there must be strong regulations or check on pumping out groundwater by farming at farm and industrial sector. There must be some penalties to the persons or industries involved, in contaminating and wastage of water. The government policies should be implemented effectively with zero tolerance.

References

Aulakh MS, Khurana MPS, Singh D (2009) Water pollution related to agricultural, industrial and urban activities, and its effects on the food chain: case studies from Punjab. J New Seeds 10 (2):112–137

Bajwa BS, Kumar S, Singh S, Sahoo SK, Tripathi RM (2015) Uranium and other heavy toxic elements distribution in the drinking water samples of SW-Punjab, India. J Rad Res Appl Sci

Blaurock-Busch E, Busch YM, Friedle A, Buerner H, Parkash C, Kaur A (2014) Comparing the metal concentration in the hair of cancer patients and healthy people living in the Malwa region of Punjab, India. Clin Med Insights Oncol 8(1)

CGWB (2009) Central Ground Water Board Ministry of Water Resources. Government of India, New Delhi

CGWB (2012) Ground Water Year Book, Ministry of Water Resources, Government of India, New Delhi

CGWB Central Ground Water Board Report (2007) Ministry of Water Resources New Delhi. http://cgwb.gov.in/District_Profile/Punjab/Bathinda.pdf. Accessed 12 Aug 2014

CGWB Central Ground Water Board Report (2011) Ministry of Water Resources New Delhi. http://www.cgwb.gov.in/documents/Dynamic%20GW%20Resources%20-2009.pdf. Accessed 14 May 2014

Central Ground Water Board (CGWB) (2014) Water Quality Issues & Challenges in Punjab. Ministry of Water Resources, Government of India

CGWB Central Ground Water Board Report (2016) Ministry of Water Resources North Western Region Chandigarh. http://cgwbchd.nic.in/qulpb.htm. Accessed 12 Jan 2016

Department of Irrigation, Government of Punjab, Canal Administration. http://www.pbirrigation.gov.in/OldVersion/canal_administration.html

Department of Soil and Water Conservation, Punjab. http://dswcpunjab.gov.in/contents/success_stories.htm

Department of Soil and Water Conservation, Punjab "Success Stories". http://dswcpunjab.gov.in/contents/success_stories.htm

ENVIS Newsletter (2014–2015) Water Resources of Punjab Vol 12. http://www.rcenetwork.org/mirror/sites/default/files/Envis%20Newsletter%20Vol%2012%20No.3.pdf
ENVIS (2015a) Issues and challenges of water sector. http://punenvis.nic.in/index2.aspx?slid=5814&mid=1&langid=1&sublinkid=995
ENVIS Punjab (2015b) Ground water resources of Punjab. http://punenvis.nic.in/index2.aspx?slid=5815&mid=1&langid=1&sublinkid=996
Hundal HS, Khurana MP (2013) An appraisal of groundwater quality for drinking and irrigation purposes in southern part of Bathinda district of Punjab, northwest India. Environ Earth Sci 70 (4):1841–1851
Hundal HS, Kumar R, Singh K, Singh D (2007) Occurrence and geochemistry of arsenic in groundwater of Punjab, Northwest India. Commun Soil Sci Plant Anal 38(17–18):2257–2277
Hundal HS, Singh K, Singh D (2009) Arsenic content in ground and canal waters of Punjab, North-West India. Environ Monit Assess 154(1–4):393–400
Indian Meteorological Department (Ministry of Earth Sciences) Southwest Monsoon (2014) End of season report Punjab. http://amssdelhi.gov.in/Nigam/MONSOON_2014_PUNJAB.pdf
Jain AK (2013) Role of water management for sustainability of ground water resources in Punjab. Dept. of Soil and water engineering, Punjab Agricultural University, Ludhiana, Punjab. In: Workshop on roadmap for sustainable development of groundwater resources in states of Punjab and Haryana. Organized by Central Ground Water Board, Chandigarh
Kochhar N, Gill GS, Tuli N, Dadwal V, Balaram V (2007) Chemical quality of groundwater in relation to incidence of cancer in parts of SW Punjab, India. Asian J Water Environ Pollut 4:107–112
Kumar R (2005) An epidemiological study of cancer cases reported from villages of Talwandi Sabo block, District Bhatinda, Punjab. PGIMER, Chandigarh & PPCB, Patiala
Mittal S, Kaur G, Vishwakarma GS (2014) Effects of environmental pesticides on the health of rural communities in the Malwa Region of Punjab, India: a review. Hum Ecol Risk Assess Int J 20(2):366–387
Sharma R (2012) A report on groundwater quality studies in Malwa region of Punjab, MUKTSAR. Int J Eng Res Appl 1(4):70–77
Singh S, Rani A, Mahajan RK, Walia TPS (2003) Analysis of uranium and its correlation with some physico-chemical properties of drinking water samples from Amritsar, Punjab. J Environ Monit 5(6):917–921
Sharma C, Mahajan A, Garg UK (2013) Assessment of arsenic in drinking water samples in south-western districts of Punjab—India. Desalin Water Treat 51(28–30):5701–5709
Singh L, Kumar R, Kumar S, Bajwa BS, Singh S (2013) Health risk assessments due to uranium contamination of drinking water in Bathinda region, Punjab state, India. Radioprotection 48 (02):191–202
SOE Punjab (State of Environment) (2007) Punjab State Council for Science & Technology, Chandigarh. http://envfor.nic.in/sites/default/files/SoE%20report%20of%20Punjab_0.pdf
Swarup R, Mishra SN, Jauhari VP (eds) (1992) Environmental water pollution and its control, vol 15. Mittal Publications
The Tribune (2015) Article Punjab facing a veritable water crisis by Sarbjit Dhaliwal

Issues and Challenges of River Health Assessment in India

M. L. Kansal

Abstract Just like human health, a river's health is desired to be measured on a continuum basis. As there is no single measure for human health, there is no single measure for river health as well. Rather, a healthy river is composed of many facets —physical, chemical, biological, and even human. Any river does not have to be a pristine river, untouched by any human development or activities, but a healthy river should be resilient and able to recover from natural and man-made disturbances. A river in the wilderness may rank the highest, but a river flowing through a major metropolitan region is not inherently unhealthy. The traits of an ecologically healthy river will have certain components that will fall within a range that allows the river to maintain its ecosystem functions. India is the second largest country in the world with a population of about 1.28 billion. 33 cities have a population more than a million and about 65 cities on the banks of some major rivers. Central Pollution Control Board (CPCB) has categorized the rivers on the basis of designated best use. Different authors/researchers have assessed the river water quality on the basis of few physical/chemical or biological parameters. However, there are no comprehensive guidelines issued at central or Indian standards level which depict the true river health. Therefore, it is felt that there is a need for comprehensive guidelines at the country level. Keeping this in mind, this study highlights the characteristics of a healthy river. Further, complexities of the river health parameters and problems in the aggregation of these quality parameters are highlighted. It is recommended that the river health of all the major rivers should be assessed on the basis of suggested guidelines. Monitoring and modeling of the river health should be carried out and plans should be made according to the deterioration level of river health. Further, river health improvement program should be prepared keeping in mind the sustainability of such initiative.

Keywords River health · Water quality parameters · Water quality index

M. L. Kansal (✉)
Department of Water Resources Development & Management, Indian Institute of Technology Roorkee, Roorkee 247667, Uttarakhand, India
e-mail: mlkfwt@iitr.ac.in

V.P. Singh et al. (eds.), *Water Resources Management*, Water Science and Technology Library 78, https://doi.org/10.1007/978-981-10-5711-3_8

Introduction

Next to air, water is the most essential requirement for survival of life. The world over, most of the civilizations have flourished along the river banks. In the beginning, these rivers were sufficient to take care of the human requirements on its bank. With the passage of time with increased concentration of population on the banks of various rivers, these have crumbled to maintain its health and freshness. With the increased anthropogenic activities and human requirements, most of the rivers passing through various cities are contaminated. This has resulted in large-scale waterborne diseases. As per one estimate, about 80% of all diseases, and over one-third of deaths in developing countries are water-associated. Further, on an average, as much as one-tenth of each person's productive time is spent in paying attention to water-related diseases.

Massive investments have been made to provide safe and adequate drinking water to both urban and rural populations of the country. The strategies revolve around the three distincts but inter-related issues: coverage of large population, sustainability of safe drinking water systems, and water quality monitoring and surveillance systems. A distinguished success has been achieved in covering the majority of her populations with potable drinking water. But, lot more is desired on sustainability and adequate supply, and on water quality issues so as to ensure safe and sufficient supply of drinking water to masses at their doorstep.

Nearly, half of the world population is living in urban areas. India has a population of about 1252 million, out of which 375 million lives in urban areas. In metropolitan cities, there are more than 225 million people. 33 cities are more than 1 million, and about 65 cities are located on the banks of one or the other river as shown in Fig. 1. Almost 50% of the population in metros live in slums which face problems related to water, sanitation and hygiene (WASH). The growth process and expansion of economic activities has led to increasing demands for diverse uses of water.

According to Brundtland Commission Report (1987), sustainable development is that development which meets the needs of the present without compromising the ability of future generations to meet their own needs. In order to have sustainable development, it should be based on equitable, bearable, and viable considerations. Thus, sustainable development is related to the three major sectors—Economic, Environmental, and the Social considerations.

Traditionally, water supply systems are designed on the basis of standard design criterion or the rules of thumb that attempt to provide hundred percent reliable water service. In real terms, the reliability of a water supply scheme continues to decrease with aging of the network and is dependent on a large number of factors. Generally, two types of reliability, viz., Mechanical and hydraulic, are studied and advocated by various researchers for urban water supply schemes. Recently, with the focus on water quality at consumer's tap level, the third type of reliability, viz., water quality, reliability has become an area of active research. This depends on the water quality of the raw water source, which in most of the cities are the rivers.

Indian River Systems and its approximate lengths are shown in Fig. 2.

S No	City	River	S No	City	River
1	Agra	Yamuna	34	Shimoga	Tunga River
2	Ahmedabad	Sabarmati	35	Bhadravathi	Bhadra
3	Allahabad	Ganges	36	Hospet	Tungabhadra
4	Ayodhya	Saryu	37	Karwar	Kali
5	Badrinath	Ganga	38	Bagalkot	Ghataprabha
6	Kolkata	Hooghly	39	Honnavar	Sharavathi
7	Cuttack	Mahanadi	40	Gwalior	Chambal
8	**New Delhi**	**Yamuna**	41	Gorakhpur	Rapti
9	Dibrugarh	Brahmaputra	42	**Lucknow**	**Gomti**
10	Ferozpur	Sutlej	43	Kanpur Cantt	Ganges
11	Guwahati	Brahmaputra	44	Shuklaganj	Ganges
12	Haridwar	Ganges	45	Chakeri	Ganges
13	Hyderabad	Musi	46	Malegaon	Girna River
14	Jabalpur	Narmada	47	Sambalpur	Mahanadi
15	Kanpur	Ganges	48	Rourkela	Brahmani
16	Kota	Chambal	49	Pune	Mula, Mutha
17	Jaunpur	Gomti	50	Daman	Daman Ganga
18	Patna	Ganges	51	Madurai	Vaigai
19	Rajahmundry	Godavari	52	Chennai	Cooum, Adyar
20	Srinagar	Jhelum	53	Coimbatore	Noyyal
21	Surat	Tapti	54	Erode	Kaveri
22	Tiruchirapalli	Cauvery	55	Tirunelveli	Thamirabarani
23	Varanasi	Ganges	56	Bharuch	Narmada
24	Vijayawada	Krishna	57	Karjat	Ulhas
25	Vadodara	Vishwamitri	58	Nashik	Godavari
26	Mathura	Yamuna	59	Mahad	Savitri
27	Auraiya	Yamuna	60	Nanded	Godavari
28	Etawah	Yamuna	61	Nellore	Pennar
29	Bangalore	Vrishabhavathi	62	Sangli	Krishna
30	Farrukhabad	Ganges	63	Karad	Krishna,Koyna
31	Fatehgarh	Ganges	64	Hajipur	Ganges
32	Kannauj	Ganges	65	Ujjain	Shipra
33	Mangalore	Netravati, Gurupura	About 10 % of Indian population in these cities		

Fig. 1 List of cities located on the banks of a river and major river basins in India

Indus (India -2880 (total 3200 km)	Indus	Indus	Indus	Indus				Penganga (676 km)	Penganga (676 km)	Godavari
Chenab (960 km)	Chenab							Godavari (1465 km)	Godavari	
Jhelum (725 km)		Chenab				NG		Manjira (724 km)		
Kishenganga (Neelum) (245 km)	Jhelum		Chenab		Indus (Pakistan)					
Ravi (720 km)	Ravi	Ravi		Panchnad (Pakistan)				Bhima (861 km)		
Satluj (1500 km)								Krishna (1400 km)	Krishna (1400 km)	Krishna (1400 km)
Beas (470 km)	Satluj	Satluj	Satluj			Total length of all major rivers > 35,000 km		Tungbhadra (531 km)	Tungbhadra (531 km)	
Yamuna (1376 km)									Hugli (260 km)	Hugli
Chambal (960 km)	Yamuna									
Betwa (480 km)	Betwa	Yamuna	Ganga	Ganga						
Ganga (2525 km)	Ganga	Ganga			Ganga	Ganga				
Gomati (900 km)	Gomati	Gomati	Gomati				Ganga	Ganga		
Ghaghra (1080 km)	Ghaghra	Ghaghra	Ghaghra	Ghaghra					Ganga	Padma (120 km)
Son (784 km)	Son	Son	Son	Son	Son					
Gandak (630 km)	Gandak	Gandak	Gandak	Gandak	Gandak	Gandak				
Kosi (730 km)	Kosi	Kosi	Kosi	Kosi	Kosi	Kosi	Kosi			
Brahmputra (2900 km)	Brahmputra	Brahmputra	Brahmputra	Brahmputra	Brahmputra	Brahmputra	Brahmputra	Brahmputra	Brahmputra	
Luni (450 km)	Sabarmati (317 km)	Mahi (583 km)	Narmada (1057 km)	Tapti (724 km)	Mahanadi (858 km)	Pennar (597 km)	Cauveri (805 km)	Vaigai (258 km)	Brahmani (480 km)	Saryu (350 km)

Fig. 2 Indian river systems. *Source* https://bh.wikipedia.org

Fig. 3 Healthy and unhealthy rivers in India

It can be observed that the Indian rivers fall under four categories—the Himalayan, rivers traversing to the Deccan Plateau, the coastal, and those draining inland as shown in Fig. 1. Himalayan rivers are perennial whose major source of water is glaciers and monsoon rains. The rivers of the peninsular plateau are mainly fed by rain and hence are of intermittent nature. The Godavari basin in the peninsula is the largest in the country. Coastal rivers flow from the peaks of the Western Ghats into the Arabian Sea and are in torrents during the rains but their flow slows down after the monsoon. Streams like the Sambhar in western Rajasthan are mainly seasonal in character, draining into the inland basins and salt lakes.

Most of these rivers are the source of water for many cities and towns. Health of these rivers varies in space and time as shown in Fig. 3.

Keeping the importance of water in mind, Government of India has launched a massive river cleaning program in mission mode. A massive National Mission on Clean Ganga has been launched for rejuvenation/restoring the wholeness of river by maintaining continuous and unpolluted flow for ecological and geological integrities. Also in the past, massive investment has been made in terms of Ganga Action Plan and Yamuna Action Plan. Further, also on social front, various sections of the society have carried out various awareness programs to highlight the importance of rivers in our lives and the need for maintaining the healthy conditions of the rivers. However, on the technical front, there is a lot more desired to study the various issues related to river health and its assessment. This study highlights the river health parameters and defines it in terms of water quality index (WQI). Further, the various procedures suggested by researchers for estimating WQI have been critically examined. Finally, it is suggested that there is a need for adopting a comprehensive procedure for assessing the river health in India. A program can be initiated to know quantify the river health in various parts of the country. Depending upon the river health condition, programs can be initiated to improve its health. This helps in taking decision with respect to investment plan and to evaluate the worth of using resources for such an activity.

River Health—Issues of Concern

Rivers provide goods and service for the people like water for drinking and eating, means of transportation, source of energy generation, and recreational opportunities. Healthy rivers are vital to protect and enhance the riverine ecosystems. Ecological assets in a country guide the rational river management policy. An organism is healthy when it performs all its functions properly with an ability to recover from normal stresses with minimal outside support.

Generally, 'health' is a term related to human health which reflects the soundness of body or mind, free from disease or ailment. It reflects a flourishing condition, well-being, vitality, or prosperity. Further, criteria of being 'healthy' change with age and gender. For example, criteria of a healthy child, healthy young person, or healthy old person as well as healthy mother or healthy father are different. Similarly, a country is healthy when a flourishing economy provides for the well-being of its citizens. Further, an environment is healthy when the supply of goods and services required by both human and nonhuman residents is sustained. Thus, 'health' is shorthand for 'good condition'. It is a property of organism or the system as a whole which reflects its 'good condition'. This property of goodness or soundness is subjective in nature and varies in space and time as well as with the context. Despite the simplicity and the breadth of this concept, different researchers have tried to define the '**river health**' in an ecological context. Since, river health perception varies from person to person, place to place and from time to time, it is very difficult to arrive at a consensus definition and characteristics of '**healthy river**'. In India also, there are no holistic or scientific guidelines to define the "**River Health or River Sickness due to Pollution**". Thus, there is a need to have an acceptable "River Policy" in a country like India which supports the second largest population in the world.

There are no universal guidelines for defining the river health. It is complex and analogous to *Human Health*. For example, the characteristics of a healthy person are clear skin, steady heartbeat that is neither too fast nor too slow, symmetrical features, blood pressure in range, clear eyes with even pupils that track together, appropriate flexibility in the joints, expected reflex responses, weight in the expected ranges—neither too high nor too low, a reasonable amount of endurance and strength, and so on, i.e., there is nothing to complain about physical and mental conditions. But it is difficult to quantify these characteristics for an individual.

Similarly, a healthy river is the one that maintains its physical, chemical, and biological structure and function; recovers after short-term natural disturbances such as floods and droughts; supports local biota; and maintains key processes, such as sediment transport, nutrient cycling, assimilation of waste products, and energy exchange. A healthy river should have a **natural flow** that varies in magnitude, frequency, duration, timing, and rate of change. It should **allow transport of sediment and nutrients**. Rocks, gravel, sand, silt, and organic debris are important components of a healthy river and contribute in creating floodplains, sandbars, riparian areas, and nourish a river's bed and channels. A healthy river in equilibrium does not allow too much erosion or excessive scouring of the riverbank and riverbed. It should have **good water quality**. A healthy river has temperature levels, dissolved oxygen content, salinity, turbidity, hardness, acidity, and alkalinity (water pH) that are all within a natural range of that river and its species. A healthy river will also have minimal amounts of pollution and toxics, such as pesticides, nitrogen, phosphate, fecal coliform, and heavy metals. A healthy river should have **strong and varied plant communities**. Native plant species provide critical habitat for fish and other riverine animals, regulate water temperatures, prevent excessive erosion of riverbanks, and can remove pollutants from river water. Vegetation as it decomposes is also an important source of nutrients and habitat. It should have **productive and diverse habitat** that can support numerous animal species. The natural movement of sediment throughout a river creates riffles, pools, side channels, and backwater areas providing both spawning and rearing habitat for many species of fish. Aquatic insects are the primary food for many riverine species. **Abundance and diversity of insect species** can be a strong indicator of river health. While the number of fish and wildlife species will vary with each river, a **diverse number of species** is often an indicator of river health. In order to have a healthy river, the **community** in the river basin should protect it through **wise management and community planning**.

In India, Central Pollution Control Board (CPCB) has tried to define the river health in terms of designated uses. They have suggested the use-based classification of rivers as mentioned in Table 1.

Water quality criteria for bathing reaches in rivers is notified by Ministry of Environment and Forests (MoEF) and is shown in Table 2.

However, these are not sufficient characteristics of river health. From the water quality point of view, various researchers and authors have considered various parameters. There are no uniform guidelines about the number of parameters to be considered toward the river health. For example, Alberta River Water Quality Index (ARWQI) (http://aep.alberta.ca/water/reports-data/alberta-river-water-quality-index.aspx) is based on the average of four subindices calculated annually for four variable groups of:

- Metals (up to 22 variables measured quarterly);
- Pesticides (17 variables measured 4 times during open-water season);
- Nutrients (6 variables measured monthly); and
- Bacteria (2 variables measured monthly).

The list of the variables considered in the ARWQI are shown in Table 3.

Table 1 Use-based classification of surface waters in India

Designated best use	Class of water	Criteria
Drinking water source without conventional treatment but after disinfection	A	1. Total coliforms organism MPN/ 100 ml shall be 50 or less 2. pH between 6.5 and 8.5 3. Dissolved oxygen 6 mg/l or more 4. Biochemical oxygen demand 5 days 20 °C 2 mg/l or less
Outdoor bathing (organized)	B	1. Total coliforms organism MPN/ 100 ml shall be 500 or less 2. pH between 6.5 and S.5 3. Dissolved oxygen 5 mg/l or more 4. Biochemical oxygen demand 5 days 20 °C 3 mg/l or less
Drinking water source after conventional treatment and disinfection	C	1. Total coliforms organism MPN/ l00 ml shall be 5000 or less 2. pH between 6 and 9 3. Dissolved oxygen 4 mg/l or more 4. Biochemical oxygen demand 5 days 20 °C 3 mg/l or less
Propagation of wildlife and fisheries	D	1. pH between 6.5 and 8.5 2. Dissolved oxygen 4 mg/l or more 3. Free ammonia (as N) 1.2 mg/1 or less
Irrigation. Industrial cooling. Controlled waste disposal	E	1. pH between 6.0 and 8.5 2. Electrical conductivity at 25 °C micro mhos/cm max. 2250 3. Sodium absorption ratio max. 26 4. Boron max. 2 mg/l

The other ways of classifying the water quality parameters are shown in Table 4a–c.

Depending upon the requirements, various countries have recommended its own list of parameters. Most common lists of parameters considered for a river health and the target, acceptable, and permissible values are as shown in Table 5.

It may be noted that the safe/desirable limits of the physical, chemical, and biological parameters are the necessary but not the sufficient conditions for the good river health. These are important from the 'river sickness' or the river pollution point of view. Besides these, Hydrological (i.e., E-flows—Minimum and Maximum, or seasonal) and Aquatic Flora and Fauna (i.e., algae, riverine vegetation, benthic macroinvertebrates like fishes are also important.

Table 2 Primary water quality criteria for bathing

Criteria	Limit	Rationale
Fecal coliform	500 (desirable) MPN/100 ml 2500 (Max. permissible)	To ensure low sewage contamination, fecal coliform and fecal streptococci are considered as they reflect the bacterial pathogenicity The desirable and permissible limits are suggested to allow for fluctuation in environmental conditions such as seasonal changes, changes in flow conditions, etc.
Fecal streptococci	100 (desirable) MPN/100 ml 500 (max. permissible)	
pH	Between 6.5 and 8.5	The range provides protection of the skin and delicate organs like eyes, nose, and ears, which are directly exposed during outdoor bathing
Dissolved oxygen (DO)	5 mg/l or more	The minimum DO concentration of 5 mg/1 ensures reasonable freedom from oxygen-consuming organic pollution immediately U/s which is necessary for preventing the production of anaerobic gases (obnoxious gases) from sediments
Biochemical oxygen demand (BOD) (3 day 27 °C)	3 mg/1 or less	The BOD of 3 mg/1 or less of the water ensures reasonable freedom from oxygen demanding pollutants and prevent production of obnoxious gases

Table 3 List of water quality parameters considered in ARWQI

Metals and ions			
Aluminum	Copper	Molybdenum	Vanadium
Arsenic	Iron	Nickel	Zinc
Beryllium	Lead	Selenium	Cyanide
Boron	Lithium	Silver	Fluoride
Cadmium	Manganese	Thallium	
Cobalt	Mercury	Uranium	
Pesticides			
2,4-D	Picloram	Bromoxynil	Chlorpyrifos
MCPP	Dicamba	Cyanazine	Imazamethabenz
MCPA	Triallate	Malathion	Diuron
Diazinon	Atrazine	Methoxychlor	Dichlorprop
Lindane			
Nutrients and related variables			
Dissolved oxygen	Total phosphorus	Nitrite–nitrogen (NO_2-N)	
pH	Total nitrogen	Ammonia nitrogen	
Bacteria			
Fecal coliforms	*Escherichia coli*		

Table 4 (a) Categories of water quality parameters, (b) categories of water quality parameters, (c) categories of water quality parameters

Category no.	Category type	List of parameters
(a)		
I	Organoleptic parameters	Turbidity, color, visible oil, tastes and odors
II	Physical	Flow, conductivity, temperature, resistivity, suspended solids, silt density index, sludge density, mixed liquor volatile suspended sludge (MLVSS), sludge volume index (SVI), suspended SVI (SSVI), diluted SVI (DSVI)
III	Inorganic chemical parameters	pH, dissolved oxygen, hardness, alkalinity, redox potential, chlorine, ozone, total dissolved solids, metal species (calcium, sodium, iron, copper, total etc.), total heavy metals, nonmetals (nitrates, phosphates, etc.), and water balance
IV	Organic chemical parameters	Biochemical oxygen demand, chemical oxygen demand, PV, TOD, TOC, TIC, volatile organic compounds, aluminide nitrogen, TKN, OKN, separable oil, total oil, halogenated methane, pesticides, PAH, detergents, and regulated substances
V	Microbiological parameters	Bacteria, *F. Coli*, *E. Coli*, total count, protozoa, algae, fungi, yeasts, molds, and viruses
(b)		
I	Organoleptic and physical	Temperature, total solid, and turbidity
II	Metals and ions	Al, Ba, B, Cu, F, Fe, Mg, Mn, SE, Ag, Zn, Cd, CN, Pb, Hg, Mo, Ni, As, and Cr
III	Nutrients and related variables	DO, nitrates (N-NH_3, N-NO_3), nitrite (N-NO_2), total phosphorus (pH)
IV	Pesticides	See IS 10500-2012
V	Bacteria	*F. Coli* and *E. Coli*, BOD5

Physical	Chemical	Toxicological	Radioactive	Biological
(c)				
Color	BOD	Cadmium	Alpha emitters	Bacteria
Turbidity	Inorganic elements	Cyanide	Beta emitters	Algae
Total dissolved solids	Mineral oil	Lead		Virus
Temperature	Anionic detergents	Mercury		Protozoa
pH	Nitrates	Arsenic		
Odor	Carbonates	Molybdenum		
Taste	Sulfates	Chromium		
	Total hardness	Nickel		
	Phenolic compounds	Trihalomethanes		
		Pesticides		
		PCB		
		PAH		

Table 5 General list of water quality parameters

S. No.	Description	Optimal value (target value) >90	Acceptable value 80–90	Permissible value 60–80
1	Temperature	20 °C	15–25	Up to 10 and 30
2	pH	7	6.5–8	6.5 and 8.5
3	Dissolved oxygen (DO) (mg/L)	7	6.00	4
4	Turbidity (NTU) (mg/L)	10 (20)	60 (50)	100 (80)
5	Conductivity (μS/cm)	50	500	1500
6	Total organic carbon (TOC) (mg/L)	0.1	10	15
7	Bacteria (MPN/100 mL)	Nil	25	50
8	Viruses (MPN/100 mL)	Nil	25	50
9	Chemical oxygen demand (COD)	10	20	30
10	Biochemical oxygen demand (BOD) (mg/L)	1	3	4
11	Metals and nonmetals (Cr, Cd, Ni, As, Hg, Na, Br...)	Nil	nil	nil
12	Total phosphates (mg/L)	0.03	0.05	0.1
13	Nitrogen compounds (mg/L) (NO_3)	3	6	10
14	Organic compounds (μg/L)	Nil	0.1	0.2
15	Total suspended solids (mg/L)	30	60	100
16	Salts (total dissolved salts)	500	1000	1500
17	Hardness (mg/L)	<60	60–100	100–300
18	SAR	<4	4–7	7–10
19	Boron	<0.4	0.4–0.7	0.7–1.0

Water Quality Index—Issues of Concern

While selecting the indicators of WQI, one should try to answer the following:

1. What are the different categories of indicators/sub-indicators which are important for defining the water quality of the river under study?
2. How to aggregate the sub-indicators and the indicators to determine the water quality subindex (WQSI) and water quality index (WQI).

Category of Indicators and Sub-indicators for River Health

Most of the indicators mentioned in the previous section can be categorized under Physical, chemical, and biological sections. Another way of categorizing the indicators and thereafter the sub-indicators are as defined in Chinese technical regulation for assessment of river health (CRAES 2012) as mentioned in Table 6.

The physical category indicator deals with hydrological and physical form stressors. Such stressors are universally recognized as an important controlling variables for the health of biota in streams. The various hydrological sub-indicators are high flow volume (HFV), low flow volume (LFV), highest monthly flow (HMF), lowest monthly flow (LMF), persistently higher flow (PHF), persistently lower flow (PLF), persistently very low (PVL) and seasonality flow shift (SFS). Each flow health indicator characterizes the degree of deviation in a specific aspect of the flow regime that is conceptually linked to ecological health.

Similarly, physical form stressor index (PFS) is based on six stressor sub-indicators: Free-flow interruption sub-indicator (FFI), Sediment transport interruption sub-indicator (STI), Longitudinal-continuity barrier sub-indicator (LoCB), Lateral-continuity barrier sub-indicator (LaCB), Bed disturbance sub-indicator (BD), and Bank stabilization sub-indicator (BS). Four of these are measured primarily in the field, and two (STI & LoCB) are desktop-measured using maps, aerial photography, and other records. The field-measured physical form stressor sub-indicators are sampled only in spring (i.e., not included in the autumn

Table 6 Quality indicators categories and sub-indicators

Indicator categories	Indicator subcategories	Measuring location	Sub-indicators	Indicators
Physical	Hydrology	Hydrology stations	HFV, LFV, HMF, PHF, PLF, PVL, SFS	FH
	Physical form stressor	Field sites, and up and downstream	FFI, STI, LaCB, BD, BS, LoCB	PFS
Water quality	Water quality represented by six parameter groups	Hydrology stations	WQP, WQC, WQM, WQn-M, WQN, WQO	WQ1
		Field sites	WQP, WQC, WQM, WQn-M, WQN, WQO	WQ2
Biological	Fish	Field sites	NA	NA
	Benthic macroinvertebrates	Field sites	EPTr, S2(OCP)w	BM
	Algae	Field sites	IBD, IPS, Chl-aper, Chl-aphy	A
	Riverine vegetation	Field sites	WR, LR, CR	VR

Source CRAES (2012)

field sampling). The desktop-measured sub-indicators are measured once per year, at a convenient time, using information collected over the sampling year (CRAES 2012).

Chemical water quality is a key attribute of aquatic ecosystem health. The characteristics of physical and chemical parameters are a response to both natural processes and the anthropogenic activities. Both are resilient to water quality but the stress of chronically poor water quality (even if for the small duration) affects the ecological health. Various chemical sub-indicators are discussed in previous section. Generally, the variables from five classes, namely, oxygen level, eutrophication, health aspects, physical characteristics, and dissolved substances, are considered for chemical water quality.

Fishes are useful biological sub-indicator of river health as they are sensitive to water quality and habitat deterioration. Benthic macroinvertebrates have generally limited mobility and its diversity ensures the good water quality or the river health as these are also very sensitive to the changes in quality. Algae are also useful indicators of river health as they are abundant in most of natural streams and they respond quickly to the changed conditions. Algae, whether benthic (periphyton) or suspended (phytoplankton), have narrow tolerance to pH, nutrients, and salinity and hence act as an indicator of river health. Riverine vegetation refers to the plants in the river channel and riparian zone including that of riverbanks, the floodplains, and its wetlands. Its diversity and longitudinal continuity speaks of good river health.

Further, different indicators are important in different stages of the river. For example, physical indicators involving hydrological and physical parameters are more important in young stage rather than chemical parameters. Similarly, chemical parameters are more important in mature stage due to increased anthropogenic activities. Further, biological and/or chemical and physical parameters become important in the old stage near the mouth of river. Also, it depends on the usage of river water. For example, CPCB has given guidelines for the designated use of river water. Similarly, MOEF has given guidelines for the use of river water for bathing purpose.

Aggregation of Sub-indicators/Indicators

As discussed previously, a large number of variables are used for assessing the physical, chemical, and biological characteristics of river water quality. The suitability of a river source for the intended purpose depends on the magnitude of these variables. Therefore, it is useful to aggregate the various variables into a single index or subindex by mathematically aggregating the same. Various researchers have suggested the different ways of aggregating the same. Tyagi et al. (2013) have reviewed the various methods suggested for WQI estimation. Some of the popular indicators are as follows:

1. National Sanitation Foundation Water Quality Index (NSFWQI) (Brown et al. 1970),
2. Canadian Council of Ministers of the Environment Water Quality Index (CCME WQI) (CCME 2001),
3. Oregon Water Quality Index (OWQI) (Cude 2001), and
4. Bhargava method.

Summary of these indicators is shown in Table 7.

Table 7 Popular water quality indices

S. No.	Method name	Formula	WQI value	Rating of quality of water
1.	NSFWQI (Brown et al. 1970)	$\mathrm{WQI} = \sum_{i=1}^{n} Q_i W_i$	95–100	Excellent
			80–94	Good
			60–79	Fair
			45–59	Marginal
			0–44	Poor
2.	Weighted arithmetic water quality index method (Brown et al. 1972)	$\mathrm{WQI} = \sum Q_i W / \sum W_i$ $Q_i = 100\left[\frac{V_i - V_o}{S_i - V_o}\right]$ $W_i = \frac{K}{S_i}$ $K = 1/\sum S_i$	0–25	Excellent (A grade)
			26–50	Good (B grade)
			51–75	Poor (C grade)
			76–100	Very Poor (D grade)
			Above 100	Unsuitable for drinking purpose (E grade)
3.	CCME WQI (CCME 2001)	$\mathrm{WQI} = 100 - \left(\frac{\sqrt{F_1^2 + F_2^2 + F_3^2}}{1.732}\right)$	95–100	Excellent
			80–94	Good
			65–79	Fair
			45–64	Marginal
			0–44	Poor
4.	OWQI (Cude 2001)	$\mathrm{WQI} = \sqrt{\frac{n}{\sum_{i=1}^{n} \frac{1}{SI_i^2}}}$	90–100	Excellent
			85–89	Good
			80–84	Fair
			60–79	Poor
			0–59	Very poor
5.	Bhargava method	$\mathrm{WQI} = \prod_{i=1}^{n} f_i(P_i)^{1/n}$	100–90	Excellent
			89–65	Good
			64–35	Acceptable
			34–11	Polluted
			<10	Severe polluted

Swamee et al. (2000) have analyzed the problems of ambiguity and eclipsing in determining the overall WQI. They have suggested a mathematical index form that avoids the problem of ambiguity and eclipsing in aggregation.

Grading Scale for River Health

The river health can be represented in different scales. For example, Table 8a, b categorizes the ten and five grades of river water quality and the river health classes. Different countries and researchers have suggested their own scales. Table 8a, b has been prepared after consulting various grading systems proposed in the literature.

Table 8 (a, b) 10- and 5-point scale for grade/class of river health

Grade category	River health score	Class	River health class	River condition(s)
(a) *Five-point scale*				
I	1	91–99	Excellent	Minimal disturbance or pristine/near pristine or in permissible range
II	0.8	71–90	Good	Healthy/mild disturbance to permissible range
III	0.6	51–70	Fair/ Satisfactory	Mild to moderately disturbed
IV	0.3	26–50	Poor	Moderately to severely disturbed/Unhealthy
V	0	0–25	Unacceptable	Critically disturbed/Nothing alive
Grade category	River health score	Class	River health class	River condition(s)
(b) *Ten-point scale*				
I	1	91–99	Outstanding	Pristine/near pristine
II	0.9	81–90	Excellent	Minimal disturbance
III	0.8	71–80	Very good	Healthy
IV	0.7	61–70	Good	Alarming but healthy
V	0.6	56–60	Above average	Alarming
VI	0.5	46–55	Average	Disturbed
VII	0.4	41–45	Below average	Disturbed and alarming
VIII	0.3	26–40	Poor	Unhealthy
IX	0.2	11–25	Very poor	Severely disturbed
X	0.1	1–10	Worse	Nothing alive

Conclusions

This study highlights the complexities involved in defining the river health, though it is easy to understand but difficult to define. Analogy has been drawn with the human health. Further, it has been highlighted that the river health perception changes from place to place and from time to time. Also, there is no universal acceptance about the parameters to be considered in defining the river health. Various categories of parameters that are useful in defining the river health have been suggested and the optimal, acceptable, and permissible ranges are discussed. Further, river health issues and challenges in a country like India are highlighted and it has been emphasized that India needs a comprehensive river health policy. This is particularly needed in view of the ambitious plan of the government of India to improve the river health of rivers like the Ganges. Also, the various methods for aggregation of parameters to define the overall water quality index (WQI) of a river have been mentioned and the limitations are discussed. A more acceptable 10 and 5 point scales are proposed for grading the water quality of a river. Overall, it emphasized that there is a need for adopting holistic approach for defining and quantifying the river health in India as well as there is a need for suggesting a comprehensive river policy for the country.

References

Brown RM, McClelland NI, Deininger RA, Ronald GT (1970) A water quality index—do we dare? Water Sewage Works 117(10):339–343

Brown RM, Mc-Cleiland NJ, Deiniger RA, Connor MFA (1972) Water quality index—crossing the physical barrier. In: Proceeding in international conference on water pollution Research, Jerusalem, vol 6, pp 787–797

Brundtland G, Khalid M, Agnelli S, Al-Athel S, Chidzero B, Fadika L, Okita S (1987) Our common future (\'Brundtland report\')

Canadian Council of Ministers of the Environment (CCME) (2001) Canadian water quality guidelines for the protection of aquatic life: CCME Water Quality Index 1.0. Technical report, Canadian Council of Ministers of the environment Winnipeg, MB, Canada. Available at http://www.ccme.ca/en/resources/canadian_environmental_quality_guidelines/calculators.htmll

Chinese Research Academy of Environmental Sciences (2012) Technical Regulations for assessment of river health

Cude CG (2001) Oregon water quality index a tool for evaluating water quality management effectiveness. J Am Water Resour Assoc 37(1):125–137

Swamee PK, Tyagi A (2000) Describing water quality with aggregate index. J Environ Eng 126 (5):451–455

Tyagi S, Sharma B, Singh P, Dobhal R (2013) Water quality assessment in terms of water quality index. Am J Water Resour 1(3):34–38. http://aep.alberta.ca/water/reports-data/alberta-river-water-quality-index.aspx

Coastal Reservoir—How to Develop Freshwater from the Sea Without Desalination

Shu-Qing Yang

Abstract Next to air, freshwater has been always considered as a key resource and central for economic development and human's basic needs. Currently, the total population is about 7 billion, and by 2050, global population is projected to be 9 billion. Additional 10 more Nile Rivers are needed, and the water demand is increasing steadily and significantly. On the other hand, more people are migrating toward coasts, the existing reservoirs' capacity is continuously being reduced due to sedimentation by 1% annually; this has intensified the water deficit. In this paper, the water deficit is discussed, and a new technology of coastal reservoir and its successful application in Shanghai are presented, which can pump freshwater from the sea without desalination. The preliminary designs of coastal reservoirs in Malaysia, India/Bangladesh, and USA show that the coastal reservoir is a feasible and effective technology in the world for the water crisis for water supply and flood disaster mitigation.

Introduction and Problem Statement

When the author was growing up in 1960s, there were only about 3 billion people on the planet. At that time, water was an abundant resource, almost none of us worries about the sustainability of water supply. As shown in Fig. 1, the world's population is increasing quickly. Now, the total population is more than 7 billion; among them more than a billion people lack adequate, clean drinking water and by 2025, it is estimated that two-thirds of the world's people will live in water-stressed condition. Irrigation currently accounts for two-thirds of water use worldwide and to meet the crop demands projected for 2025, additional 192 cubic miles of water could be required—a volume nearly equivalent to the annual flow of the Nile 10 times over. Therefore, water shortage is a major socioeconomic problem society

S.-Q. Yang (✉)
School of Civil, Mining and Environmental Engineering, Centre for Coastal Reservoir Research, University of Wollongong, Wollongong, NSW 2522, Australia
e-mail: shuqing@uow.edu.au

V. P. Singh et al. (eds.), *Water Resources Management*, Water Science and Technology Library 78, https://doi.org/10.1007/978-981-10-5711-3_9

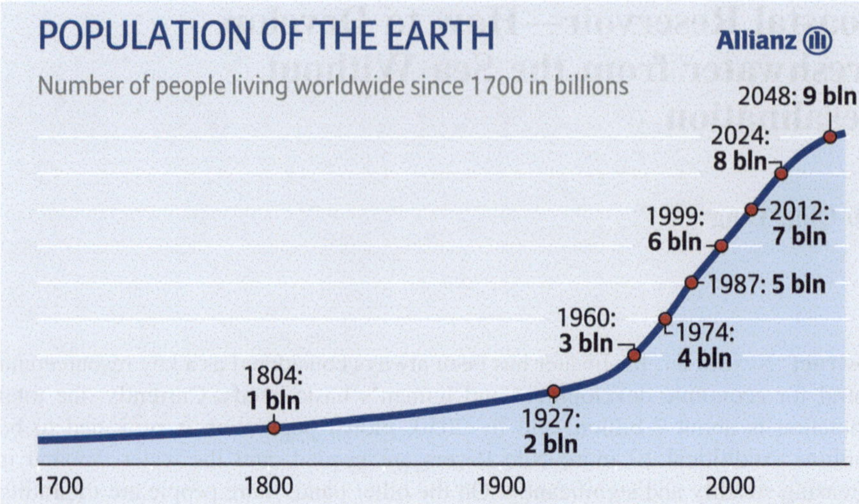

Fig. 1 The growth of world population after industrial revolution. *Source* United Nations World Population prospects, Deutsche Stiftung Weltbevölkerung. For further information please visit: knowledge.allianz.com

facing today and will soon become one of the major constraints for future economic development. Severe water shortages also strike particular regions of water-rich countries, such as the U.S., India/Bangladesh, Malaysia, and China. Thus obtaining an adequate supply of freshwater has consequently been the focus of human ingenuity and a greater challenge.

Before the industrial revolution in 1760s, the world population was almost uniformly distributed over the arable lands, and the coastal areas had almost the same population density as the inland areas. The industrial revolution cuts off the direct connection of people and land just like the umbilical cord. Consequently, urbanization and industrialization appear in the world. The deltaic or coastal areas have fertile lands for agricultural production and fisheries resources, natural waterways for navigation, and the relative flat topography for economic activities. It is natural that more people are attracted to the coastal and deltaic area. Consequently, the coastal areas have higher and higher population density. In the world, almost all megacities appear in the coastal areas as shown in Fig. 2, or the densest population in the world lives the megacities in the coastal area. Table 1 shows the trend of population in urban areas due to urbanization, and it can be seen that about 70% of total population will live in urbans; most of these urbans are located in coastal areas or the estuaries where the water is unstable in quantity and quality as a lower reach of rivers near seas generally suffers very serious disasters related to water, because it must bear the consequences when the upper reach of a river draws too much water, or discharges too much flood/wastewater into the river. It is certain that disputes would be raised between the upper or lower reaches over water. Someone predicts that different from the twentieth century for Petroleum,

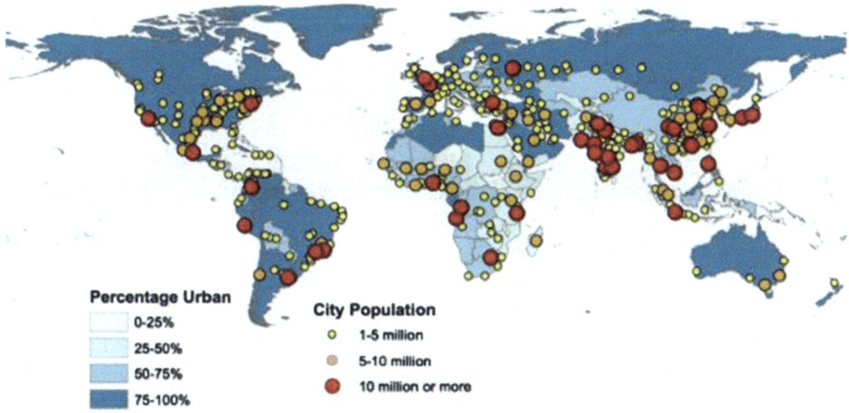

Fig. 2 The megacities in the world—almost all of them appear in the coastal lines

Table 1 Urban population growth from 1950 to 2050

Year	Urban population	Total population	% of urban population
1950	748	2515	29.7
1955	850	2752	30.9
1960	985	3024	32.6
1965	1189	3330	35.7
1970	1325	3636	36.4
1975	1495	4044	37.0
1980	1733	4451	38.9
1985	1937	4791	40.4
1990	2277	5369	42.4
1995	2583	5777	44.7
2000	2820	6117	46.1
2005	3228	6558	49.2
2010	3568	6898	51.7
2015	4010	7374	54.4
2020	4316	7714	55.9
2025	4757	8189	58.1
2030	5063	8461	59.8
2035	5471	8835	61.9
2040	5743	9005	63.8
2045	6117	9379	65.2
2050	6456	9684	66.7

wars might be broken out for water in the twenty-first century. The big challenge for water supply includes the water demand as being increased significantly, and also the dismatch of dams with the population centers intensifies the water crisis.

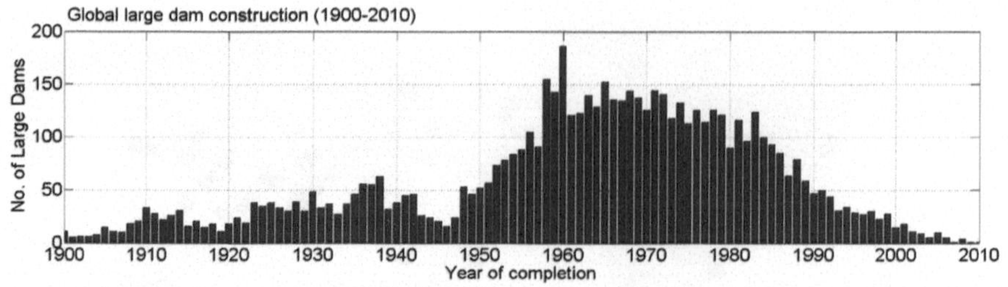

Fig. 3 The global large dam number constructed in the past 100 years

Currently, almost all water supplies for urbans come from inland reservoirs which are constructed in upstream gorge sites. Figure 3 shows the large dams constructed worldwide, which is defined by the dam height (>15 m). There exist about 40,000 large dams worldwide (WCD 2000), and 300 major dams—giant sand bags are at least 150 m in height. It can be seen that after 2010 there are no more new dams constructed because (1) almost all sites with ideal hydro-social and geotechnical combinations have been dammed; (2) many negative impacts on the ecosystem by the inland dams have been found; and (3) the most important fact is the public's opposition that forces the suspension of new dams.

Everything in the universe has its life span. The expected life span of concrete and steel structures for any infrastructures is about 100–200 years. The dam's lifespan also depends on its sedimentation rate. For reservoirs, the worst enemy for its long-term sustainable use is sedimentation (USBR 2006). The total sediment loss in the world is estimated to be 13.5×10^9 ton/annum, about 75% of it is trapped, retained, and stored in the lakes, reservoirs, and river systems (Shen 1999; Batuca and Jordaan 2000). Consequently, the silting process is reducing the storage capacity of the world's reservoirs by more than 1% per year (James and Chanson 1999), often with a rapid filling period. So we are losing our existing reservoir capacity rapidly. Considering the structure's lifespan and sedimentation, it is reasonable to assume that on average the lifespan of a dam is about 150 years. Based on this, the total dam number and reservoirs' storage capacity are shown in Fig. 4, which clearly indicates that from 2000 to 2060, the world has the highest number of large dams and also has the highest reservoir storage capacity. The calculation results show that after 2060, the dam's number and storage capacity starts to decrease, and this decrease will last to 2150, by that time all reservoirs will disappear and the water infrastructures lose their function. Therefore, Fig. 4 posts a biggest challenge to the water resources community: where should we drink when no replacement of the aging dams is available?

Coastal Reservoir Strategy and Shanghai's Successful Application: The traditional ways of water resources development may not be able to fully accommodate the increasing demand, especially in the coastal regions. To meet the water demand in the regions, technologies like desalination of seawater, wastewater recycling, etc. have been proposed and applied worldwide. The large-scale desalination facilities remain expensive and is used only in wealthy areas, it also

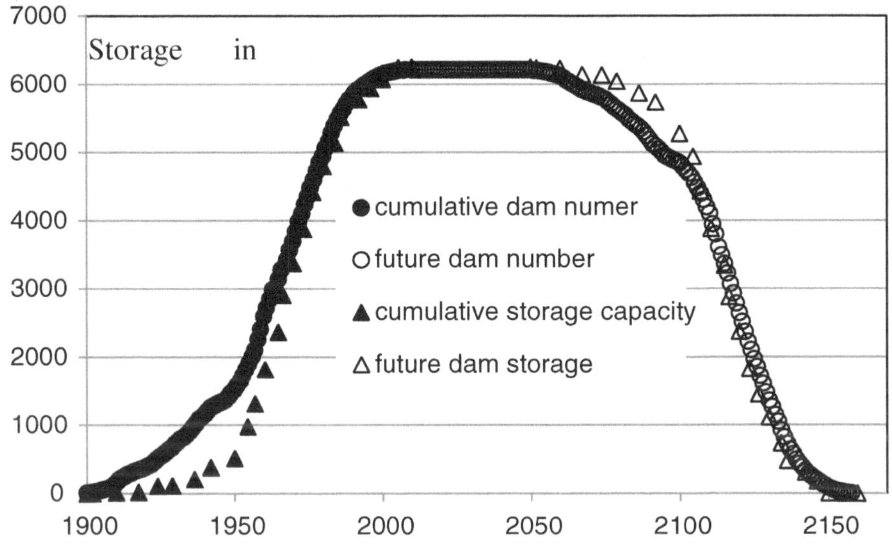

Fig. 4 The cumulative dam number and storage capacity in the world (dam number from Fig. 3 for the period of 1900–2010), and the future prediction based on the assumption of 150 years dam's life span

provides less than 0.2% of global withdrawals and cannot be used economically for large-scale agricultural development. Obviously, the strategy of desalination cannot fully solve the water stress in large scale.

Because the overall freshwater from rivers into the ocean represents globally on an average 42,000 km^3 per year, six times more than current total water supply, it is believable that this portion of freshwater will be innovatively developed in the next several decades. The central challenge for the development is to keep river water from mixing with seawater and wastewater, by noticing that the human society uses one-sixth of available water resource, the remaining is the runoff lost to the sea, and most of them using the floodwater which cannot be developed in a convenient way. Different from the existing inland reservoirs, Yang (2001, 2002, 2003, 2004, 2005a) claimed that the world's water crisis can be effectively solved using the technology of coastal reservoir that is a freshwater reservoir in seawater by developing the runoff to the sea. Based on Fig. 4, he predicts that after 100 years, almost everyone needs to drink the water from a coastal reservoir.

What a coastal reservoir needs is an impermeable barrier between the fresh river water and the salty sea water. Yang (2001) outlined SPP (3 guidelines) for the successful construction of a coastal reservoir. The first guideline of SPP is to separate freshwater and seawater by a barrier, and, to separate clean water from polluted river water by an intake regulation. Next is protection, i.e., to protect the collected fresh water against polluted river water and external pollution. Last is prevention, meaning the successful prevention of saltwater intrusion into the stored freshwater. Different from inland reservoirs that can only collect water from a small portion of a catchment, a coastal reservoir has the potential to collect every single drop of water from a catchment. A coastal reservoir can be classified into various categories, in terms of location, barrage, water quality, etc. Existing freshwater lakes or lagoons on the shore can be regarded as special or natural coastal reservoirs. The main differences between coastal reservoirs and on-land reservoirs are summarized in Table 2, which shows that biggest difference is the water pressure; for existing inland reservoirs, the wall should be very strong to sustain the very high water pressure; this is one of the reasons why the construction cost is very high. The water pressure for a coastal reservoir is relatively small (less than 10 m) between the inside and outside water levels. As the sea is the biggest reservoir in the world, the coastal reservoir can be constructed in any place in the sea, even offshore if it is required. The biggest challenge for a successful coastal reservoir is its pollution control and prevention; this is because the coastal reservoir also has the potential to collect all pollutant generated from a catchment. The SPP strategy is effective to collect good quality water, because the intake gates will be closed when poor quality water or first flush appears in the river mouth; the polluted water then will flow to the sea bypassing the coastal reservoir. The water in the reservoir could be improved in its quality if a wetland pretreatment is allowed.

The so-called coastal reservoir is a new strategy to the global water crisis in systematical point of view. But in practice, there exist many coastal reservoirs as listed in Table 3, where the original designers may not know that their solution for the local thirst can be extended to other places. Its technical feasibility can be realized simply by the ocean or coastal engineers relative to the inland reservoir's designers and managers. One of the famous coastal reservoirs is Zuiderzee in Holland, which was originally created for land reclamation using 33 km dam. The seawater was turned into freshwater due to dilution by river water in 5 years after the embankment was enclosed. Recently, more and more coastal reservoirs were constructed as listed in Table 3; among them, Hong Kong coastal reservoir may be the first one used for drinking water supply. It is interesting to note that after its operation in 50–80 years, so far no significant environmental impacts by these coastal reservoirs have been reported. Of course, these primary and simple designs for coastal reservoirs should be improved in order to improve the water quality and reduce potential impacts.

Table 4 compares different strategies for half billion m^3/year of water being supplied to a middle-size city like Australian Brisbane. Yang (2015) calculated the

Table 2 Differences between on-land reservoirs and coastal reservoirs

Item	Inland reservoir	Coastal reservoir
Dam site	Limited (gorge area)	Unlimited (in the sea)
Dam design	High pressure	Low pressure but with wave surge
Seepage	By pressure difference	By density difference
Pollutant	Land-based	Land-based + seawater
Emigrant cost	High	No
Water supply	By gravity	By pump

Table 3 Existing coastal reservoirs in the world

Coastal reservoir	Catchment area (1000 km^2)	Length of dike (km)	Storage capacity (billion m^3)	Construction cost (billion US$)	Constructed year
Holland	170	33	5.6		1937
Qingcaosha Shanghai	1800	43	0.55	2.8	2011
Saemanguem S. Korea	3320	33	0.53	2.1	2006
Marina Bay, Singapore	1130	0.35	0.04	1.5	2008
Plover Cove, Hong Kong	459	2	0.23	0.007	1968

Table 4 Comparison of environmental costs for different solutions to coastal water stress

	Inland dams	Desalination	Recycled wastewater	Coastal reservoir
Energy used ($\times 10^9$ kWh)	0	2.0	1.0	0
Greenhouse emission (CO_2) in million ton	0	3.33	1.67	0
Construction cost (U$ in billion)	11.42	9.28	10	2.8
Maintenance/ operation cost	Low	High	High	low
Impacts on ecosystem	Loss of biodiversity, beach erosion, etc.	Brine result in loss of marine biodiversity and saline fish may be either extinct, threatened or endangered	Low	No negative impacts on ecosystem
Life span	100	20 years	20 years	Infinity
Sustainability	The damage on ecosystem is not remediable	The damage on ecosystem is remediable	The damage on ecosystem	Sustainable

gas emission for a desalination plant if so much freshwater is produced, and the results show that the CO_2 is equivalent to the amount of gas generated by all cars in Australia. The strategy of wastewater reuse needs a very high energy cost similar to the desalination. The inland dam has very high construction cost induced by relocating people for their settlement and flooding the land. Consequently, Table 4 shows that the strategy of coastal reservoir is the best option for the coastal communities as it has little impacts to the ecosystem as shown in Hong Kong and Holland.

Shanghai's Application: Shanghai is situated in the delta of the Yangtze River in the middle portion of the Chinese coast. It is one of the largest megacities in the world with total area of 6340 km^2 and population 24 million in 2013. The Yangtze River is the longest river in Asia and the third-longest in the world; it drains one-fifth of the land area of China and the river basin is home to one-third of China's population. Its annual runoff stands at 951.3×10^9 m^3, about 52% of the national total. The Tai Lake in the west of Shanghai is the third largest freshwater lake in China; its surface area covers 2238 km^2, its volume is 4.66×10^9 m^3 (Yang and Liu 2010), and on average the lake discharges about 70–80% of its water, or 10.6×10^9 m^3/year to the Huangpu River, the mother river of Shanghai. The total water availability for Shanghai from the Yangtze and the Tai Lake is about 962×10^9 m^3 per year. This is a strikingly water-rich city surrounded with abundant water resources.

In 1996, a conference organized by the UN Center for human settlements predicted that Shanghai would be one of the dozen cities in the world with the most severe water crisis (N'Dow 1996). Shanghai is notorious for the crisis induced by pollution even though there is plenty of water for its residents. To solve this problem, Shanghai invited many international experts to discuss the possible solution. It is natural that the strategies like desalination, wastewater recycle, etc., listed in Table 4 were suggested by these experts. The author was also luckily invited to give a seminar to about 500 engineers and government officers on July 26, 2005. Based on this presentation, the author published his paper and claimed that "Completely Solve Water shortage Problem in Shanghai by Coastal Reservoirs" (Yang 2005b).

In 2005, the author proposed that Shanghai could have the best quality of water if the technology of coastal reservoir was used. The scheme proposed is shown in Fig. 5 in which the intake is located at the head of Chongming Island, whose wetland will be used to pretreat the river water from the intake; and finally the treated water will be stored in the coastal reservoir located at northern Chongming Island.

In December 2005, the author was informed that Shanghai government decided to construct the coastal reservoir in the south of Chongming Island, i.e., Qingcaosha in Fig. 5 in order to saving the construction cost mainly incurred by the underground pipes. The author decided to make his final effort to change the plan by flying back from Korea, and in the meeting with the chief engineer of Shanghai water authority; the following disadvantages were identified for their plan:

(1) The Qingcaosha Reservoir is vulnerable to "too much" water. As shown in Fig. 5, the reservoir is located in the main passage of floodwater. If a thousand year flood appeared in the Yangtze River, the reservoir together with the newly formed island could be swept into the sea by its torrents. But the reservoir marked in red in Fig. 5 is safe as it is located in a small branch of the river.
(2) The Qingcaosha Reservoir is vulnerable to "too little" water. Currently, the interface of freshwater and seawater oscillates around the yellow dot or Qingcaosha, i.e., during low tide, the reservoir is surrounded by freshwater and

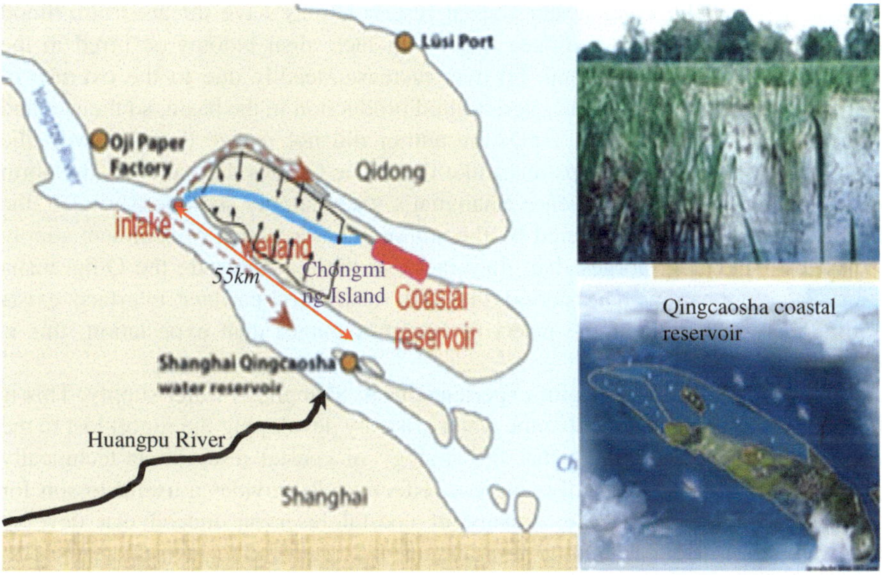

Fig. 5 The coastal reservoir proposed by Yang (2005a, b) in the northern Chongming Island (*red bar*), and its intake is located at the head of Chongming Island (*red dot*), the *arrows* in the Chongming Island represent the wetland pretreatment, the *thick blue line* is the perforate underground pipe to infiltrate the wetland treated water that is stored in the coastal reservoir marked by *red bars*. The *yellow dot* is the constructed coastal reservoir in 2010, and the detail is shown in the *left-hand side* near the *bottom*

the reservoir's intake can take water from the river; during high tide, seawater surrounds the reservoir, and the gates will be closed to prevent seawater pollution. But this seawater/freshwater interface is jointly decided by the river water and tidal flow. In an extremely dry year, seawater intrusion will push the interface to upstream of the reservoir, which may lead to the failure of freshwater intake. In the author's proposal, the intake locates at 55 km upstream of Qingcaosha; thus it is safe in dry seasons.

(3) The Qingcaosha Reservoir is vulnerable to "too dirty" water. It can be seen that the yellow dot is located at the intersection point of Yangtze River's main flow and the Huangpu River that carries all pollutants generated by 23 million Shanghai people. The tidal flow oscillates the plume of Huangpu river up and down, and finally the pollutants will enter the reservoir, resulting in algal blooms. But the author's proposal avoids the pollutants from Huangpu River and the river water can be pretreated by the wetland.

Unfortunately, the chief engineer rejected the writer's proposal, and decided to construct the Qingcaosha reservoir in 2008, the construction was completed in 2010. Half year later, the reservoir was diluted into a freshwater reservoir. Now, this coastal reservoir successfully supplies 70% of treated water needed by Shanghai; the remaining water comes still from the Huangpu River.

In long term, the Qingcaosha coastal reservoir may have threats from "flood water", "polluted water", and "sea water". In fact, algal blooms occurred in the reservoir and in future TN and TP may increase steadily due to the overuse of fertilizer and pesticide for higher agricultural production in the basin, so the wetland pretreatment is necessary. In 2005, the author did not realize that the hydraulic engineering projects in the upstream like the Three Gorges dam and South–North Water Diversion Project influence Shanghai's water supply so seriously; also, the sea level rise was not considered by the author. The reduction of sediment supply caused by the Three Gorges dam may increase the risk to erode the Qingcaosha reservoir's foundation. The period that the freshwater/seawater interface exists continuously upstream of the reservoir becomes longer than expectation; this is another threat.

Anyway, there is some useful experience from Shanghai's water supply. This is the first time to completely solve the water crisis by developing the runoff lost to the sea for a megacity. It proves that the strategy of coastal reservoir is technically feasible, environmental friendly, and cost-effective. It provides a useful lesson for the world to follow that the strategy of coastal reservoir indeed can develop freshwater from the sea without desalination. It is the time to investigate the feasibility of this strategy in other places in the world.

Applications of Coastal Reservoir in the World: This paper advocates the technology of coastal reservoirs to provide sufficient freshwater to the global population centers that are most likely located on the coastal/deltaic areas. This paper will extend the successful example in Hong Kong, Singapore, and Shanghai to other places in the world, and discuss the feasibility to pump freshwater from the sea without desalination. The ideal sites for these coastal reservoirs are suggested, and its feasibility is preliminarily investigated in terms of water availability, water demand, cost-effectiveness, and environmental sustainability.

Malaysia: In the world, many islands have highest population density and exists great water stress. One of the examples is the Langkawi Island in Malaysia (see Fig. 6), which is one of the famous tourist attractions for its sandy beach. It is located in the northwestern part of Peninsular Malaysia. It has a land area of 464 km^2 (similar to Singapore) with population of about 100,000 in 2010. The urbanized area concentrates at the west and south of the island near the Langkawi Airport, Kuah, the largest town at the south.

Currently, Malaysia is developing the Langkawi as another Singapore, but its progress toward this target has been hindered by the lack of freshwater in the island as there is no a big river on the island, even the annual rainfall is about 2500 mm. It is infeasible to transport freshwater from other places using ships and pipes over the seabed due to the high cost. The proposal of desalination plant has been also rejected by the developer as the high energy cost cannot make the industry profitable. It is predicted that the water demand in Langkawi will increase to 107 MLD by 2020 and 128 MLD by 2030. With the current design capacity of 87.3 MLD from the inland reservoirs listed in Fig. 6, Langkawi will face water deficit by 2020 if no further upgrading works or additional water source are explored. Of the existing water sources in Langkawi Island, there are three inland reservoirs created,

Fig. 6 Location of inland reservoirs in Langkawi and the largest catchment in the island

Empangan Malut, Padang Saga, and Padang Mat Sirat bunded storage as marked by yellow pins in Fig. 1. Their catchment sizes are only 3.4–12 km^2 only. This is an example to show that the inland reservoirs can only be used to catch rainwater from a small portion of the catchment.

Currently, a breakwater structure was constructed at the shoreline of Langkawi near Sg Melaka estuary to protect the Langkawi port and the airport. This existing breakwater structure shall be an ideal location to construct a coastal reservoir with further study to be conducted to explore its feasibility and the economic viability (refer Fig. 7).

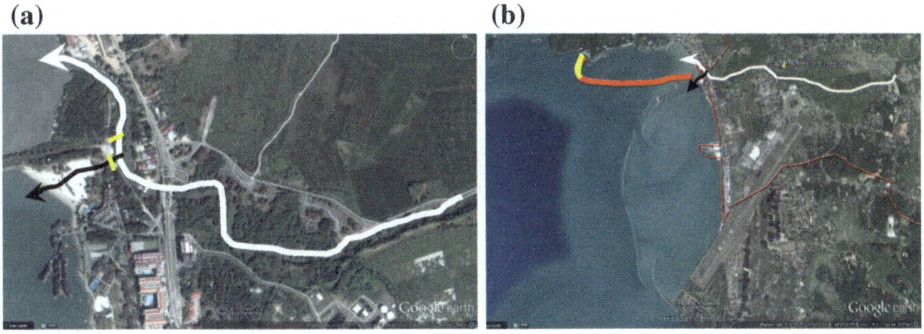

Fig. 7 Design of coastal reservoir for Langkawi Island's biggest catchment. **a** Water intake design (route for poor quality water like first flush water = *black line with arrow*, clean water route = *white line*; **b** barrage of coastal reservoir (*red line*), *yellow lines* = inlet/outlet)

The design for the coastal reservoir is shown in Fig. 7, where the clean water and polluted water are separated as shown in Fig. 7a, where a gate at its entrance is installed to regulate the incoming water. If the measuring sensor shows that the incoming water is polluted like the first flush flows, then the gates toward the coastal reservoir are closed, and the gates to the sea will be opened; thus the flow will be diverted to the sea as specified by the black line; vice versa, the clean water will flow to the coastal reservoir as specified by the white line in Fig. 7a. The coastal reservoir is shown in Fig. 7b and the barrage is indicated by the yellow line, which is similar to the existing breakwater; hydraulic gates or inflatable rubber dams will be installed as part of the yellow to regular the water level in the coastal reservoir and to keep water in motion inside the reservoir, and thus the chance like algal blooms is limited.

For the coastal reservoir design, it is very important that the barrage is high enough to sustain surfs in extreme weather condition. This coast is relatively calm as there is no typhoon. It is suggested that the dike height is similar to the level of the nearby airport, and the width of the barrage could be as wide as a highway, and thus this barrage can be used for road transport.

(1) The water quality is the crucial part for the successful development of good quality water by the technology of coastal reservoir. It is simple to dilute the initial seawater in the reservoir to the drinkable level as the incoming freshwater is stored and the reservoir water is discharged to the ocean by opening the gates/deflating the rubber dams during low tidal. This dilution process generally takes half year to 5 years dependent on the ratio of inflow rate to the reservoir capacity. The seawater intrusion from the groundwater is negligible as the foundation is impermeable and the water level in the reservoir is higher than that in the sea during most of the time. The risk for pollution by overtopping surges is very low as the airport has never been flooded by the seawater after its construction.
(2) The contamination by pollutants from the catchment and the prevention of algal blooms in the reservoir are also very important. It is necessary to have a better catchment management strategy to reduce the pollutants; also the intake gates are important to prevent the first flush's pollution as shown in Fig. 7a. To prevent algal blooms, it is suggested that special hydraulic models are to be conducted for the design of hydraulic gates/rubber dams in the barrage, which drives the reservoir water in motion.

The catchment marked in red is about one-fourth of the island, i.e., about 120 km^2, and the rainfall is about 2.5 m/year; if the evaporation loss is about 50%, then the runoff to the sea is about (= 120 * 2.5/2 = 150 million m^3/year). If 20% of the incoming water is the polluted water and should be discharged to the sea, then the developable water will be 20 million m^3/year or 330 MLD. This is about 3 times of water demand in 2030. Therefore, it can be seen that using the technology of coastal reservoir, the island has sufficient freshwater for its development, and its quality is comparable with the existing dams.

India and Bangladesh: India lies roughly between 10°E–93°E and 7°N–36°N with an area of about 3.29 million km^2, and population of 1.3 billion in 2014. The topography of India consists of the lofty and high Himalayan mountains in the north, the plain valleys of the Gang, Indus, and Brahmaputra in the south of the Himalaya, and large peninsular hills and plateau, and some narrow coastal plains. The rainfall in this country mainly depends on the southwest monsoon, the average annual rainfall over the country is about 1250 mm, but the spatial variation is very high, from less than 100 mm in the western deserts to 11,000 mm in parts of the northeast. The total freshwater resource is 1950 billion m^3 or 1950 km^3, of which 690 km^3/year is considered as utilizable surface sources, and 432 km^3/year for utilizable groundwater sources. Currently, the total water usage is about 550 km^3/year. It is estimated by the National Commission (1999) that the annual withdrawal requirements, by 2050, could be between 973 and 1180 km^3. The major basin/basin groups in India are listed in Table 5. India has 2% of the world's geographical area, 16% of the population, and only 4% of freshwater resources.

The Ganges–Brahmaputra–Meghna (GMB) basin now has 550 million people, more than North America and Western Europe combined. It also has the largest population of the world's poor mainly caused by water disasters. This region has the potential to be rich if its natural resources could be harnessed efficiently, sustainably, and rationally. There are four countries in the basin: Bangladesh, Bhutan, India, and Nepal. The river begins in the Himalayas in Nepal, and then flows to India and finally reaches Bangladesh and drains into the Bay of Bengal.

The basin's average rainfall is more than 1500 mm/year (Nepal 1520 mm, India 1300 mm, Bangladesh 2400 mm, Bhutan 2000–5000 mm per year). 75–80% of the annual rainfall occurs in 3–4 months, and most of the water is received in a period of 30–40 days. The GBM basin is the third largest river in the world in terms of discharge, only being exceeded by the Congo and the Amazon river systems (Chowdhury and Ward 2004). The ultimate utilizable surface water resources in India have been estimated as 4% of the available water resources of the Brahmaputra and Meghna, or the maximum utilizable surface water resources in India will be 2419 km^3/year. The river carries abundant water into Bangladesh that is enough to submerge the whole country under more than 9 m of water every year, of which only 7% is generated from this low-lying country. Every year from June to

Table 5 Indian water resources and its spatial distribution

River basin	% of the country's area	% of national water resources
Brahmaputra and Meghna arms of GBM basin	7.2	34.7
Ganga arm of the GBM basin	26.2	26.9
Subarnarekha, Brahmani, Baitarani, and Mahanadi	5.9	10.5
Narmada	3.0	2.4
West-flowing river south of TAPI	3.4	10.3
All other river basins	54.3	15.2

September, one-third of the country is vulnerable to flood. But water shortages prevail during the winter with drought or drought-like conditions. During November–May, the mighty river in the world no longer reaches the sea as the upstream diversions do not leave enough water to the downstream rivers after mid-1970s when the Farakka barrage was constructed. The seasonal water deficit poses a bottleneck for Bangladesh's development, and the seawater intrusion has also caused a rapid advance of a saline front across the western portion of the river delta, threatening agricultural production. The problems raised in 1971 war have remained unsolved completely; the agreement between India and Bangladesh expired in 1988 for sharing the dry-season flow from the Ganges. In March 1998, the government of the Netherlands convened a Ganges forum for experts and decision-makers to collaborate for the basin's sustainable development. The critical problem is how to provide Bangladesh with sufficient water in the lean season, especially from January to April. Only 5% of the total annual flow of the river would be needed to solve the downstream water scarcity during the lean season. Certainly, this requires a new way to store water in the catchment to be used when the flow diminished.

The annual discharge of the Brahmaputra river basin from Bhutan to India 78 km^3 and from China to India is 165.40 km^3. The annual discharge of the Ganges river basin from China to Nepal is 12.0 km^3. The annual flow of the Brahmaputra river basin from India to Bangladesh is 537.2 km^3. All flows in Nepal drain into the Ganges river with an annual flow of 210.2 km^3 to India. The annual flow of the Ganges basin from India to Bangladesh is 525.0 km^3. The annual flow of the Meghna river basin from India to Bangladesh is 48.36 km^3. This gives a total annual GBM river basin inflow into Bangladesh from India of 1110.6 km^3; this is larger than the Yangtze River in China. Based on flood records, the drought period starts in October, the river flow is minimum between March and April and is maximum between August and the last week of September (Parua 2001). About 138,700 m^3/s of water flows into the Bay of Bengal during floods through a single outlet of the GBM river. This is the largest in the world for a single outlet to the sea and exceeds even that the Amazon discharge into the sea by about 1.5 times (Parua 2001).

Historical floods in Bangladesh are perhaps unavoidable phenomenon, which witnessed the visit of nearly half a dozen floods in every century. For example, the floods of 1998 and 2007 were catastrophic, leading to widespread destruction, and high loss of life. To use the surplus water for irrigation as well as to reduce the flood losses, the Bangladesh government developed some projects for flood control. Some major projects are Ganges-Kobadak irrigation project (G-K Project), Dhaka-Narayanganj-Demra (DND) Project, etc. These control and drainage projects heavily depend on dredging, embankments, polder, and gravity drainage. Despite huge amounts of investment in flood control and drainage projects, the benefits have been less than satisfactory.

The factors for causing floods are (a) general low topography of the country with a congested river network system, (b) very high rainfall in the upstream country or in the mainland, (c) snow melt in the Himalayas and glacial displacement,

(d) synchronization of flood waves from one river on the other, and (e) tidal and wind effects on slowing down the river outflow (backwater effect).

To solve the problem, we proposed a coastal reservoir as shown in Fig. 8 highlighted by red/yellow lines, and problems like the flood disasters, water shortage, and seawater intrusion can be integrally solved, because of the following:

(1) The huge coastal reservoir can provide sufficient freshwater for Bangladesh's economic development. This is a water-rich region, but the poverty is caused by the water-related disasters. Once the coastal reservoir is constructed, the country will have sufficient freshwater for its development during lean seasons. The best quality water will be stored in the reservoir enclosed by the yellow line and long red line.
(2) This huge coastal reservoir can greatly mitigate flood disasters during flood period (i.e., 30–40 days) whose gates/inflatable dams will regulate the water in the following way: similar to the reservoir in Fig. 7b, the coastal reservoir has inflatable rubber dams to regulate the water level (yellow line). During low tides, the gates along the red line will be opened, and the excessive reservoir water will be released to the sea. During high tides, the gates will be closed and the water level in the reservoir will be at the low level, and thus a large storage capacity is ready for the imminent peak flood waves.

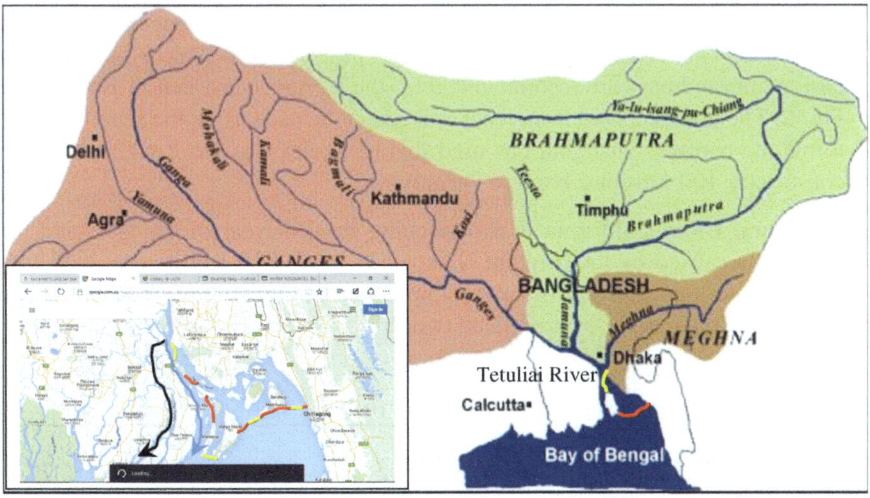

Fig. 8 The catchment of the Ganges–Brahmaputra–Meghna (GBM) basin and the proposed coastal reservoir (*red* = barrage and *yellow lines* = inlet/outlets). For flood disaster mitigation: discharging the reservoir water to the sea during low tide and waiting for peaks of flood waves. For water supply in lean season: using the stored floodwater in the reservoir. For reduction of seawater intrusion into the soil: discharging the environmental flow (e.g., Farakka project) in dry periods through the Tetulia River. *Black line* with arrow is the flow passage all the time; the *blue line* with arrow is the flow passage only for safe flood water in flood period, the disastrous water will flow to the reservoir which is empty prior to the arrival of peak flood

(3) The safe flood can be defined as the floods whose water level below 6.2 m above PWD at Bhairab Bazar of Meghna River, 13 m at Hardinge Bridge of Ganges River, and 19.5 m at Bahadurabad of Brahmaputra. If a safe flood appears, all inlet/outlets of the reservoir will be closed, thus the reservoir's water level is the lowest, and the reservoir's capacity is not occupied by the safe flood water, but remains at the lowest tidal level. Once the river water level is higher than the safe flood level, the gates/dams at the inlet will be opened, and thus the upstream water level will be lowered quickly. Once the rivers' water level is lower than the defined safe water level, the reservoirs' inlet will be closed, and the outlet of the reservoir will release the excessive reservoir water during low tides in order to prepare for the next flood waves. The average tidal height in the Bay of Bengal is about 1–3 m, if the reservoir's area is assumed to be 2500 km^2, the reservoir capacity for flood water would be 2500 * 2.5 = 6.2 km^3, the largest reservoir in the world. Therefore, it is understandable that the coastal reservoir can effectively mitigate the flood disasters, and it is enough for the country's development.

(4) The small environmental flows will be forced to enter the sea through the Tetulia River, and therefore the seawater intrusion can be alleviated because the huge coastal reservoir is also effective to stop the seawater intrusion to the agricultural lands.

It is important to protect the reservoir's water quality; to achieve so, it is important that pollutants from first flushes will bypass the reservoir as shown in Fig. 7a. For this purpose, the Tetulia River will be the main outlet for the pollutants. This coastal reservoir is a huge project, to reduce its construction cost; giant sand bags are recommended for the barrage construction as shown in Fig. 9. The dredged sand from local seabed can be used to fill the sand bags and forms the barrage. Shanghai has used this technology in the Yangtze River to construct a 100-km-long barrage cheaply.

California, USA: California is experiencing its fifth year of drought, and the economic loss caused by the drought is expected to hit \$2.74 billion in 2015 (see Fig. 10a) due to the reduction of water supplies in the agriculture, industry, urban water management, and the environment; two-thirds of the loss is from the agriculture

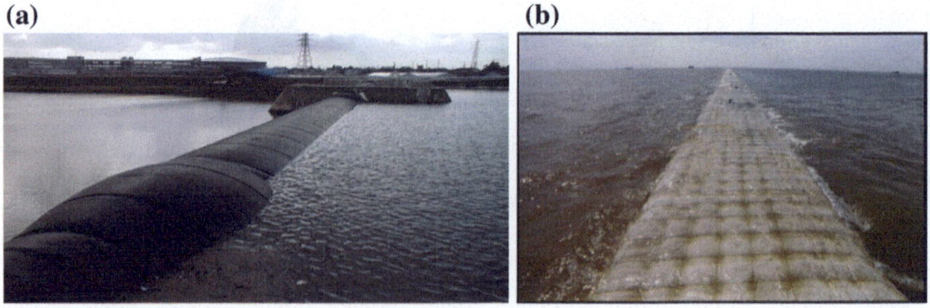

Fig. 9 Inflatable rubber dam (**a**) and giant sand bag (**b**) for the barrage construction

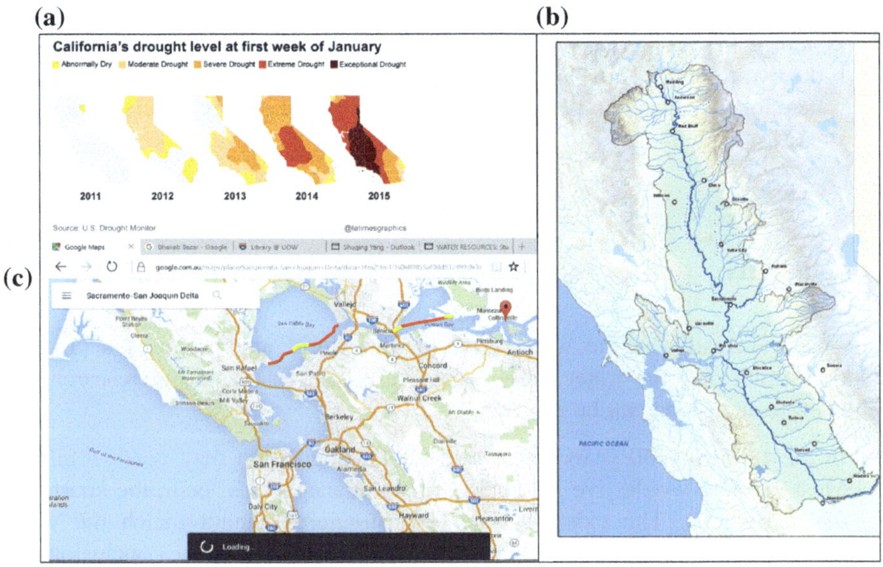

Fig. 10 California drought (**a**), map of the (**b**), and the possible coastal reservoirs by barrages (*red lines*) and the gates (*yellow lines*), in dry periods the wastewater will flow to the sea

and industry. In 2015, farmers had to follow half a million acres of land due to difficulty obtaining water. To address these challenges, California has taken a series of measures, for example, a $7.5 billion water bond was approved to fund investments in water recycling, groundwater cleanup, water storage, and desalination.

The Sacramento and San Joaquin Basin includes two major watersheds—the Sacramento River on the north and the San Joaquin River on the south—plus the Tulare Lake Basin. These rivers play a key role in California's development; six of the top 10 agricultural counties in the state use the water. It also waters farms, homes, and industry in California's Central Valley as well as major urban centers like the San Francisco Bay Area. To solve the water crisis, this paper provides the following conceptual design for the freshwater storage as shown in Fig. 10c:

(1) The barrages are constructed in the San Pablo Bay and Grizzly Bay as the red lines shown in Fig. 10c. It is expected that the upstream of Point Pinole would be freshwater, as the channel width is very narrow in this section, and thus high flow can stop the seawater intrusion. The narrow channel at the Point Pinole will not cause flood disasters, because there are gates installed along the barrage of San Pablo marked in yellow. If needed, they will be opened to increase the discharge for flood disaster mitigation.
(2) Similar to the gates' operation in other coastal reservoirs proposed for Malaysia and Bangladesh, the gates will be closed all the time in the long dry period. The water level inside the reservoirs will be lowest just before the arrival of flood waves. The reservoirs' intakes will be opened and the peak flood water will be

stored; this mitigates the flood disasters and the best quality water in the river system is kept for the dry periods.

(3) If the two reservoirs area is 300 km^2 and the average depth 2 m is assumed, the water storage capacity will be 600 million m^3; it will greatly alleviate the water stress in this area.

Conclusions

This paper investigates the water crisis in the world and proposes a possible solution to alleviate the water stress coastal reservoir. Based on this research, the following conclusions can be drawn:

(1) In the following 30 years, the world's total population may reach 9 billion. Most of them would live in urbans, cities, or megacities, or in the coastal or deltaic regions. The existing water infrastructures cannot match the water demand in these regions. In other words, this paper supports the conclusion that this world is experiencing water crisis, but this paper further highlights that the most severe water shortage will occur in the coastal/deltaic regions. International society needs to pay more attention and investment to the coastal areas.

(2) Currently, the world's water supply heavily depends on inland reservoirs which are aging quickly due to sedimentation and others. The inland reservoirs have played important roles for urbanization and industrialization, but the practice shows that the inland reservoirs generate many negative impacts on the ecosystem, thus no new large dams have or will be constructed. This will intensify the world's water stress. The appearance of coastal reservoir—a freshwater reservoir in the seawater can provide sufficient freshwater for the future.

(3) The coastal reservoirs are effective to provide clean water to the coastal communities. In its design, it is important to bypass the polluted wastewater, thus the high-quality water can be collected and stored in the coastal reservoirs. This can be seen from the example provided for Malaysia's Langkawi Island. For other small islands or coastal catchment, these innovative ways to develop runoff to the sea can be used similarly.

(4) The coastal reservoir can develop freshwater from the sea without desalination; it also can mitigate the flood disasters. The example is provided here for Bangladesh. To achieve this, it is required that the reservoir's water level is lowered prior to the arrival of peak flood waves. The coastal reservoir also can prevent the seawater intrusion.

(5) Coastal reservoir also includes the conversion of lagoon or coastal bays into freshwater sources. An example is provided for California's San Francisco region.

The strategy of coastal reservoir has the potential to ease the global water crisis, i.e., too much (floods), too little (droughts) and too dirty (pollution). In the future, more detail research is needed to validate its application worldwide. Most importantly, the UN, World Bank, and governments should lead the revolution of water resources development.

Acknowledgements The partial support provided by the National Science Foundation of China (51239001) is gratefully acknowledged.

References

Batuca DG, Jordaan JM (2000) Silting and desilting of reservoirs. Balkema Publishers, Rotterdam, The Netherlands

Chowdhury R, Ward N (2004) Hydro-meteorological variability in the greater Ganges-Brahmaputra-Meghna basins

James DP, Chanson H (1999) A study of extreme reservoir siltation in Australia. Water 99 Joint Congress-Australia, pp 1109–1114

National Commission (1999) For integrated water resources development. Ministry of the Water Resources, Government of India

N'Dow (1996) Water crisis to strike most developing world cities by 2010 Istanbul: Habitat II. Available at http://www.un.org/Conferences/habitat/unchs/press/

Parua PK (2001) Flood management in Ganga-Brahmaputra-Meghna basin: some aspects of regional co-operation

Shen HW (1999) Flushing sediment through reservoirs. J Hydraul Res IAHR 37(6):743–757

U.S. Bureau of Reclamation (USBR) (2006) Erosion and sedimentation manual. U.S. Department of the Interior, Bureau of Reclamation, Technical Service Center, Sedimentation and River Hydraulics Group, Denver, Colorado

World Commissions on Dam (WCD) (2000) Dams and development. Earthscan Publication Ltd., London

Yang SQ (2002) How to alleviate global water stress, case study about China water diversion scheme. In: Proceeding of 13th congress of Asia and Pacific division of international association for hydraulic research (IAHR). World Scientific Press

Yang SQ et al (2001) Coastal reservoirs. Patent No. 200504653-7 (in Singapore)

Yang SQ (2003) Potential water resources in Singapore. J of Water Supply Res Technol Int Water Assoc 52(6):425–434

Yang SQ (2004) Global and China's water crisis and its solutions in the 21st century. Tianjin University Press. ISBN 7-5618-2069-0 (in Chinese)

Yang SQ (2005a) Storage and isolation of clean freshwater in the sea. Patent No. ZL200510016106.1 (in China)

Yang SQ (2005b) Completely solve water shortage problem in Shanghai by coastal reservoirs. http://www.paper.edu.cn/html/releasepaper/2005/08/130/. China Science and Technology Online Papers

Yang SQ (2015) A preliminary feasibility study for a backup water supply-coastal reservoir in Southeast Queensland, Australia. J Water Supply-Res Technol 64(4):470–480

Yang SQ, Liu PW (2010) Strategy of pollution prevention in Taihu Lake and its effects analysis. J Great Lakes Res 36(1):150–158

Leakage Detection Studies for Water Supply Systems—A Review

P. Darsana and K. Varija

Abstract This paper summarises the recent studies done for the leakage detection, more like a general review than specifically discussing any area of study. This is an analysis of the leakage detection methods with the objective to identify the current state-of-the-art in the field. Numerous publications are cited in an attempt to comprehend the recent trends in the considered area. From the leakage detection studies the main conclusion can be drawn as many of the approaches are data hungry, which are difficult or expensive to obtain. Since the pressure measuring sensors and other equipments are expensive the need for a method with minimum number of attributes which can be measured easily in the field has increased. Many of the studies have been done for the virtual data and thus its field application is to be studied further and specifically for the area which we are considering the particular method. Finally, the review paper ends with a conclusion that there are lots of studies to be undertaken for the betterment of the leakage detection methods.

Introduction

Each year more than 32 billion m^3 of treated water is lost through leakage from distribution networks and 16 billion m^3 per year is delivered to customers but not invoiced because of theft, poor metering, or corruption. Proper understanding of pressure leakage relationship is essential to planning pressure management and leakage reduction and water loss reduction (Choia et al. 2014) programme. Leakage assessment is crucial in understanding the performance of current infrastructure and determining effective ways to manage and reduce quantified levels of leakage. It is generally accepted that leaks are more sensitive to pressure than described by the traditional fixed area Orifice Equation, highlighted by empirically derived leakage exponent values greater than 0.5 from both laboratory tests and field investigations.

P. Darsana (✉) · K. Varija
Department of Applied Mechanics, Indian Institute of Technology Madras, Chennai, India
e-mail: darhupc@gmail.com

V. P. Singh et al. (eds.), *Water Resources Management*, Water Science and Technology Library 78, https://doi.org/10.1007/978-981-10-5711-3_10

This is predominantly due to the dynamic nature of the leak area. Various studies have focussed on the development of leakage models accounting for the observed pressure-dependent leakage. Research on leakage detection includes numerical and physical studies aimed at quantifying the structural deformations of leaks in order to determine an associated leakage exponent, as well as field work aimed at characterising the total system leakage behaviour. The majority of these studies assume a one-to-one relationship between the system pressure and the resulting leakage flow rate. Leaks in commonly used plastic pipes, such as polyethylene, have been shown to result in a more complex relationship between pressure and leakage flow rate due to the inherent material properties, namely viscoelasticity showed that linear-elastic leak area models produced 'good approximations' of the leakage response of cracks in viscoelastic pipes over a two-day period. When fitting to experimental data, the cumulative error increases with increasing time.

If high leakage is so harmful, why is more not done to reduce it? The answer is complex and relates to a lack of awareness, knowledge, and priority. It is clear that lowering leakage will have a cost to locate and eliminate the bursts which can be offset by the saving in costs, both in the reduced production and the consequential scope to optimise it, through pumping at night when the electricity tariff is lower. By considering the social significance of water resources and hence water supply system, it should be continuously improved and updated based on specific plans and utmost care should be taken to maintain the stability and safety of the water supply (Yoo et al. 2014). More method to facilitate more reliable determination of the rehabilitation priority order for water pipes by taking into account the pipes' hydraulic importance should be given more importance.

Background

Breakages and leakages in pipes lead to the higher energy consumption to meet the demand. Thus leakage managements require more consideration both technologically and financially. Leakage management can be broadly classified as shown in Fig 1.

The first step of leakage management is the leakage assessment devoted to the entity of losses completely ignoring their position and location. Typically, this step is done by evaluating the overall water balance assessment (Almandoz et al. 2005; AHL 2006).

Fig. 1 Leakage management steps

The second step of leakage management process is mainly devoted to identify the possible area where a loss occurs and may be performed through analytical- or equipment-based techniques or again a combination of both of these together also can be used. The major equipment-based methods are leakage reflection method (LRM), inverse transient analysis (ITA) (Liggett and Chen 1994), impulse response analysis (IRA), transient dumping method (TDM), and frequency response method (FRM).

The final steps of leakage analysis, that is, leakage localisation and rehabilitation, became more efficient in the last 30 years making more new ways to reduce the uncertainty related to the leakage management and rehabilitation processes. For example, a Decisional Support System (Rogers 2014), which automatically evaluates the severity of the leakage, and assesses how much can be recovered, predict the future trend so as to quantify the total potential saving and compare it with the likely cost of the intervention that has been developed as a part of EU-funded PALM (pump and leakage management) project recently completed in central Italy.

Recent Studies on Leakage Detection

Pudar and Liggett (1992) research paper regarding the leak detection method paved path for a whole new era. In the research paper, a new leak detection method based on pressure measurement was proposed and it was called inverse problem. Liggett and Chen (1994) was the first to propose using Inverse Transient Analysis for leak detection. Pudar and Liggett (1992) works proved that inverse analysis can be used for leak detection under steady-state conditions. Brunone (1999) modified the idea of using transient analysis-based technique and used for leak detection in outfall pipes. According to Pudar and Liggett (1992), nodal leaks become additional demands at the nodes, the locations and quantities being unknown; in the usual problem—the *forward* problem—the demands and the characteristics of the system are known and have to find the pressures and how the flow separate. The characteristics of the system and the demands are known but some parameters (unaccounted nodal outflows, leaks) are unknown in the *inverse* problem. If the known parameters are extended (i.e., pressure measurements), the leaks can be found out according to the theoretical explanation. An inverse programme is unlikely to provide the definitive results that would replace more conventional methods and also it can serve as a supplement to leak surveys. Inverse transient analysis (Covas et al. 2005) is further refined by simulating scenarios of various transient severity and leak sizes. Based on the results of the simulation experiment, a timid conclusion was made that the inverse transient analysis method may be an effective tool for detecting leaks of intermediate to large size. The inverse transient analysis proposed by the Liggett and Chen (1994) has been refined modified by many others over the past years to optimise the objective function and hydraulic model. Covas and Ramos (2011) used inverse transient analysis for leakage detection and location in pipe networks. Leaks location in pipe networks is identified by using observed pressure data, collected during the occurrence of transient events, and by the minimisation of the difference between observed and calculated parameters. This approach is presented theoretically and employed in a software tool. Though the leak flow was around one-third of total flow at the inlet of the system, the method was able to detect leaks. The method still has to be tested for smaller leaks to assess the minimum detectable leak. Nevertheless, the method might not able to locate accurately small leaks.

Leakage assessment can be divided into two broad categories. Top-down and bottom-up leakage assessment methods. Of this bottom-up approaches are relatively desirable since they are less data hungry. The bottom-up real loss assessment can be done in two ways. One is 24-hour zone measurement (HZM) and Minimum night flow analysis (MNF). MNF in urban areas usually takes place during morning hours. Several studies have been done on this topic. Recently, Fox et al. (2015) proposed traditional leakage assessment based on minimum night flow (MNF) analysis for calculating the total real losses from a district metered areas (DMA) based on a small amount of input data, irrespective of the specific pipe material found to be effective. Though previously by considering the shortcomings of MNF, a new method FAVAD (Lambert 2001) means fixed and variable area discharges. This concept was implemented in a recent study by Deyi et al. (in real network). The study applied both conventional equation and FAVAD concept to the network. Networks with leakage exponent value greater than 1.5 showed greater anomalies.

Leakage assessment studies are basically carried out to quantify total losses including, if possible, real and apparent losses. This lead to the development of leakage detection methods with the aim to detect and pinpoint leaks. A Bayesian system identification methodology is proposed for leakage detection in water pipe networks by Poulakis et al. (2003). The unavoidable uncertainties in measurement and modelling errors have been properly handled. Boulos and Altman (1991) work to develop an algorithm for optimization of nonlinear distribution network based on previous studies also become base for the development of a Bayesian system. Estimates of the most probable leakage events (magnitude and location of leakage) and the uncertainties in such estimates have been distinctly calculated using this method. The optimised model (Qi et al. 2014) combining genetic algorithm and Bayesian decision theory makes good use of the calculated values and measured values of all pressure monitoring points to locate leakage.

Akensla et al. (2008) introduced an unsupervised method in which it is not necessary to have information on all the leaks by using a self-organising map (SOM) for leak detection. The study experiments showed an excellent ability to detect leakages in a water distribution network. The results strongly suggest that combining existing flow data with a leak function to model leakages is a valid method for solving the leakage detection problem. Fuzzy numbers are used to model uncertainties of the independent parameters in the work done by Islam et al. (2011). The developed method can be used to measure the degree of leakage as well its severity in terms of the index of leakage propensity (ILP) based on monitored pressure and flow. The developed model has been implemented on water distribution system (WDS) and from the analysis, it can be seen that it can detect the leakage and diagnose the exact location of the leakage.

An analysis is done by Colombo and Karney (2002) to point up the role of total system demand, leak location, and topological complexity. A thorough analysis of the relation between water loss and energy cost has been done. The impact of leaks on water age is also evaluated through simulation and via a dimensionless expression relating leak size and location to residence time. Performance indicators which consider the seasonal variation as well as the pressure variation have been introduced later in the field of leakage analysis. The Infrastructure Leakage Index (ILI) is an outstanding indicator of physical losses developed by IWA, which takes into account how the network is managed. The IWA and the American Water Works Association (AWWA) Water Loss Control Committee both recommend this indicator. In the cases with low NRW, ILI is found useful in networks the ILI can help to identify which areas can be reduced further (Winarni 2009). Reducing the uncertainty associated with the Infrastructure Leakage Index will lead to the betterment of results (Babic et al. 2014).

The case study was done by Ristovski (2011) which implemented two of four well-known IWA WLTF recommended strategies: Active Leakage Control and Pressure Management in particular DMA "Lisiche" in Republic of Macedonia. Active leakage control (ALC), which is a cost-effective and efficient method for leakage management, is now an internationally accepted and well-established technique.

Major issues faced by the urban water supply systems are in different aspects. The rusting of the pipeline (Annus and Vassiljev 2014) is one among them. Thus comes the importance of leakage management. A graph-based analysis proposed by (Candelieria et al. 2014) is to improve leakage management in water distribution networks. Leakage management process in urban WDN was improved and developed by the application of graph-based analysis which leads to an effective computational leakage localization approach. This approach is based on a combination of simulation of different leaks, in terms of location and severity, and the graph-based clustering analysis of the pressure and flow variations resulting from each simulation run (leakage scenario). In particular, spectral clustering, in its pure and approximated version, has been investigated according to its widely proved quality. Spectral clustering approximations have been considered to deal the critical issue of Big Data associated with the analytical task. Finally, the possibility to combine the graph-based analysis for leakage localization with more traditional machine learning techniques (i.e., regression) for the estimation of leak severity allows us to implement a workflow able to further improve localization and support WDN managers in defining a suitable investigation and intervention plan.

Sala and Kołakowski (2014) presented results of research to implement a system for detection of leaks and optimal control of flow in water distribution networks (WDNs) based on pressure measurements, where the leak detection system is supported by an electronic system for remote transfer of measurement data using GSM protocols. The Virtual Distortion Method (VDM) presented in Holnicki-Szulc (2008) is served as the basis of analytical formulation of the leak identification problem. Major developments made are as follows: first, the problem has been formulated for the Hazen-Williams constitutive law, which was subsequently linearized. Second, the ability of the leak identification algorithm to track the position of leak precisely along network branch was added. Third, the practical aspect of the problem, i.e., taking measurements of water head levels at hydrants instead of nodes, was considered. These solutions are possible, thanks to appropriate modifications of the influence matrix.

A methodology for the estimation of area leakages in virtual areas of a water network proposed by Adachi et al. (2014) helps to prioritise leak surveys for the areas. The leakage estimation was done by minimising the difference between the measured and the hydraulic model estimated pressure and flow rates within the areas. The accuracy was evaluated by a simulation study, where the effect of head measurement error was considered by a Monte-Carlo-type method and mean absolute error of area leakage was less than 2% of the system flow. Requirement of comparatively lesser parameters than many of the previous studies and lower computational load and increased stability of the optimization process are the main advantages of this method, where in some cases it becomes very difficult to calculate the accurate value of leakage due to the absence of proper data. The proposed method can be applied to fewer sensors for the estimation after the calibration of the hydraulic model. Compared to the previous studies this method can only prioritise virtual areas and cannot identify possible locations of leak hotspots. However, the information provided by the method could be useful to accelerate the leakage reduction process.

The use of a particle filter (PF)-based technique for the detection of leaks in water pipelines proposed by Anjana et al. (2015) was applied to a real-time network in Mandya, Karnataka, India. Automation of WDS (with water quantity and quality sensors) has helped the water authorities around the world to get near real-time data of important system parameters. Various models developed on the basis of Extended Kalman Filter (EKF), Nonlinear Kalman Filter, etc., have been widely used for leak detection in pipelines. Also, online burst detection using extended Kalman filter (Okeya et al. 2014) and hydraulic modelling found to be effective to detect bursts in a reliable manner within a district metering area under assumed test scenario. These models are applied to flow, pressure, and acoustic signals from the system. But the main disadvantage of most of these models is that they require the nonlinear pipeline model to be linearized. In this study, a particle filter based technique was proposed for state estimation in a supply-based water distribution system. PF algorithms can be used for state estimation (flow estimation) in supply-based tree water distribution networks. But further study is required to calculate its applicability in a looped network. Since flow metres are generally more expensive than pressure sensors, further study is required to optimise the ratio of flow measurements to pressure measurements required for any system, since the use of many flow measurements will increase the cost of system instrumentation.

Challenges of urban water distribution network identification and the measures of leakage control are one of most serious issues which water industry face. In theory of blind source separation (Gao et al. 2014), which is applied to calculate the leakage of water distribution networks, special leakage experimental platform of water distribution networks is established. Fast independent component analysis method (Fast ICA) was used to divide the physical leakage from total node flow, which used for reducing physical leakage. Separation of water consumption and water leakage from water network outlet flow was done with fast independent component analysis algorithm; calculation of correlation coefficient was undertaken for results separation. Comon (1994) proposed independent component analysis (ICA) theory and suggested the theoretical model of ICA. ICA is basically a method of blind source separation theory, which can search inner statistically independent and non-Gaussian source signals from multidimensional signal. Results lead to the conclusion that it is feasible and effective to apply blind source separation theory, to separate water leakage.

Quality of the leakage identification procedure is particularly improved according to the quality of information available from the real network (Bort et al. 2014); for these reasons, it has been developed a methodology in which positioning of pressure sensors into the hydraulic network is done with high efficiency. Three different scenarios are suggested such that first one is a condition in which the pressure sensors are displayed according to empirical equations, second a case in which only sensitivity analysis of the network on the position of the leakage is made, and finally a condition in which a correlation analysis of the pressure measurement nodes is done afterward the sensitivity analysis of the network.

Low and negative pressure events have the potential as they result in intrusion of pollutants: negative pressures create a suction effect inside the pipe and the contaminant intrusion through pipe leaks. This study (Fontanazza et al. 2015) presents a different aspect of leakage analysis. Paper presents the results of experimental tests on the intrusion of contaminant through pipe cracks in water distribution network resulting from low/negative pressures. Also, the energy cost impact of leakage is another aspect, where various parameters are needed to be addressed. An analysis was done by Colombo and Karney (2002) to point up the role of total system demand, leak location, and topological complexity. A thorough analysis of the relation between water loss and energy cost has been done. The impact of leaks on water age is also evaluated through simulation and via a dimensionless expression relating leak size and location to residence time.

Discussion

From the leakage detection studies, the main conclusion can be drawn is that many of the approaches are data hungry. For accurate results we have to go for many control parameters. Since the pressure measuring sensors and other equipments are expensive, the need for a method with minimum number of attributes which can be measured easily in the field has increased. Many of the studies have been done for the virtual data and thus its field application is to be studied further and specifically for the area which we are considering the particular method. As much as leakage detection and localization methods are considered and research works are done, similarly the rehabilitation methods are also need to be addressed. As earliest of the leakage detection methods which are mainly equipment based on nature, Pudar and Liggett (1992) presented an article in which how pressure measurements can be used to detect leakages is analysed.

Hydraulic transient flow is very difficult to measure due to the lack of instruments or flow metres which react instantaneously to these transients. Early studies (Wiggert et al. 1968) on fluid transients using leak detection lead to the recent developments. By analysing the study done by Liggett and Chen (1994), it can be observed that though they were limited by the technology of their time, optimistic suggestion regarding the efficiency has been put forward by coupling it to unsteady state analysis. As studied by Brunone (1999) the precision of transient analysis was further confirmed with accurate results. This method was found to be appropriate for outfall pipe analysis. Covas and Ramos (2001) had implemented the inverse transient methodology by using a software tool. According to their conclusion inverse transient analysis method was able to detect leak, which was around two-third of the total flow. Its applicability when the leaks are small in size is still need to be explored in detail. Leakage response of cracks in viscoelastic pipes can be estimated accurately by the leakage exponent methodology by MNF analysis.

Very few leakage detection methodologies are based on probabilistic approaches. Bayesian system approaches are one among them. Errors are associated with the calculation of leakages so to find out the more probable answer Bayesian concept is used. The only problem is that it requires great computer efficiency. The uncertainties of the water distribution network can be modelled by using fuzzy numbers. Only when a number of pressure sensors are limited and multiple leakages are there, the computational efficiency should be higher. Several other ANN-based methods are also recently used to estimate the leakage points accurately. They are one among the most promising online models. The model developed by Islam et al. (2011) can be integrated into the EPANET software, which increases its capacity to be used for leakage detection. The successful development of these methods will lead to the development leakage control measures, rehabilitation methods, etc. Above all the sampling and designing procedure for the locating position of pressure and flow sensors along network should be developed such that leakage location will be more feasible and reliable.

References

Adachi S, Takahashi S, Kurisu H, Tadokoro H (2014) Estimating area leakage in water networks based on hydraulic model and asset information. Elsvier Procedia Eng 89:278–285

Aksela KM, Aksela M, Vahala R (2009) Leakage detection in a real distribution network using a SOM. Urban Water J 6(4):279–289

Almandoz J, Cabrera E, Arregui F, Cabrera E Jr, Cobacho R (2005) Leakage assessment through water distribution network simulation. J Water Resour Plann Manage 131(6):458–466. doi:10.1061/(ASCE)0733-9496

Anjana GR, Sheetal Kumar KR, Mohan Kumar MS, Amrutur B (2015) A particle filter based leak detection technique for water distribution systems. Elsvier Procedia Eng 119:28–34
Annus I, Vassiljev A (2014) Different approaches for calibration of an operational water distribution system containing old pipes. Elsvier Procedia Eng 119(2015):526–534
Babic B, Stanic M, Prodanovic D, Dzodanovic B, Dukic A (2014) Reducing uncertainity of infrastructure leakage index—a case study. Procedia Eng 89:1577–1584
Boulos P, Altman T (1991) A graph-theoretic approach to exhibit nonlinear pipe network optimization. Appl Math Model 15:459–466
Brunone B (1999) Transient test-based technique for leak detection in outfall pipes. J Water Resour Plann Manage
Candelieria A, Contib D, Archettia F (2014) A graph based analysis of leak localization in urban water networks. Elsvier Procedia Eng 70:228–237
Chiara CM, Notaro V, Puleo V, Nicolosi P, Freni G (2015) Contaminant intrusion through leaks in water distribution system: experimental analysis. Procedia Eng 119:426–433 Elsevier
Choia YJ, Ahn JC, Im HT, Koo A (2014) Best management practices for water loss control in seoul. Elsvier Procedia Eng 89(2014):1585–1593
Colombo AF, Karney BW (2002) Energy and costs of leaky pipes: toward comprehensive picture. J Water Resour Plann Manage 128(6):441–450. doi:10.1061/(ASCE)0733-9496
Comon P (1994) Independent component analysis, A new concept? Elsvier Signal Process 36:287–314
Covas D, Ramos H, Hydraulic transients used for leakage detection in water distribution systems BHR group 2001 water pipeline systems
Covas D, Ramos H, de Almeida AB (2005) Stand wave difference method for leak detection in pipeline systems. J Hydraul Eng 131(12):1106–1116
Deyi M, van Zyl J, Shepherd M (2014) Applying the FAVAD concept and leakage number to real networks: a case study in Kwadabeka, South Africa. Elsvier Procedia Eng 89:1537–1544
Ferrantea M, Brunone D, Meniconi S, Capponi C, Massari C (2014) The leak law: from local to global scale. Elsvier Procedia Eng 70:651–659
Fox S, Collins R, Boxall J (2015) Traditional leakage models for leakage modelling: effective or not? Procedia Eng 119:35–42 Elsvier
Gao J, Qi S, Wu W, Li D, Ruan T, Chen L, Shi T, Zheng C, Zhuang Y (2014) Study on leakage rate in water distribution network using fast independent component analysis. Procedia Eng 89:934–941
Giorgio Bort CM, Righetti M, Bertola P (2014) Methodology for leakage isolation using pressure sensitivity and correlation analysis in water distribution systems. Elsvier Procedia Eng 89:1561–1568
Karney B, Khani D, Halfawy MR, Hunaidi O A simulation study on using inverse transient analysis for leak detection in water distribution networks. J Water Manage Model
Lambert A (2001) What do we know about pressure: leakage relationships in distribution systems? In: Proceedings of IWA conference 'system approach to leakage control and water distribution system management, Brno, Czech Republic
Liggett J, Chen L (1994) Inverse transient analysis in pipe networks. J Hydraul Eng 120(8):934–955. doi:10.1061/(ASCE)0733-9429
Okeya I, Kapelan Z, Hutton C, Naga D (2014) Online burst detection in a water distribution system using the Kalman filter and hydraulic modelling. Elsvier Procedia Eng 89:418–427
Poulakis Z, Valougeorgis D, Papadimitriou C (2003) Leakage detection in water pipe networks using a Bayesian probabilistic framework. Probabilistic Eng Mech 18:315–327
Pudar RS, Liggett JA (1992) Leaks in pipe networks. J Hydraul Eng ASCE 118(7), 1031e1046
Qi S, Wu W, Gao J, Qiao M Tu W, Wang J (2014) Research on optimized leakage locating model on water distribution system. Procedia Eng 89:934–941
Ristovski B (2011) Pressure management and active leakage control in particular DMA (Lisiche) in the city of Skopje. FYROM Water Utility J 2:45–49
Rogers D (2014) Leaking water networks: an economic and environmental disaster. Elsevier Procedia Eng 70:1421–1429

Sala D, Kołakowski P (2014) Detection of leaks in a small-scale water distribution network based on pressure data—experimental verification. Elsvier Procedia Eng 70:1460–1469
Wiggert DC (1968) Unsteady flows in lines with distributed leakage. ASCE J Hydraul Div 94 (HY1):143–161
Winarni W (2009) Infrastructure leakage index (ILI) as water losses indicator. Civ Eng Dimension 11(2):126–134
Yoo DG, Kang D, Jun H, Kim JH (2014) Rehabilitation priority determination of water pipes based on hydraulic importance. Water 6

Quantification of Water Footprint of National Capital Territory (NCT) of Delhi, India

Deepa Chalisgaonkar, Sharad K. Jain, M. K. Nema and P. K. Mishra

Abstract Overexploitation of freshwater caused by the increasing population and urbanization over the past few decades has resulted in water scarcity in National Capital Territory (NCT) of Delhi. Water footprint (WF) is a useful tool to better understand the linkages between humanity's activities and their growing pressure on the freshwater resources. It indicates the amount of water required to produce all the goods and services by the individual or community, or geographic area directly and indirectly (virtual water). It is measured in terms of water volumes consumed (evaporated) and/or polluted per unit of time. Many studies have been carried out pertaining to the water requirement of NCT of Delhi. The present study is focused on the assessment of the WF of NCT of Delhi. The total WF of NCT of Delhi for the year 2010 has been assessed as 15,926 MCM per annum. Out of this, 6530, 780 and 865 MCM per annum have been contributed from domestic, agriculture, and industrial sectors, respectively, and rest (7751 MCM per annum) is as virtual water import. Certain assumptions were made due to non-availability of some of the data. With the availability of more data, the assessment can be improved. Quantification of WF of NCT of Delhi indicates that domestic sector is the major water consumer. Sector-wise results can facilitate authorities to develop improved management policies, action plans, and strategies for better management of freshwater resources.

Keywords Water footprint · NCT-delhi · Virtual water · Grey water

Introduction

The increasing population and overexploitation of surface and groundwater over the past few decades have resulted in the water scarcity in NCT of Delhi. Efforts are continuously being made by various organizations to develop strategies to meet out the water scarcities. From public as well as private sectors different initiatives have

D. Chalisgaonkar (✉) · S. K. Jain · M. K. Nema · P. K. Mishra
National Institute of Hydrology, Roorkee 247667, India
e-mail: deepa_c@yahoo.com

V. P. Singh et al. (eds.), *Water Resources Management*, Water Science and Technology Library 78, https://doi.org/10.1007/978-981-10-5711-3_11

been launched (e.g. Reduce–Reuse–Recycle, rainwater harvesting, automated irrigation system, anti-boring laws for groundwater extraction, water pricing, etc.) to prevent and mitigate water scarcity for developing water sustainability. In the recent past, few water indicators also have been developed in this regard. Some indicators which are considered to best represent the overall status of water resources of a region are virtual water, water neutrality, water mark, water debt, water audit, peaking water and water footprint. Out of these indicators, virtual water and water footprint have been worked out in this study.

Virtual water: The water required for the production of a commodity, goods or services is called virtual water (VW). The virtual water concept is primarily used for measuring the distribution/transportation of water through trade between states/regions/countries. The virtual water of a product may vary from place to place because of the difference in climatic conditions. The concept of virtual water helps us to realize how much water is needed to produce different goods and services. In semiarid and arid areas, knowing the virtual water content of a goods or service can be useful towards determining how best to use the available scarce water. It can be computed by

$$\text{VW Content} = \text{Direct VW} + \text{Imported VW} - \text{Exported VW}. \quad (1)$$

Water footprint (WF): The WF is a consumption-based indicator of water use that looks at both direct and indirect water use of a consumer or a producer or a region. The WF of an individual, community or business is defined as the total volume of all kinds of water that is used to produce the goods and services consumed by the individual or community or produced by the business. It is measured in terms of water volumes consumed (evaporated) and/or polluted. It can be calculated for any well-defined group of consumers (e.g. an individual, family, city or nation) or producers (e.g. an organization). For example, the WF of a cotton T-shirt is 2600 litres and covers the water used from cotton plantation until the T-shirt is put on the stores' shelves. Information about WF of products, countries and companies will help to understand how a more sustainable and equitable use of freshwater can be achieved. The WF of an area is primarily assessed in three important sectors, i.e. domestic, agriculture and industrial by considering the direct as well as indirect water consumptions.

Domestic water footprint: The domestic WF of a person is the amount of water he/she uses in and around his/her house throughout the day. This includes the water he/she uses directly (i.e. from a tap) as well as the water he/she uses indirectly (food he/she eats and to dilute the water he/she pollutes, etc.). One may not drink, feel or see this **virtual water**, but it makes up the majority of his water footprint. One of the main factors determining the water footprint of the different regions is the consumption habits of its people both through direct water consumption and virtual water consumption due to a strong demand for industrial products and a diet high in meat. Figure 1 shows the water needed, either used or polluted, to make common consumer goods.

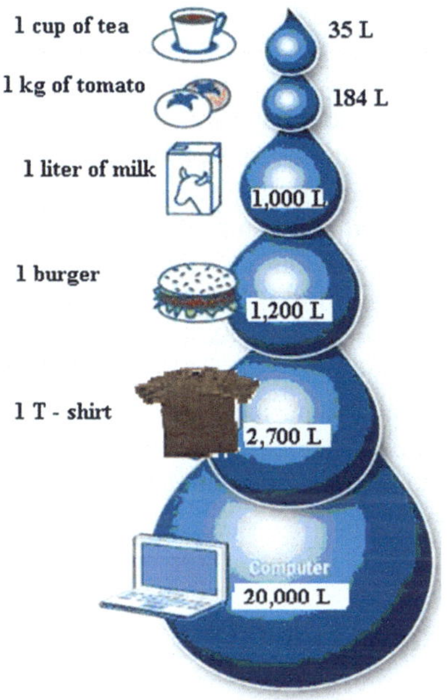

Fig. 1 Water needed (used or polluted) to make common consumer goods (*Source* https://fosteringeducationandenvironmentfordevelopment.files.wordpress.com/2012/11/water-footprint-analysis-for-water-management-education.pdf)

Agriculture water footprint: Agricultural sector (including forest and shrub area) is a major water-consuming sector. The agricultural WF corresponds with the water use in the agricultural sector (i.e. in the form of crop evapotranspiration (ET) or water pollution). The agricultural water footprint measures the volume of ET or water use of a crop per unit mass of yield. Comparing water footprints of different management practices in agriculture can help in evaluating drought tolerance, water use efficiency, effective use of rainfall and the significance of irrigation.

Industrial water footprint: The industrial sector needs an appropriate quantity and quality of water for consumption as an elementary ingredient for various processes and operations. Although many industries have high shares of effluents leading towards pollution to hydro-ecosystems, risking water reserves for sustaining long-run operations. Industrial WF is defined as the total volume of freshwater used directly or indirectly to run and support a business (rather industrial or commercial). It is an indicator of water use for industry quantifying the total amount of water volumes consumed, their destinations, flows standards and volumes of water pollution flowing back to an environmental system.

As the WF is the volume of freshwater used to produce the product, measured over the full supply chain, all the three sectoral WFs, i.e. domestic, agriculture and industrial WFs, are further subdivided into three components given below:

Blue water footprint: It is the volume of surface water and groundwater consumed (i.e. evaporated or incorporated into the product) during production processes (also includes irrigation water used for crop growth).

Green water footprint: It is the volume of rainwater consumed (i.e. evaporated or incorporated into the product) by the product. In other words, it is the consumptive use of rainwater for crop growth (Evapotranspiration).

Grey water footprint: It is the amount of freshwater required to mix pollutants and maintain water quality according to agreed water quality standards. In other words, it is the water required to dilute the water pollution that is caused by the production of the commodity to acceptable levels.

The present study estimates the water footprints of NCT of Delhi from consumption perspective by quantifying the green, blue and grey water footprints for all the major water-consuming sectors (domestic, agriculture and industrial) within NCT of Delhi. The results of the study could be useful in quantifying the overall water scenario of NCT of Delhi.

Water Footprint: Status Quo

The conceptual tools necessary to compare, quantify and price the water consumption and trade have emerged within the last 15 years. In 2004, the first global study including water footprint and virtual water trade analysis was published (Chapagain and Hoekstra 2004), containing virtual water flows per country related to international trade of crop, livestock and industrial products for the period 1997–2001, and the water footprint of national consumption for almost every country in the world. A refined analysis was presented in the book "Globalization of Water" (Hoekstra and Chapagain 2008), in which specific case studies were discussed in detail. Water footprints were presented versus water scarcity, self-sufficiency and water import dependency of various countries (Water Footprint and Corporate Water Accounting for Resources Efficiency 2011; World Economic Forum Water Initiative Draft 2009). Figures presented demonstrate the nature of water as a geopolitical resource and urge decision-makers to give priority to this resource in their political agendas. Several water footprint assessment studies have been undertaken since 2004 at different geographical levels. Holcomb (2010) quantified water footprint for growing crops sustainably in Northwest India. In 2011, a collaborative study on water footprint was done by Korea International Cooperation and UNEP for analysing corporate water accounting for resource efficiency. Later Fulton et al. (2014) studied the impact of outcomes of the water footprint of California State on policy relevance changes concluding with a comparison with results at the national level. This reveals that calculating water footprints helps to find out the global nature of freshwater. The water scarcity can only be measured locally, but accounting for the water footprints of products means that the amounts of water being "traded" can be monitored.

Keeping this in view, the major objective of the present study is to quantify the various components of the water footprint of National Capital Territory (NCT) of

Delhi, India for all the three important sectors, i.e. domestic, agriculture and industrial by considering the direct as well as indirect water consumptions. This study also estimates the virtual water import to the NCT of Delhi.

Study Area

The India's capital city Delhi has a specific position in the Indian institutional system. Although about 75% of the total geographical area (1483 km^2) is urbanized, the urban agglomeration of Delhi extends its limits out of the NCT, with satellite towns like Gurgaon, Noida, Faridabad and Ghaziabad, growing in the immediate vicinity of Delhi in the neighbouring states of Haryana and Uttar Pradesh. The location of NCT of Delhi is shown in Fig. 2.

The climate of the NCT of Delhi region is semiarid in nature. About 87% of the annual rainfall is received during the monsoon months, viz. June to September. The average annual rainfall of the Delhi of NCT is about 800 mm. NCT of Delhi's population grew at an annual growth rate greater than 4% during the last decade. Although the NCT of Delhi is said to be 75% urban, the 25% rural area contributes a variety of livestock presence at Delhi. The travel and transport demands of NCT of Delhi are increasing with the growth of population and economic activities.

The agricultural activities in NCT of Delhi include (i) growing of field crops, fruits, seeds and vegetables and (ii) management green area and forest plantations.

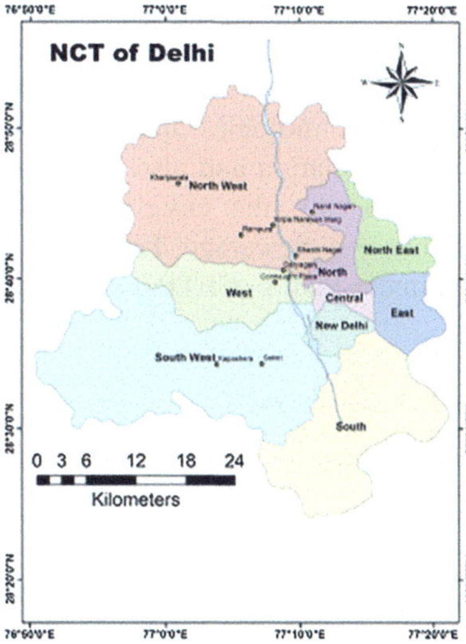

Fig. 2 Location and administrative boundaries of NCT of Delhi water management in NCT of Delhi

As per forest department of NCT of Delhi, the tree cover area is 229.6 km^2 consisting of forest area of 85 km^2. There are more than 18,000 parks and gardens in NCT of Delhi spread in about 8000 ha in various locations. Delhi has 29 planned industrial areas and 5 multi-storeyed factory complexes. In addition, 22 non-confirming industrial clusters have been notified for development. All the major activities in NCT of Delhi are based on electricity. Delhi being a city state with diminishing rural areas and agricultural activities, the thrust on energy front in Delhi is mainly to have uninterrupted power supply and to take care of increasing power demand. In order to meet this demand of electricity, NCT of Delhi has state-owned power generation facilities and also imports from the neighbouring states.

Delhi Jal Board (DJB) is responsible for procurement and treatment of allocated raw water to NCT of Delhi. Because of small size of the NCT, the DJB relies heavily on surface resources which are mobilized outside the NCT. Table 1 shows that the three major Himalayan rivers, viz. Ganga, Yamuna and Sutlej, are primary source of surface water that is being made available to NCT of Delhi through the canals and directly from river (in case of Yamuna) which accounts to 85% of the total water supply source and rest 15% is being met out from the groundwater that is being explored through Renny Wells and Tube Wells of DJB. The groundwater, the only resource available to fill the gap between drinking water requirement of the NCT and the raw water available, is in a very critical condition as the pace of groundwater recharge is far behind the pace of ground water draft.

The DJB has around 1.48 million domestic water supply connections. The data related to NCT of Delhi indicate that Yamuna River is the most prominent source which not only enables supply of drinking water but also serves as a sink for waste of Delhi. This dumping of waste has tremendously increased the pollution level in the Yamuna water in the region of Delhi. The water supply need and projected capacities show a huge gap. Thus it is an indication to search for alternatives sources, technologies and approach in order to be able to meet the demands of future population and also to safeguard the river Yamuna. With the growth of the city of Delhi, the environmental concerns have assumed greater importance. Twenty-two Sewerage Treatment Plants (STPs) and Thirteen Common Effluent Treatment Plants (CETPs) are operational in various areas of NCT of Delhi.

Table 1 Supply water resources of Delhi Jal Board in March 2011

S. No.	Resources	Quantity (MGD)
A.	**Surface water sources**	
1	Yamuna River	310
2	Ganga River	240
3	Satluj River (Bhakra Storage)	140
	Subtotal (A)	**690**
B.	**Groundwater sources**	
1	Renny Wells/Tube Wells	115
	Total (A + B)	**805**

Source Delhi Jal Board (2010)

Methodology

The methodology used in this study is largely based on earlier studies supported by water footprint network (www.waterfootprint.org), which already calculated international virtual water flows and water footprints of nations (Hoekstra and Chapagain 2007) and comprehensive introductions to water footprints (Hoekstra 2008; Hoekstra et al. 2011; Kampman 2007; Hoekstra and Mesfin 2012). In the present study, these previous methodologies are integrated as per the data availability. The study of existing water management schemes of water supply and treatment of NCT of Delhi reveals that sources of water in NCT of Delhi, consumption of water for different purposes and processing of water at the treatment plant influence the water use. The water footprints thus consist of two components: consumptive water use and wastewater pollution. The current study is focused on the quantification of consumptive water use. The impact of water pollution has been assessed by quantifying the dilution water volumes required to dilute waste flows to such extent that the quality of the water remains below agreed water quality standards. The WF of the NCT of Delhi has been computed based on the available data of direct (real) as well as indirect (virtual) water consumption for the period 2006–2010. The related data have been collected from various sources, published reports from various departments of government of NCT of Delhi, from important websites and other reports. The virtual water content-related data are available at country level not at NCT of Delhi level, so these were used for NCT of Delhi as well. The data which were not available have been assumed with justification (Shaban and Sharma 2007).

Assessment of Domestic Water Footprint

As the study area is an urban-dominated area, the green component of the domestic sector is negligible. The blue component of domestic WF has been calculated by first calculating the total volume of water available for human consumption out of the total volume of water supplied by DJB. This has been done by subtracting the water requirements of all the animals (livestock) and vehicles (transport) in NCT of Delhi from the total volume of water supplied by DJB. The blue water footprint was then computed by dividing this figure by the human population in NCT of Delhi. As the computation of grey water footprint includes the amount of freshwater required for mixing pollutants and maintaining water quality according to agreed water quality standards, the water quality data of the outlets of sewage treatment plants (STP) for the year 2010 of NCT of Delhi have been used. The water quality criterion of Central Pollution Control Board ('C' Class water) has been taken as the water quality standards for the computation of dilution water requirement. The 'Dissolved Oxygen' (DO) parameter for the STPs is not available, so it has been ignored. The grey water footprint has been taken as the highest of the dilution water

Table 2 Computation of water consumption by human population

Total domestic water supply	37.68 Lakh kL per day
Livestock water consumption	0.22 Lakh kL per day
Domestic vehicle water consumption	0.48 Lakh kL per day
Water supply available for human population	36.98 Lakh kL per day

Table 3 Water consumption per capita per day

Year	Population (lakhs)	Computed LPCD	Actual LPCD[a]	BIS recommended LPCD
2011	167.53	220.78	99.35	135

[a]Assuming 'Unaccounted for Water' is ranging from 40 to 50%

requirement of each of the components. Different steps of computation are presented below.

Water requirements (litres/day) of the livestock in NCT of Delhi have been computed. The water requirements (litres/day) for the domestic vehicles in NCT of Delhi have also been computed. Based on the above data, the water available for the human population has been computed as shown in Table 2.

Based on the Census 2011 and DJB data, the available water supply for both urban and rural settlements, the domestic water consumption was calculated as 220.79 litres per capita per day (LPCD), but due to significant transmission losses as well as 'unaccounted for water (UFW)' factors the water consumption per capita was found to be in the order of 99.35 LPCD which is far below the standard set by Central Public Health & Environmental Engineering Organization (CPHEEO), i.e. 220 LPCD as well as it does not meet the standard set Bureau of Indian Standard (BIS) i.e. 135 LPCD. The details are shown in Table 3.

Grey component of the domestic water footprint was calculated using ratio method. Based on the available water quality data from the various STP outlets of NCT of Delhi, the grey component of domestic water footprint was estimated to be 5155 MCM per Annum (843 LPCD). The blue and grey WF added together give domestic water footprint.

Assessment of Agricultural Water Footprint

The blue component of Agriculture WF is negligible since NCT is an urban-dominated region. The green component in crop water use (CWU, m^3/ha) has been calculated by accumulation of daily evapotranspiration (ET, mm/ day) over the complete growing period:

$$CWU_{green} = 10 \times \sum_{d=1}^{lp} ET_{green}, \text{ volume/area,} \quad (2)$$

where *lp* is the crop growing period.

The green component has been assessed by computing crop water requirements (CWR) for different crops in NCT of Delhi by estimating evapotranspiration rates for a specific crop, in the specific climate of NCT of Delhi. This computation has been done using the CROPWAT 8.0 and CLIMWAT 2.0 software of Food and Agriculture Organization (FAO) of the United Nations. The CROPWAT uses precipitation data, crop growth inputs and general soil data to calculate crop water requirements. The crop water requirements and irrigation requirements for all the crops of NCT of Delhi added together are presented in Table 4.

Computation of grey water footprint is a different segment of quantifying available water resources. While studying the agricultural sector, it is important to include the grey water in the total water footprint because non-point pollutants such as fertilizers, pesticides and insecticides have a significant impact on the water demands of a farm. Often times, enforced water quality standards require pollutants to be diluted by freshwater to attain certain ambient legal levels. The grey component in water footprint of growing a crop or tree has been calculated by using the formula given below:

$$WF_{proc,grey} = [(\alpha \times AR)/(C_{max} - C_{nat})]/Y, \text{ m}^3/\text{ton,} \quad (3)$$

Table 4 Crop water requirement and irrigation requirement for all crop types

Crop type	ET_c mm/annum	Eff rain mm/annum	Irr. req. mm/annum	Area in ha	CWR million litre/annum
Wheat	303.7	34.8	269.5	20,135	61,150
Rice	807.1	470.2	608.5	6848	55,270
Barley	345.8	107	292.5	70	242
Millet	351.4	348	143.1	1531	5380
Maize	470.5	426.1	187.8	40	188
Sorghum/Jowar	414.4	426.1	160.9	3341	13,845
Pulses	250.4	35.6	213.2	8	20
Potato	354.3	36.8	315.9	51	181
Sugarcane	2031.4	515.3	1522	3	61
Perennial grass	1192	515.3	750.1	8000	95,360
Perennial shrubs	1182.6	515.3	858.7	29,958	354,283
Forest	1771.5	515.3	1302.2	8500	150,578

where

AR Chemical application rate to the field per hectare (kg/ha)
α Leaching runoff fraction
C_{max} Maximum acceptable concentration (kg/m^3)
C_{nat} Natural concentration for the pollutant considered (kg/m^3)
Y Crop yield (tonne per hectare).

The grey water component was calculated based on the application of nitrogen (N) fertilizer to NCT of Delhi crop fields. Lacking Delhi-specific data, several assumptions regarding fertilizer use and transport has been made based on Hoekstra's Manual. The leaching fraction (quantity of N that reaches water bodies) was assumed to be 10% of the applied fertilizer rate. Due to unavailable local ambient N water quality standards, the United States Environmental Protection Agency (USEPA) recommendation (maximum of 10 mg of nitrate per L of water) was used (Hoekstra et al. 2009). The natural concentration of N in the receiving water body was assumed to be zero. Only the nitrogen fertilizer use was incorporated into the grey water footprint, because the grey component is expressed as a dilution water requirement. This means only the most critical pollutant with the greatest application rate need be considered (Hoekstra et al. 2009). By the addition of the total crop water requirement and grey component, total agriculture WF had been calculated.

Assessment of Industrial Water Footprint

Being an urban area, the green component of the industrial WF was assumed almost negligible. The blue component of industrial WF has been assessed by calculating the total volume of supplied by DJB for commercial and industrial purposes. As per Delhi Electricity Regulatory Authority about 7000 million units are locally generated in various power plants in Delhi. It has been reported in few of the industrial reports that production of one unit of thermal power requires around 3 litres of water and in case of hydropower the water requirement is 17 litres per unit. The water consumption for electricity generation has also been considered.

The water quality data of the outlets of common effluent sewage treatment plants (CETP) of NCT of Delhi have been used for the computation of grey water footprint in the similar way as it has been done in the case of domestic grey water footprint. The blue and grey WF have been added to get the final value of industrial water footprint of NCT of Delhi.

Assessment of Virtual Water

In the present study, for the assessment of agriculture products including livestock-based products, the data related to import of these products have been obtained from Delhi Agricultural Marketing Board. Virtual water contents for the crops from water footprint network have been considered for computation of virtual water import (Kumar and Jain 2007).

NCT of Delhi heavily depends on the neighbouring states for fulfilling the electricity requirements. It imports around 17,764 million units/year (66%) of its total electricity requirements from other states. Therefore, the assessment of the virtual water import in terms of electricity has also been included in the study. Crude oil is often attributed as the "Mother of all Commodities" because of its importance in the manufacturing of a wide variety of materials. In NCT of Delhi crude oil in various forms like LPG, petrol, diesel, aviation fuel, etc., is procured from other states. Therefore, the assessment of the virtual water import in terms of crude oil has also been included in the study.

Results and Discussion

In the present study an attempt has been made to quantify the overall water footprint of NCT of Delhi including direct and indirect water consumptions. The direct WF assessment has been done for domestic use (Blue and Grey), agricultural use (Green and Grey) and industrial use (Blue and Grey). To account for the indirect water uses, the concepts of virtual water import and grey water have also been applied. The various results of the study have been given in Tables 5, 6, 7, and 8 and discussed in further sections below.

Table 5 Domestic water footprint of NCT of Delhi

Details	Quantity		MCM
Water supply of DJB	13,754.00	Lakh kL/annum	1375.40
Livestock water requirement	79.43	Lakh kL/annum	7.94
Water requirement for domestic vehicles	173.68	Lakh kL/annum	17.36
Water supply for human population	13,500.89	Lakh kL/annum	1350.09
BIS recommended water supply (LPCD)	135.00	LPCD	–
Computed LPCD	220.78	LPCD	–
Actual LPCD (assuming UFW as 45%)	99.35	LPCD	–
Dilution water requirement	51,551.17	Lakh kL/annum	5155.11
Grey water footprint	843.00	LPCD	–
Domestic water footprint (MCM/Annum)			**6530.51**

Table 6 Agriculture water footprint of NCT of Delhi per season

Crop type	CWR million litre	CWR MCM	Grey WF (m^3/tonne)	Grey WF MCM	Footprint MCM
Wheat	61,150	61.15	329.49	30.47	91.62
Rice	55,270	55.27	354.07	10.10	65.36
Barley	242	0.242	177.3	0.04	0.28
Millet	5380	5.38	265.96	0.79	6.17
Maize	188	0.19	56.92	0.06	0.25
Sorghum/ Jowar	13,845	13.85	51.12	1.62	15.46
Pulses	20	0.02	568.18	0.04	0.06
Potato	181	0.19	2	0.04	0.23
Sugarcane	61	0.06	11.28	0.003	0.06
Perennial grass	95,360	95.36	NA	NA	95.36
Perennial shrubs	354,283	354.28	NA	NA	354.28
Forest	150,578	150.58	NA	NA	150.568
Agriculture water footprint		**736.56**		**43.161**	**779.72**

Table 7 Industrial water footprint of NCT of Delhi

Details	Quantity		Qty/Annum
Water supply of DJB	3.45	Lakh kL/day	125.92 MCM
Water consumption for generation of electricity	–	–	20.7 MCM
Grey water footprint	19.87	Lakh kL/day	718.21 MCM
Industrial water footprint	5.51	Lakh kL/day	**864.83 MCM**

Table 8 Virtual water import

Electricity purchased from other states in annually	25,823 million units
Virtual water import due to electricity import assuming hydropower as the main source (15 L per kW)	387.35 MCM
Virtual water import due to crude oil (4080 thousand tones) annually	326.30 MCM
Virtual water import due to import of crop-based products annually	5138.7 MCM
Virtual water import due to import of animal-based products annually	1898.43 MCM
Total virtual water import	**7750.78 MCM**

Note Based on the data collected from Annual report of Directorate of Animal Husbandry (2007), Delhi Statistical Hand Book (2012), Economic Survey of Delhi (2012–2013), Statistical Abstract of Delhi (2012), Tourism Survey in the State of Delhi-Annual Final Report (2009–10)

Domestic Water Footprint

Analysis and results pertaining to the domestic water footprint of NCT of Delhi are given in Table 5. It can be seen that DJB supplies about 220.78 LPCD of water which is about 163% more than the BIS recommended norms of water supply, i.e.

135 LPCD. But because of 'unaccounted for water (UFW)' ranging from 40 to 50% a citizen of NCT of Delhi is eventually getting 99.35 LPCD which is 74% of the recommended BIS norm. Domestic WF of NCT of Delhi comes of to be 6530.51 MCM per annum (4.4 MCM/km^2/annum) which also includes 5155.11 MCM per annum of indirect water consumption in the form of dilution water required for bringing the output of STPs to an acceptable level. The grey water footprint is 78% of the total domestic water footprint of NCT of Delhi.

Agriculture Water Footprint

The agriculture sector is the primary water user sector in general, but being an urban setup the NCT of Delhi does not uses much of the water for its agricultural practices. The agriculture water footprint for NCT of Delhi has been computed as 779.72 MCM per annum. As shown in Table 6 it has about 5% of grey water component. It also shows that a significant part of the agriculture water component of NCT of Delhi (76%) is utilized by perennial grass, perennial shrubs and forests and very less part goes for the production of food grains and vegetables.

Industrial Water Footprint

Being an urbanized residential area, there are very less industries in NCT of Delhi. The industrial water footprint of NCT of Delhi has been assessed as 864.83 MCM per annum (Table 7) including 718.21 MCM per annum of indirect water consumption in the form dilution water required for bringing the output of CETPs to an acceptable level. It is also assessed that 20.7 MCM/annum water is used for generation of electricity in its various thermal power plants. The grey water footprint is 83% of the total industrial water footprint of NCT of Delhi.

The sum of domestic, agriculture and industrial WF of NCT of Delhi is 8175 MCM per annum including the grey component.

Virtual Water

In addition to the earlier discussed water consumptions, a huge quantity of virtual water is also being transferred to/from NCT of Delhi in the form of various goods, commodities, products, etc. Based on the available data, the indirect WF has been assessed due to virtual water transfer of agricultural products (crop-based), animal-based products, petroleum products and electricity as 7750.78 MCM per annum as shown in Table 8.

Total Water Footprint

The total WF of NCT of Delhi has been assessed as 15,926 MCM per annum which is due to water consumption of 41% in domestic sector, 5% in agriculture sector, 5% in industrial sector and 49% because of virtual water transfer.

Conclusion

The present study has been done to have insight of water consumption in NCT of Delhi, India in the context of WF. The WFs have been assessed for domestic, agriculture and industrial uses of water in the NCT of Delhi. Water consumed per annum by various types of crops, grass, shrubs and forests in various forms as well as water consumed by other various activities including virtual water import have been assessed in the study. The total WF of NCT of Delhi has been assessed as 15,926 MCM per annum which is due to water consumption of 41% in domestic sector, 5% in agriculture sector, 5% in industrial sector and 49% because of virtual water transfers. The WF comes out to be approximately 950 m^3/capita/annum which is almost equal to the National WF average. In addition to this, the water is brought in Delhi through unauthorized tankers. The WF of the average global consumer is 1,385 m^3/capita/annum. The average consumer in the United States has a WF of 2,842 m^3/capita/annum, whereas the average citizens in China and India have WFs of 1071 and 1089 m^3/capita/annum, respectively.

The present assessment has been carried out based on the best possible available data sets related to virtual water transfer as the data of all the commodities were not available. Moreover, huge amount of water is utilized in NCT of Delhi which is brought in via tankers but there is no complete record of these tankers. With the availability of more data, the WF of NCT of Delhi is expected to increase. The results of the study will enable authorities to develop improved strategies by quantifying the freshwater demands of all the three major sectors, viz. domestic, agriculture and industrial. It will also help the authorities to identify the sectors/ commodity with large water footprint so that a check can be made in order to optimize the resources. The study reveals that the grey water footprint in all the sectors is the major shareholder which implies that water is facing huge environmental problems in NCT of Delhi and the efforts should be taken to minimize the grey water footprint.

Limitations

There is still no consensus with regards the standardized approach to be employed for the assessment of WF. The physical properties of water such as its spatial and temporal availability, along with its many different sources and types or "colours", make it exceptionally difficult to value across varying contexts or locations. One of the main factors determining the domestic water footprint is the consumption habits of its people both through direct water consumption and virtual water consumption. There are quite a number of practical issues that one will encounter when carrying out a water footprint assessment. As the water footprint studies are data intensive in nature and many a times all of the required data are not available, a major issue is to handle the lack of requisite data.

References

Annual report of Directorate of Animal Husbandry (2007) GNCT of Delhi

Chapagain AK, Hoekstra AY (2004) Water footprints of nations, research report series no 16. UNESCHO-IHE, Delft, The Netherlands

Delhi Statistical Hand Book (2012) Directorate of Economics and Statistics, Government of NCT of Delhi

Economic Survey of Delhi (2012–2013) Planning Commission of NCT, Delhi

Fulton J, Cooley H, Gleick PH (2014) Water footprint outcomes and policy options change with scale considered: evidence from California. Water Resour Manage 28(11):3637–3649

Hoekstra AY (2008) A comprehensive introduction to water footprints

Hoekstra AY, Chapagain AK (2007) Water footprints of nations: water use by people as a function of their consumption pattern. Water Resour Manage 21(1):35–48

Hoekstra AY, Chapagain AK (2008) Globalization of water: sharing the planet's freshwater resources. Wiley

Hoekstra AY, Mesfin MM (2012) The water footprint of humanity. PNAS 109(9):3232–3237

Hoekstra AY, Chapagain AK, Aldaya MM, Mekonnen MM (2009) Water footprint manual: state of the art 2009. Water footprint network. Enschede, The Netherlands

Hoekstra AY, Chapagain AK, Aldaya MM, Mekonnen MM (2011) The water footprint assessment manual: setting the global standard. Earthscan, London, UK. Water footprint Network. http://www.waterfootprint.org/?page=files/FAQ

Holcomb MK (2010) Quantifying the water footprint: growing crops sustainably in northwest India. School of Environment and Natural Resources, The Ohio State University

Kampman DA (2007) The water footprint of India. University of Twente, Enschede, The Netherlands

Kumar V, Jain SK (2007) Status of virtual water trade from India. Curr Sci 93(8):1093–1099

Shaban A, Sharma RN (2007) Water consumption patterns in domestic households in major cities. Econ Polit Wkly 42:2190–2197

Statistical Abstract of Delhi (2012) Directorate of Economics and Statistics, Government of NCT of Delhi

Tourism Survey in the State of Delhi-Annual Final Report, 2009–10, Ministry of Tourism, Government of India

Water Footprint and Corporate Water Accounting for Resources Efficiency, UNEP (2011)

World Economic Forum Water Initiative Draft (2009)

www.dpcc.delhigovt.nic.in. Accessed on 30 Jan 2014

Salt Water Intrusion and Water Security Issues of Coastal Community: Case of Thane District (Maharashtra)

M. D. Omprakash and Nivedita Gadikar

Abstract Coastal belt of India is prone to the problem of salt water intrusion. This exerts a great pressure on the coastal residents, who need to look for alternate sources of fresh water to meet their agricultural and domestic needs. Present study is an effort to understand the pattern of saltwater intrusion/inundation and the water security issues in the coastal belt of Thane district (Maharashtra). Groundwater salinity has been observed and well proven by laboratory analysis of water samples collected from groundwater wells and spatially analyzed data to understand the underlying causes. Physical, chemical, and biological parameters of the sampled water helped to draw correlation between various issues related to water security of the coastal community. Salinity could be due to inundation or induced, which is because of heavy drawing of groundwater. Various other factors like industrial waste disposal, local salt/clay pan may also be the other reasons of groundwater salinity. Spatial analysis of well location with respect to the distance reveals that there is not much significant relationship existing between distance and individual physicochemical parameters. However, the low lying area in the vicinity of creeks/ back water (rivers) has impacted the low depth wells. The low depth well (Veera) in the vicinity to creeks has been induced by saline water. This induced effect is due to the failure in functioning of tide control measures such as gates (Vasant bunds) made at local level. Due to the high salinity, cropping pattern has been changed and now farmers have restricted themselves to salt-resistant crops. Analysis of well depth with respect to elevation and other physicochemical parameters depicts significant relationships that infers to the over abstraction of groundwater. The study identified critical issues like inundation of sea water through creeps/back water (river) especially in the low laying regions, which can be avoided with the help of tidal regulation with gate operated bunds and subsurface bunds to avoid excess flow of groundwater to sea and reverse flow due to lowered hydraulic head.

Keywords Salt water intrusion · Groundwater · Water security

M. D. Omprakash (✉) · N. Gadikar
Indian Institute of Forest Management, Bhopal, India
e-mail: omprakashiifm@gmail.com

V. P. Singh et al. (eds.), *Water Resources Management*, Water Science and Technology Library 78, https://doi.org/10.1007/978-981-10-5711-3_12

Introduction

Globally more than 80% of the urban population lives in the coastal belt. With growing population and increasing trade, today the scenic landforms and unique ecosystems are degrading at an alarming rate (CGWB 1992). High population density along the banks of major rivers and coastal areas are also attributed to easy availability of water. In coastal areas, groundwater is usually considered to be the safest source of water for drinking and other purposes. The semi consolidates and unconsolidated sediments along the coastline have helped mankind to go in for deeper groundwater exploration. However, as the exploration advances toward deeper horizon, the problem of groundwater quality degradation becomes conspicuous. High density of population and changes in sedentary lifestyle of local people demand more water which has put the coastal aquifers under stress. When groundwater is pumped from the aquifers that are in hydraulic connection with the sea, the gradients that are set up may induce a flow of salt water from the sea toward that coastal aquifer. The term used for this phenomenon is 'Seawater Intrusion'. Because of peculiar coastal topography, the east coast of India already has evidences of seawater intrusion (Sharma 2008). Maharashtra State has the longest west coastal line and some district like Thane, Raigad, Ratnagiri, Sindhudurg are vulnerable to such issues of water security. The present paper is based on a study conducted to systematically understand this problem and to identify possible measures to address water security issues of the local community in coastal belt of Thane district.

Methodology

Both primary and secondary data were collected from the field and concerned department/organizations. Probability and non-probability sampling have been used under mixed method of study design. With the help of secondary data and primary information gathered through informal conversations with key persons in coastal belt of Thane district, the issue of groundwater problem had been confirmed. Hence stratified purposeful sampling was used. Groundwater wells were located in the field through consultation. The water was drawn from the well using rope and bucket method and was collected in 500 ml sterilized PET bottles. To assess the groundwater quality 100 groundwater samples were collected during March–April, 2013. All the samples were marked with sample number, location, date and parameters to be analyzed.

The bottles were submitted at the laboratory for the analysis of various chemical and biological parameters such as Ca, Mg, Cl and coliform. Physical parameters such as temperature, pH, TDS and EC were checked with the help of digital combo meter (HANNA 98, 129) and the readings were noted down. Remotely sensed data and maps of the district formed the basis for identification of samples and demarcation of land use of the sampled villages. Google Earth, image which had a high spatial resolution was used to develop understanding on the land use and land cover of the identified villages and surroundings. Based on the geo-coordinates (latitude and longitude) of the sampled wells, spatial data layers have been retrieved using GIS software. Using various groundwater attributes proximity analysis and trend assessment has been carried out. A questionnaire was designed by categorizing it into three sections, viz., part 1, primarily focused on the efficiency of groundwater usage by the local cultivators in the study area, part 2, was designed to extract the information about the developmental scenario of the study area and part 3, intended to gather information based on the knowledge of the respondents. All the responses were carefully noted and tabulated. The data collected from individual interviews was cross checked with the help of focus group discussions and the secondary data collected from the official websites of different departments of Thane district.

Discussion

To understand the relation existing between the variables described in the present study different statistical tools were used. Correlation being the most common statistical method was used to infer the relationships and the degree of strength of the relationship between these variables. Pearson correlation has been applied to understand the relation with the help of correlation coefficient. Pearson correlation coefficient is nothing but a value which summarizes large set of data representing the degree of linear association between two measured variables.

Correlation: The relationship between different variables such as well depth, distance of well from the coastline, and physicochemical and biological parameters such as temperature, pH, TDS, EC, Ca, Mg, Cl, and coliforms has been discussed here. Factors that can influence the inference have been taken into consideration. The results of Pearson correlation applied on the above-mentioned parameters and the output table (Table 1) has been furnished and the interpretations of these outputs have been discussed to understand the pattern of seawater intrusion.

Table 1 Correlation between well depth and physicochemical parameters

Parameters	Well depth	Temp	pH	TDS	EC	Ca	Mg	Cl
Well depth	1							
Temp	0.003	1						
pH	−0.136	0.33*	1					
TDS	−0.148*	0.21*	0.14	1				
EC	−0.199*	0.20*	0.17*	0.85*	1			
Ca	−0.006	0.18*	−0.06	0.49*	0.29*	1		
Mg	−0.218*	−0.02	−0.07	0.41*	0.31*	0.48*	1	
Cl	−0.210*	0.07	0.02	0.69*	0.65*	0.40*	0.56*	1

*Correlation coefficient value is significant at 0.05 level of significance

Table 2 Correlation between well distance from coastline and physicochemical parameters

Parameters	Well distance	Temp	pH	TDS	EC	Ca	Mg	Cl
Correlation coefficient	1	0.223*	0.030	0.164	0.084	0.175	0.061	0.077

*Correlation coefficient value is significant at 0.05 level of significance

Table 3 Correlation between electrical conductivity and motor pump run (h)

Correlations			
		EC	Pump run (h)
EC	Pearson correlation	1	0.038
	Sig. (2-tailed)		0.709
	N	100	100
Pump run (h)	Pearson correlation	0.038	1
	Sig. (2-tailed)	0.709	
	N	100	100

Correlation between well depth and temperature showed a positive significant value of Pearson correlation which indicates that there is a weak but positive correlation between well depth and temperature. This interprets that as the well depth increases the temperature also increases. pH and well depth were found to be negatively correlated although the correlation was not significant. This infers an inverse relationship between these two variables. Well depth and other physico-chemical parameters such as EC, Mg, and Cl showed a significant negative correlation with the significant values 0.047 ($p < 0.05$), 0.029 ($p < 0.05$), 0.036 ($p < 0.05$) respectively. Whereas it was seen that there is no significant relationship between calcium concentration in the groundwater and well depth. Usually TDS, EC, Cl values increase with increase in well depth. This negative correlation here suggests its seawater inundation rather than intrusion. This may be because of the creeks present in the study area (Table 2).

Apart from temperature, all other parameters are very weakly correlated with the distance of well from coastline. But the correlation is positive which shows as the distance of the well increases the temperature, TDS, EC, pH, Ca, Mg, and Cl has been found to increase wherein temperature is significantly correlated with significant value 0.026 ($p < 0.05$). This may be because as the distance increases still the effect or influence of creeks remains on low lying areas (Table 3).

The relationship between the number of hours motor pump is operated to water the field or to be used for other purposes and electrical conductivity was checked to understand whether EC value is high due to overdraft. No significant relation was found between these two variables. But the correlation is positive, interpreting the direct proportion between these variables.

Spatial relationship: Certain variables were analyzed using spatial analysis tools (GIS) to understand the relationship between well depth, elevation, location of well, distance from coastline, trend of changing electrical conductivity with distance, elevation, etc. (Appendix 1). The sampled wells from Kelve village are distributed over the range of 800 m to 3 km from the coastline, sampled wells from Mahim village are distributed over the range of 200 m to 4 km from coastline. But majority distribution is within the belt of 200 m to 1.6 km. Sampled wells from Shirgaon are clustered within the coastal zone from 400 m to 1.6 km. Shirgaon being located very near to the coastline the sample collected from this site showed specific alkalinity (pH) but low TDS and EC values. Local people related this with shallow dug wells (Veera) in the close vicinity of coastline. Due to raised water table, they are able to get fresh water from these shallow dug wells but this has induced the salinity in pockets as small portion of creeks exist in the area. The relationship between distances of sampled well and EC describes that in Kelve the relationship is inverse that means, EC is decreasing with the distance, which is a natural process. But in case of Mahim, as the distance from coastline increases the value of TDS also increases, because of the nearby creeks and existence of salt pans. Moreover the similar results were found for Vasai also. In case of Dahanu tehsil, inverse relationship between distance and EC was noted. Dewale, small village clustered within the distribution range of 1.2–1.6 km from the coastline, showed relatively low values of all other parameters except pH, as the distance is comparatively more from the coastline. But even if the values of EC and TDS are within permissible limit, they are significantly more than highest desirable limit. Hence local people were not using this water for drinking purpose. Whereas Bhuigaon shows the farthest distance that is 3–4 km from coastline, among all sampling points, still the value of EC was high. This is due to the influence of creek in low lying areas. Sampled wells from all the three villages of the tehsil are distributed within the range of 200–600 m from coastline. In this case, the reason of high values of TDS and EC are not only because the region is in such a close vicinity of the coastline but also the other factors such as well depth, over abstraction of groundwater for horticulture, influence of creeks, etc., have a significant impact on seawater intrusion.

There is a direct relationship between effective well depth and EC in Palghar tehsil, which depicts that EC increases with increased well depth. However two sample wells from Mahim village showed an inverse relationship between these variables and showed influence of creeks and saltpans of the area. In Vasai tehsil, at most of the places there is an inverse relationship between the parameters. This can be inferred to seawater intrusion. In the same manner Dahanu tehsil showed a significant positive relationship between well depth and EC, especially in case of Chikhale village of Dahanu tehsil. Another significant relationship exists between EC and number of hours the motor pump run to draw water from the well. Vasai tehsil shows a direct relationship between these two variables. It clearly shows that as the pumping hour's increase the corresponding EC value was found to be high, thus indicating the induced effect of aquifers.

Conclusion

Groundwater salinity has been observed and proven by the laboratory analysis conducted during the study. Salinity could be due to inundation or could be induced, which is because of heavy groundwater drawing. Various other factors like industrial waste disposal, salt/clay pan may also be the other reasons of groundwater salinity. In Thane district there is possibility of both, but the major reason is inundation or ingress of sea water due to the presence of creeks.

Analysis of well location with respect to the distance reveals that there is not much significant relationship existing between distance and each physicochemical parameter. However, the low lying area in the vicinity of creeks/back water (rivers) has affected low depth wells. The low depth wells (Veera) in the vicinity to creeks is induced with saline water, which has been confirmed both by the villagers as well as from the physicochemical analysis of well water samples. This induced effect is due to the failure in functioning of tide control measures such as gates (Vasant bunds) made at local level. The local fishing community tampers the bund which allows the tidal current to enter into the inland. There are low lying areas, with shallow dug wells for meeting the irrigation demand, they end up receiving highly mineralized water due to such unregulated fishing activities. Due to the high salinity, cropping pattern has been changed and now farmers are more restricted to crops like onion, red chili, and millet grains, which are salt resistant and can be raised without much soil constraints. Analysis of well depth with respect to elevation and other physicochemical parameters depicts a significant relationship which clearly infers the over abstraction of groundwater. This would be the reason of elevated value of electrical conductivity found in the groundwater samples. Deepening of wells to meet the agriculture and horticulture irrigation demand is most common in the area. Long duration of pumping and irrigation along with deep wells are showing high EC. This could be due to the deeper aquifer, which has lost hydrograph balance (reverse hydrograph), and has changed its flow pattern, i.e., sea to land. TDS is another parameter justifying the inference drawn on the basis of EC and excess drawing of groundwater from the deep wells. High TDS is directly proportional to the well depth, i.e., deep and shallow wells. Higher values of Mg, Ca, and Cl are an indicator of salinity. These parameters indicate groundwater salinity, which is mainly due to salt water intrusion/inundation and there is less possibility of the reason being urban and industrial pollution. However, further investigation of creek water samples is needed to confirm the fact. Coliform count is an indicator to confirm the salt water intrusion in the aquifers and flooding of creek water. Deep well has reported less counts of coliform, whereas shallow well which are in the vicinity of surface-inundated water has shown high coliform count. This could be due to the contact of human fecal material with inundated water. During the rainy season these wells submerge and contaminate the shallow low lying wells in the vicinity. This is a serious problem, though these wells are used only for irrigation purpose.

Water security issue in this region is highly influenced by the quality of the water. Because there is shallow water level and still the high amount of salinity; it has restricted the use of water for drinking as well as raising staple food crop. High input cost to meet the drinking water and selection of crop are the two major issues before the sustainable water management in the region. Purification measures like RO or other filtration devices are some of the few options left with the community to meet the drinking water demand. To some extent this problem is being addressed by the local administration by providing potable piped water supply to the households, this indicates that there is clear cut shift of source of drinking water due to the salinity of groundwater in the region.

There have been some attempts by the local government to combat the water salinity problem by constructing bunds (Vasant Bandhara Scheme), awareness program for wise water use, etc. However the bunds require overhead cost to maintain, which is not available and the status has remained same. Therefore, a designated water institution should be there to manage the village assets and ensure its functioning. Strict implementation of Coastal Regulation Zone (CRZ) Act is another issue before the institution. This would create some conflict among the fishing and non-fishing communities (cultivators) which needs to be addressed by local government (Panchayat). This study has identified critical issues like inundation of sea water through creeks/back water (river) especially in the low lying regions. This can be avoided by giving way to tidal regulation with gate operated bunds and subsurface bunds to avoid excess flow of groundwater to sea and reverse flow due to cone of depression. Further the cone of depression (hydrograph) can be regulated by region-specific rainwater harvesting measures. Community is totally aware of the importance and possibility of rainwater harvesting in the region. However, the community is reluctant in spending individually for these interventions, which they feel, are expensive at household level/scale. Integrated Watershed Management approach would be useful in addressing these issues.

Acknowledgements The authors are grateful to all the respondents of Thane district for participating in the survey and sharing valuable information. The authors are thankful to PCB, Maharashtra for helping us to carry out the water sample analysis.

Appendix 1: Spatial Relations

See Figs. 1, 2, 3, and 4.

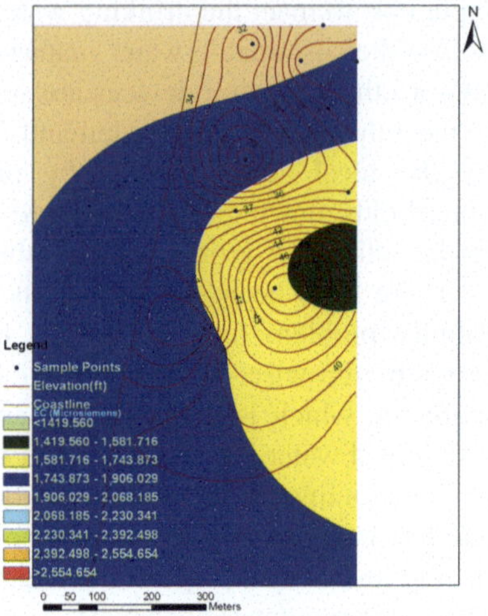

Fig. 1 Relation between EC and elevation, Bordi

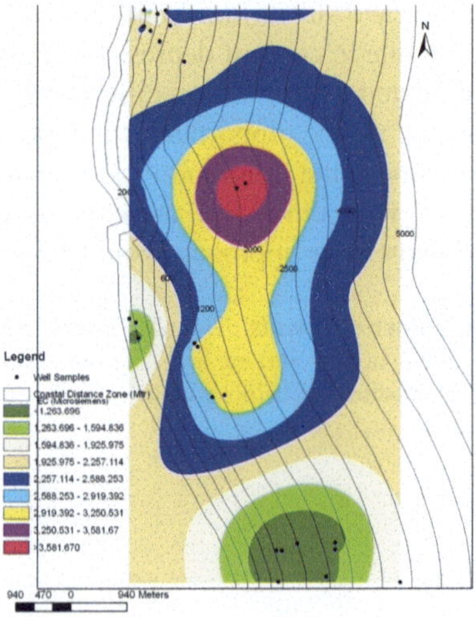

Fig. 2 Relation between the samples wells distance from coastline and EC, Palghar

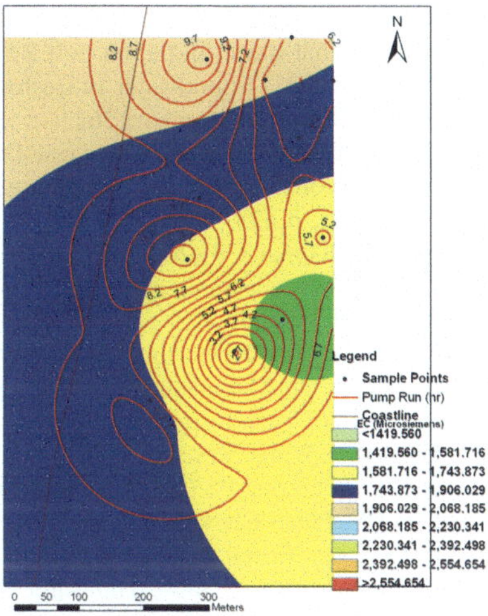

Fig. 3 Relation between EC and pump run (h), Bordi

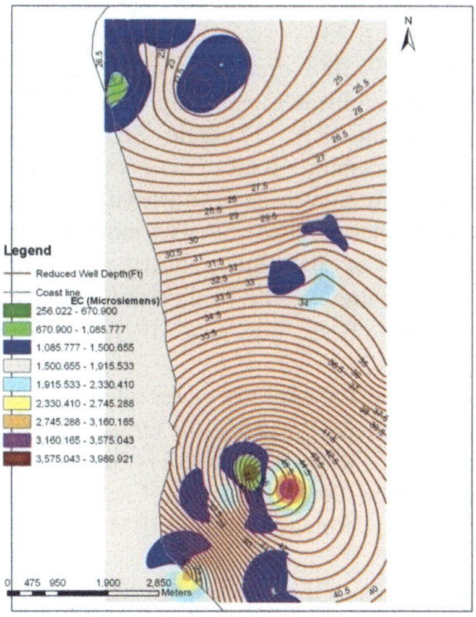

Fig. 4 Relation between well depth (reduced to AMSL) and EC, Vasai

Bibliography

Anawara HM, Akaib J, Mostofac KMG, Safiullahd S, Tareqd SM (2002) Arsenic poisoning in ground water health risk and geochemical sources in Bangladesh. Environ Int 27:597–604

Anbazhagan J, Nair AM (2003) Geographic information system and groundwater quality mapping in Panvel Basin, Maharashtra, India. Environ Geol. doi:10.1007/s00254-003-0932-09

Asian Development Water Outlook (2007) Achieving water security for Asia, Asian Development Bank

Bagade SP (1995) Groundwater potential and the causes of salinity in Vasai-Virar area, Thane District, Maharashtra, Unpub Report, CGWB, 12

Baghure GR,Mirgane SR (2009) Heavy metal concentration in groundwater and soils of Thane region, Maharashtra, India, un-pub Report, CGWB
Barlow PM (2003) Groundwater in freshwater-saltwater environments of the Atlantic coasts, US Geological survey, circular 1262
Bear J, Cheng AH-D, Shaul S, Ouazar D, Herrena I (eds) (1999) Saltwater intrusion in coastal aquifers—concepts, methods and practices. Kluwer Academic Publishers, Dordrecht, The Netherlands, p 625
Bhosale DD, Kumar CP (2009) Simulation of seawater intrusion in Ernakulam Coast. Retrieved from www.academia.edu/171255/Simulation_ofSeawater_intrusion_in_Ernakulam_Coast). Cited on 4 May 2013
CGWB Report (1992) Ground water potential and the causes of salinity in Vasai-Virar Area, Thane District, Maharashtra. A report published by Central Ground Water Board in 1992
CGWB (2010) Groundwater quality in shallow aquifers of India. Central Ground Water Board, Ministry of Water Resources, GoI
Cooper DR, Schindler PS (2001) Business research methods, 7th edn. McGraw-Hill, Irwin, Boston
Creswell JW, Plano Clark VL, Guttman M, Hanson W (2003) Advanced mixed methods research designs. In: Tashakkori A, Teddlie C (eds) Handbook of mixed methods in social and behavioral research. Sage, Thousand Oak, CA, pp 209–240
Das A, Dutta B (2001) Simulation of seawater intrusion in coastal aquifers: some typical responces, Sadhana, vol 26, part 4, pp 317–352
De AK (2006) Environmental chemistry, 6th edn. New Age International Publishers, New Delhi, India, p 232
DTE 1999–2000 droughts. Article appeared in Down to Earth, 15 Jan 2000. Retrieved from http://www.rainwaterharvesting.org/crisis/First_visit.htm on 10 May 2013
Duraiswami RA, Thingale SS, Pawar NJ, Meshram DC (2000) Coastal environment of Maharashtra: issues vis-à-vis development. Prof. KB Powar Commemoration volume, contribution to environmental geosciences, pp 59–73
Effectiveness and social/environmental impacts of irrigation projects: a review. In: Annual report 1988 of the International Institute for Land Reclamation and Improvement (ILRI), Wageningen, The Netherlands, pp 18–34
Gall MD, Borg WR, Gall JP (1996) Educational research: an introduction. Longman, White Plains, NY
Guidelines for drinking water quality (1996) Health criteria and other supportive information, 2nd ed, vol 2. World Health Organization, Geneva
Gupta SC, Kapoor VK (1984) Fundamentals of applied statistics, 3rd edn. Sultan Chand, Delhi
GWQS/09/2006-2007 Ground water quality series. Published by Central Pollution Control Board
IBM Corp. Released (2011) IBM SPSS Statistics for Windows, Version 20.0 Armonk, NY: IBM Corp
International Organization for Standardization (1985) Water quality—determination of electrical conductivity, Geneva
IS 3025 (Part 40, Part 32 and Part 46) (2009) Indian standard: water and waste water methods of sampling and test (physical and chemical) Part 40 Calcium, Part 46 Magnesium, Part 32 Chlorides
Lasserre Cortez S (2006) A mixed methods examination of professional development through whole faculty study groups. Unpublished Doctoral dissertation, Loisiana State University, Baton Rouge
Maharashtra Pollution Control Board Report (2005) Environmental status of Thane Region
Mc Namara C (2009) General guidelines for conducting interviews. Retrieved from http://managementhelp.org/evaluatn/interview.htm. Accessed on Mar 2013
National Groundwater Association (2001) 601 Dempsey Road, Westerville, Ohio 614-898-7791. Retrieved from http://www.h2O-ngwa.org/pubaff/bacq_a.html. Accessed on 11 Mar 2013
Naval Bajpai (2010) Business Statistics. Pearson, Delhi
NEERI (1991) Manual on water and waste water analysis. National Environmental Engineering Research Institute, Nagpur

Nowroozi AA, Stephen BH, Henderson P (1999) Saltwater intrusion into the freshwater aquifer in the eastern shore of Virginia: a reconnaissance electrical resistivity survey. J Appl Geophys 42:1–22
Parasnis I, Samar VJ, Fischer SD (2005) Deaf college student's attitude towards radical/ethnic diversity, campus climate and role models. Am Ann Deaf 150(1):47–58
Patton MQ (2002) Qualitative research and evaluation methods, 3rd edn. Sage, Thousand Oaks, CA
Paul Wyrwoll P (2012) India's groundwater crisis, discussion paper 1228. Australian National University, Australia
Ranjan P (2007) Effect of climate change and land use change on saltwater intrusion. Retrieved from http://www.eoearth.org/article/effect_of_climate_chnage_and_land_use_change_on_saltwater_intrusion. Accessed on 9 Nov 2012
Reilly TE, Goodman AS (1985) Quantitative analysis of saltwater-freshwater relationships in groundwater systems—a historical perspective. J Hydrol 80:125–160
Reilly TE, Goodman AS (1987) Analysis of saltwater upcoming beneath a pumping well. J Hydrol 89(3–4):169–204
Salinity ingress and combating measures http://agri.gujarat.gov.in/hods/dire_agriculture/irrigation_salinity_ingress.htm
Sastry KV, Rathee P (1998) Physico-chemical and microbiological characteristics of water of village Kanneli (dist. Rohtak), Haryana. Proc Acad Biol 7(1):103–108
Shah T (2011) Innovations in groundwater management: examples from India. International Water Management Institute. Retrieved from http://rosenberg.ucanr.org/documents/argentina/TusharShahFinal.pdf. Cited on 18 Apr 2013
Shamoo AE, Resnik BR (2003) Responsible conduct of research. Oxford University Press
Sharma SK (2008) Influence of seawater ingress: a case study from east coast aquifers from India. Presented in 20th seawater intrusion meeting, at Florida, USA, 23–27 June 2008
Singh SI, Mosley LM (2003) Trace metal levels in drinking water on Viti Levu, Fiji Islands. S Pac J Nat Sci 21:31–34
Soni AK, Pujari PR (2010) Groundwater vis-à-vis seawater intrusion analysis for a part of limestone track of Gujarat coast, India. J Water Resour Port 2:462–468
Soni AK, Pujari PR (2012) Seawater intrusion studies for coastal aquifers: some ponts to ponder. Open Hydrol J 2012:24–30
Suseela P, Visalakshy KP (2006) Subsurface dykes—case study for its suitability in Kerala. J Agric Eng 43(2)
Teddlie C (2005) Methodological issues related to causal studies of leadership: a mixed method perspective. Educ Manage Adm Leadersh USA 33(2):211–217
TERI (2010) A coastal policy newsletter, published by TERI, 2010
Umrikar BN, Devi NM (2011) Groundwater quality assessment for Guhaghar coastal area, Maharashtra, India. J Environ Res Dev 5(4)
US EPA (2004) Field sampling guidance document # 1220-groundwater well sampling 2004, US. Environment Protection Agency, Region 9 Laboratory, Richmond, California
van Hoorn JW, van Alphen JG (2006) Salinity control. In: Ritzema HP (ed) Drainage principles and applications, Publication 16. International Institute for Land Reclamation and Improvement (ILRI), Wageningen, The Netherlands, pp 533–600. ISBN 90-70754-33-9
Werner AD, Bakker M, Vincent EA, Vandenbohede A, Lu C, Ataei B, Simmons CT, Barry DA (2012) Seawater intrusion processes, investigation and management: recent advances and future challenges. Adv Water Resour. Retrieved from journal homepage: www.elsevier.com/locate/advwaters
World Bank (2010) Deep wells and prudence: towards pragmatic action for addressing groundwater overexploitation in India' an article published in World Bank 2010 report

Study of Suction Versus Water Content of Soil of Turamdih Area Mixed with Bentonite and Its Implication on the Liner Property of Tailing Dam: A Case Study of East Singhbhum, Jharkhand, Eastern India

Brahm Deo Yadav, Sunil Kumar Gupta and Sahendra Singh

Abstract The pressure difference between air and water component in soil void is a key variable in the analysis of hydromechanical behavior of unsaturated soils. Capillary action is directly related to the free energy of the pore water in a soil and can be used to classify the relative swelling potential of expansive soil. Lower hydraulic conductivity requirement of compact clay liner (CCL) is fulfilled through addition of bentonite to the locally available soil material. Thus, the nature of CCL becomes expansive. It is clear from the test data that higher the suction pressure, lower is the hydraulic conductivity and higher is the swelling. The behavior of alternate swelling and shrinking is harmful in other case, but in the area of linear application it is always saturated under wet tailings. So, in between high swelling and low hydraulic conductivity, suction gives positive response towards liner design in uranium ore tailing dams. The relationship between the affinity of soil to retain water and suction can be measured based on the filter paper technique of total section. The obtained value of total suction was thereafter used to estimate the expansiveness of soils. Compacted soils have been widely used as landfill barriers because of many favorable characteristics such as low permeability and high swelling. Compacted clay liner made of different alternatives are normally unsaturated and therefore, suction can be used as a behavior indicator in addition to generally used factor such as water content, dry density, void ratio hydraulic conductivity, etc. This study is mainly focused on investigating suction characteristics of CCL mixtures. Suction was measured using filter paper method for these combinations. The laboratory results were analyzed to provide on understanding of

B. D. Yadav · S. K. Gupta
Department of Environmental Science and Engineering, Indian Institute of Technology (Indian School of Mines), Dhanbad 826004, Jharkhand, India

S. Singh (✉)
Department of Applied Geology, Indian Institute of Technology (Indian School of Mines), Dhanbad 826004, Jharkhand, India
e-mail: sahendrasingh02@gmail.com

V. P. Singh et al. (eds.), *Water Resources Management*, Water Science and Technology Library 78, https://doi.org/10.1007/978-981-10-5711-3_13

the suction concept. It was found that suction depends primarily on the water content and the bentonite content of the mixture.

Keywords Suction · Compacted clay liner · Landfill barrier · Hydration Hydraulic conductivity · Swelling · Filter paper

Introduction

In landfill applications of compacted clay, one of the criteria commonly used for the performance of landfill liner is the coefficient of permeability of the liner. The presence of clay material is, therefore, important. The addition of bentonite to granular material can change the performance of a highly permeable material and transform it to a material suitable for use as an engineered barrier for landfill. This type of mixture is often referred to as a bentonite enhanced mixture with low percentage of bentonite. On dry mass basis, addition of 01% sodium-type bentonite is sufficient to reduce the permeability of material up to several degree of magnitude (Stewart et al. 1999). A further increase in the percentage of bentonite may not lead to a decrease in the coefficient of permeability of the mixture (Studds et al. 1998). For a less active bentonite such as a calcium type bentonite, a higher percentage of bentonite, with higher field compaction density, is required to achieve the same performance as a sodium-type bentonite–local soil compacted mixtures.

Compacted bentonite soil mixtures are normally unsaturated and, therefore, suction can be used as a behavioral indicator in addition to generally used factors such as water content, dry density, or void ratio. The presence of suction in compacted bentonite soil mixture is often associated with collapse behavior of compacted mixtures. However there is also a correlation between suction, collapse behavior, and the coefficient of permeability of the mixtures. A mixture, which is compacted at moisture content lesser than its optimum value, has a relatively low degree of saturation and high suction and tends to exhibit collapse under specific over burden pressures because of its metastable flocculated structures. Therefore, the mixture can have a relatively high coefficient of permeability. If the same mixture gets compacted close to its optimum water content, maximum dry density with some compaction energy has a higher degree of saturation and lower suction. The mixture undergoes swelling when wetted at the same load and has a lower coefficient of permeability. Strength and stiffness of a compacted bentonite soil mixture generally increases with increase in suction. The objective of this study is to investigate the suction characteristics of bentonite soil mixtures containing different proportions of bentonite.

Assessments of optimum bentonite content in bentonite soil mixtures in relation to the application in landfill are not within the scope of this paper. Although compacted mixtures may undergo changes in suction in the field due to environmentally induced wetting and drying cycles. The initial suction plays an important role in the behavior of the compacted mixtures (Fredlund 1979; Mckeen 1981, 1992; Dineen et al. 1999; Likos and Lu 2003).

Theoretical Background of Suction

The powerful molecular and physicochemical forces acting at the boundary between the soil particle and the soil water lying above the water table causes the water to be either drawn up into the empty void spaces or held there without drainage following infiltration from the surface. Thus, the attractive force that the soil exerts on the water is termed as soil suction and it is regarded as tensile hydraulic stress in a saturated piezometer with a porous filter placed in intimate contact with water in the soil. The magnitude of attractive force that the soil above water table exerts on water is governed by the size of void in a manner similar to the capillary dia. The smaller the void, harder is to remove the water. Meniscus formed between adjacent particles of soil by soil suction, creates a normal force between the particles, which bonds them. Thus soil suction can enhance the stability of earth structure. However, soil suction also provides an attractive force for free water, which can result in loss of stability in loosely compacted soil or swelling in densely compacted soil.

Total suction using Kelvin equation, derived from ideal gas law is given as

$$h_t = RT/V \ln(P/P_o)$$

R Universal gas constant
T Absolute temperature
V Molecular volume of water
P/P_o Relative humidity
P Partial pressure of pore water vapor
P_o Saturation pressure of water vapor

At a reference temperature of 25 °C, the following relation exists, h_t = 13.7.182 ln (P/P_o):

The suction is calculated either as $\log_{10}$ (kPa) or pF.
The pF is represented by $\log_{10}$ (suction in cm of water).
The two systems are approximately related by log (kPa) = pF − 1 (Bulut et al. 2001).

The osmotic suction of electrolyte solutions, which are usually employed in the calibration of filter papers, can be calculated using the relationship between osmotic coefficient and osmotic suction. Osmotic coefficient can be obtained from the following relationship

$$\phi = -\rho_w/(vmw)\ln(P/P_o)$$

where

ϕ	Osmotic coefficient
v	Number of ions from one molecule of soil 2 for NaCl, Kcl, NH_4Cl and 3 for Na_2SO_4, $CaCl_2$, Na_2 S_2O_3, etc.
m	Molality
w	Molecular mass of water
ρ_w	density of water
P/P_o	Relative humility also known as activity of water (9w) combining with Kelvin equation
ho	$-vRTm\ \phi$.

Soil suction is a microscopic property that indicates the intensity of free energy level of water that the soil attracts (Fredlund and Rahardjo 1993; Bulut et al. 2001; Ridley et al. 2003; Rao and Shivananda 2005; Sreedeep and Singh 2006). Soil suction comprises two components—Osmotic and capillary (Matric) suction. The suction due to capillary nature, texture, and adsorptive forces of unsaturated soils and which varies with change in moisture content of the soils is called matric suction. The osmotic suction is a result of the presence of dissolved salts in the pore fluid. The relation between different types and suction (Chen 1988) are:

$h_t = h_o + h_m$
h_o = Osmotic suction = $U_a - U_w$
U_a = Pore air pressure
U_w = Pore water pressure
H_m = Matric suction
H_t = Total suction.

For expansive soil matric suction is dominant, while for nonexpansive soil osmotic suction can be generated by saturating the soil with salt solution.

Hydration forces play an important role in controlling the suction characteristics especially matric component in dry condition, because of unhydrated exchangeable cations near the clay surface. An increase in water content satisfies these forces and increase the interlayer separation distance to about three mono layers of water molecules (about 10 Å), as a result crystalline swelling occurs (Yong 1999). The hydration forces provide an additional driving force for water in a similar manner to capillary forces and osmotic suction. Besides the hydration forces, the other contributing forces arise from van der Waals force fields. Both the hydration and van der Waals forces are operative at a short range from clay particles and are called sorptive forces. These sorptive forces dominate the matric component of suction; the presence of water menisci or capillary action is not necessary for soil to have matric suction (Young 1999).

The increase in moisture content is usually associated with the decrease in suction and vice versa (Sreedeep and Singh 2006). Conversely, soil volume decreases as the soil suction increases and vice versa. An increase in suction will remove the absorbed water from the soils. When the moisture content of the clay soils is reduced the clay shrinks causing downward movement. On the other hand, decrease in suction triggers the entry of water molecules between the clay layers, thus inducing the swelling of soil (Lucian 2009). Ultimately, hydraulic conductivity and swelling characteristics is inverse to the suction.

Expansive potential using suction values

Using matric suction values, PI, and estimated over burden pressure relation was employed to estimate the expansiveness (Brackley 1980)

$$\text{Swell } (\%) = \text{PI} - 10 \, \log_{10}(\text{S}/\text{P})$$

where

S Soil suction at the center of layer
P Overburden plus foundation stress at that depta.

Methods Used

There are four different methods for determining soil suction namely,

- Filter paper (FP) Technique
- Psychrometer (PSY) Technique
- Dew point sensor (DP) Technique
- Chilled mirror hygrometer (CMH) Technique.

Out of the four methods, the simple and cheap favorable method to conduct the suction test over a wide range of suction is by the use of filter paper in accordance with ASTM D 5298. Only the FP technique is used to measure total and matric suction in both field and in laboratory. The other three techniques only measures total suction. In the FP method, the soil specimen and filter paper are brought to equilibrium either in contact (for matric suction measurement), or in a noncontact (for total suction measurement) method in a constant temperature environment. This view is explained in Fig. 1. After equilibrium is established between the filter paper and soil, the water content of the filter paper disc is measured. Then, by using a filter paper calibration curve of water content versus suction, the corresponding suction value is found from the curve. The filter paper method is an indirect method of measuring soil suction. Therefore, a calibration curve should be constructed using pressure plate apparatus for suction less than 1500 kPa or be adopted. In this study, calibration curve for whatman No 42 filter paper disks in ASTM D 5298-94 are adopted.

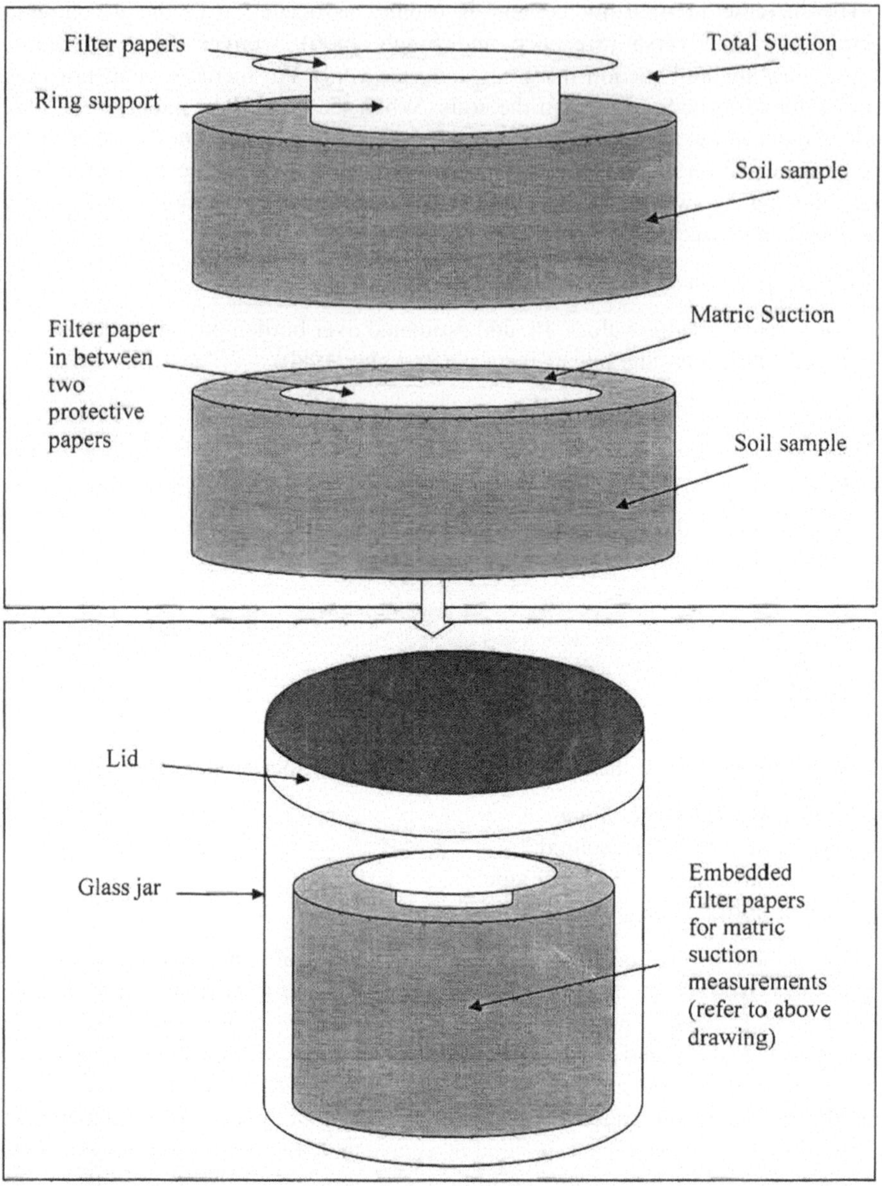

Fig. 1 Contact and noncontact filter paper methods for measuring total matric suction

Apparatus required for soil suction measurements:

Filter papers; the ash-free quantitative Schleicher & Shuell No. 589 Whatman No. 42-type filter papers.

(i) Sealed containers; glass jar with lids.
(ii) Small aluminum cans; the cans with lids are used as carriers for filter papers during moisture content determination.
(iii) A balance with accuracy to the nearest 0.0001 gm is used for moisture content determination.
(iv) An oven; for determining moisture contents of the filter papers by leaving them in it for 24 h at 105 ± 5 °C temperature in the aluminum moisture cans (standard test method for water content determination of soils).
(v) A temperature room; a controlled temperature room in which the temperature fluctuations are kept below ±1 °C (used for equilibrium period).
(vi) An aluminum block; the block is used as heat sink to cool the aluminum cans for about 20 s after removing them from the oven.

In addition, latex gloves, tweezers, plastic tapes, plastic bags, ice-chests, PVC Ring, scissors, and a knife are used to setup the test.

Soil Total Suction Measurement

Glass jar that are between 250 and 500 ml volume size are readily available in the market can be easily adopted for suction measurement. Glass Jars, especially, 3.5–4 inch diameter can contain the 3 inch diameter Shelby tube samples very nicely.

Experimental Procedure

1. At least 75% by volume of a glass jar is filled up with the soil; the smaller the empty space remaining in the glass jar, the smaller the time period that the filter paper and the soil system requires to come to equilibrium.
2. A ring-type support, which has diameter smaller than filter paper diameter and about 1–2 cm in height, is put on top of the soil to provide a noncontact system between the filter paper and the soil. Care must be taken when selecting the support material; materials that can corrode should be avoided, plastic or glass-type materials are much better for this job.
3. Two filter papers one on top of the other are inserted on the ring using tweezers. The filter papers should not touch the soil, the inside wall of the jar, and underneath the lid in any way.
4. Then, the glass jar lid is sealed very tightly with plastic tape.

5. Steps 1–4 are repeated for every soil sample.
6. After that, the glass jar is put into the ice-chests in a controlled temperature room for equilibrium.

The suggested equilibrium period is at least one week (ASTM D 5298). After the equilibrium time, the procedure for the filter paper water content measurement can be as follows:

1. Before removing the glass jar containers from the temperature room, all aluminum cans that are used for moisture content measurements are weighed to the nearest 0.0001 gm accuracy and recorded.
2. After that, all measurements are carried out by two persons. While one person is opening the sealed glass jar, the other is putting the filter paper into the aluminum can very quickly in a few seconds using tweezers.
3. Then, the weights of each can with wet filter paper inside are taken very quickly.
4. Steps 2 and 3 are fallowed for every glass jar. Then, all cans are put into the oven with the lids half-open to allow evaporation. All filter papers are kept at 105 ± 5 °C temperature inside the oven for at least 10 h.
5. Before taking measurements on the dried filter papers, the cans are closed with their lids and allowed to equilibrate for about 5 min. Then, a can is removed from the oven and put on an aluminum block (heat sink) for about 20 s to cool down; the aluminum block functions as a heat sink and expedites the cooling of the can. After that, the can with the dry filter paper inside is weighed very quickly. The dry filter paper is taken from the can and the cooled can is weighed again in a few seconds.
6. Step 5 is repeated for every can.

After obtaining all of the filter paper water contents an appropriate calibration curve is employed to get total suction values of the soil sample.

Soil Matric Suction Measurements

It is similar to the total suction measurements but a good intimate contact should be provided between the filter paper and the soil for matric suction measurements. Both matric and total suction measurements can be performed on the same soil sample in a glass jar as shown in Fig. 1.

Experimental Procedure

1. A filter paper is sandwiched between two larger size protective filter papers, so either a filter paper is cut to a smaller diameter and sandwiched between two 5.5 cm papers or bigger diameter filter papers are used as protective.

2. Then, these sandwiched filter papers are inserted into the soil sample in a very good contact manner. An intimate contact between the filter paper and the soil is very important.
3. After that, the soil sample with embedded filter papers is put into the glass jar container. The glass container is sealed up very tightly with plastic tape.
4. Step 1–3 are repeated for every soil sample.
5. The prepared containers are put into ice-chests in a controlled temperature room for equilibrium.

The suggested equilibrium period is 3–5 days (ASTM D 5298). However, if both matric and total suction measurements are performed on the same sample in the glass jar, then the final equilibrating time will be at least 7 days of total suction equilibrating period (Figs. 2 and 3).

The procedure for the filter paper water content measurements at the end of the equilibration is exactly same as that of outlined for the total suction water content measurements. After obtaining all the filter paper water contents, the appropriate calibration curve may be employed to get the matric suction values of the soil sample.

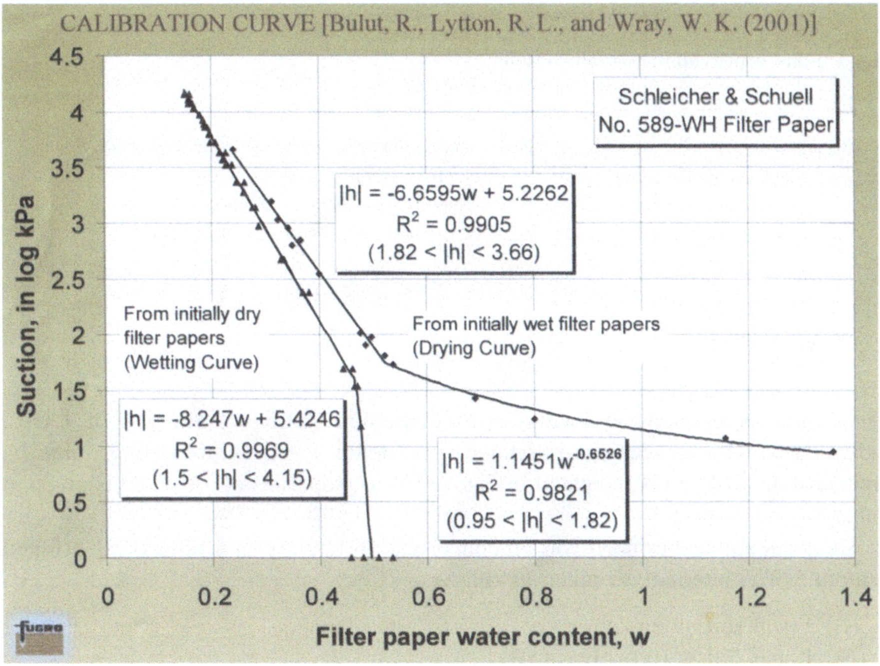

Fig. 2 Calibration curve for suction pressure measurement

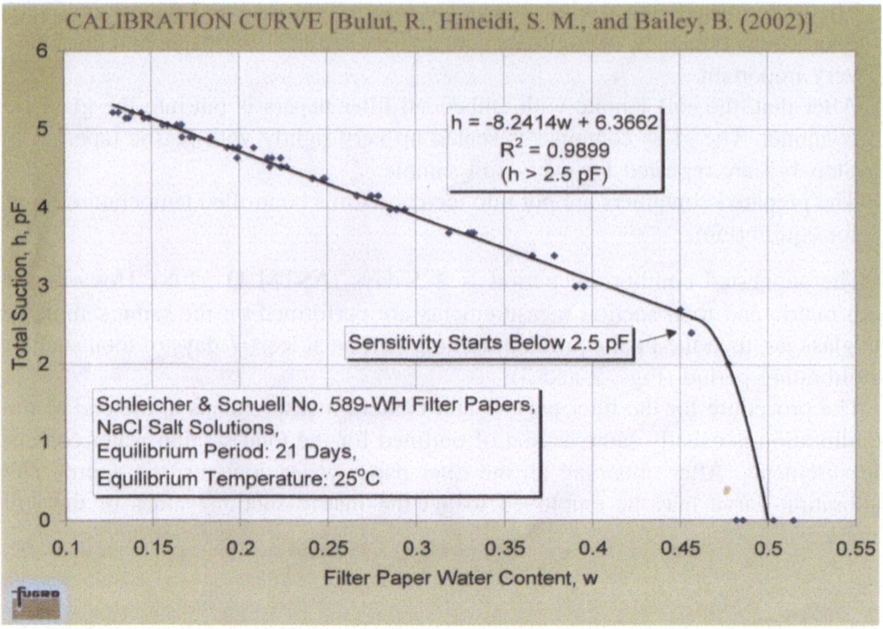

Fig. 3 Filter paper calibration relationship

Filter Paper	$\log_{10}$ |suction in kPa|	pF = $\log_{10}$ |suction in cm of water|
Schneider & Schuell No.—589—WH	$\|h\| = 5.4246–8.247\ \omega$ $R^2 = 0.9969$ $1.5 < \|h\| < 4.15$	$\|h\| = 6.3662–8.2414\ \omega$ $R^2 = 0.9899$ $\|h\| > 2.5$ pF

Test Materials

Tests were performed on local soil of Turamdih tailings dam site of UCIL, Jaduguda which has been used as liner construction for Uranium tailings dam. Its grain size distribution is indicated in Fig. 4. Basic properties of local soil is given in Table 1.

The local soil as per need was amended with different proportions of fly ash and bentonite. The alternatives taken as follows:

(i) Local soil alone (A)
(ii) Local soil + 10% Flyash (B)
(iii) Local soil + 20% Flyash (C)
(iv) Local soil + 10% Bentonite (D)
(v) Local soil + 20% Bentonite (E)

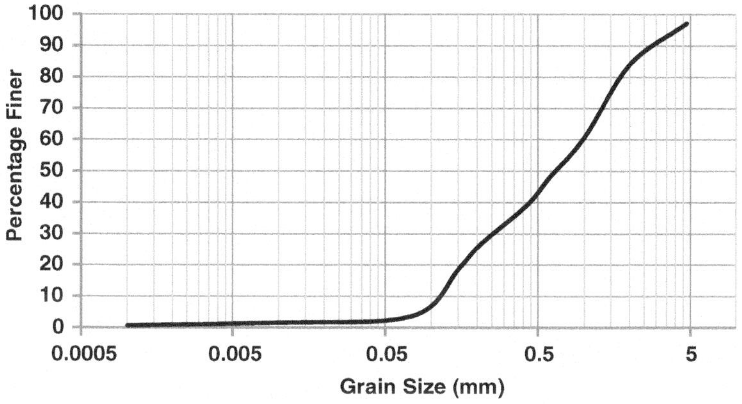

Fig. 4 Grain size distribution curve of UCIL soil

Table 1 Basic properties of local soil

LL (%)	28.86
PL (%)	20.32
SL (%)	16.19
PI (%)	8.54
Specific gravity	2.47
OMC (standard Proctor) (%)	15.25
Dry density (kN/m^3)	18.75
OMC (modified Proctor) (%)	14.85
Dry density (kN/m^3)	20.35
USCS classification	SW-SM
Clay (%)	1.61
Gravel (%)	2.82
Silt (%)	2.11
Sand (%)	93.46
Coefficient of uniformity (C_u)	9.357
Coefficient of contraction (C_C)	0.211
Cohesion (C) in kN/m^2	43.921 kN/m^2
Angle of Internal Friction (ϕ)	23°7′

(vi) Local soil + 10% Flyash + 10% Bentonite (F)
(vii) Local soil + 20% Flyash + 10% Bentonite (G)
(viii) Local soil + 20% Flyash + 20% Bentonite (H)

The soil materials were sieved to avoid the presence of coarse grains (max size 4.75 mm) then mixed at initial water content, and an adequate amount of water was subsequently added using a water sprayer to reach the target water content. The composition of bentonite/and or flyash in the mixture was prepared to obtain different materials with different degree of plasticity, and in which suction is expected to differ. The mixture were subsequently cured in two layered plastic bags for 2 weeks to allow the hydric equilibrium to establish. The different compaction curves were obtained by dynamically compacting the mixture using the standard proctor method following the ASTM D 698-91. The total suction measurement using the FP technique was performed on standard proctor sample having 102 mm in diameter and height. The Whatman 42 filter paper was used in all 23.35 mm tests.

Result and Discussion

(i) The total suction of different alternatives of soil, flyash, and bentonite mixture is primarily a function of water content and bentonite content. There is a lesser dependency on fly ash.

(ii) Pore geometry or fabric ultimately void ratio have no apparent effect on total suction, and thus an insignificant contribution of the capillary component of suction.

(iii) Redistribution of water occurs after compaction because of a difference in total suction in the different levels of the pores, so total suction of specimen in the compacted state does not represent the true suction at equilibrium.

(iv) The in contact filter paper technique appears to measure the capillary matric suction component. The component due to the action of sorptive forces is not measured using this technique. This technique should only be used to measure a capillary suction component of less than 1500 kPa.

(v) The noncontact filter paper technique measures total suction but due to its long equilibration time, it is very difficult for measuring as compacted total suction of different samples.

(vi) The results of total suction for different sample at their OMC, shows that as the suction increases, hydraulic conductivity decreases and volumetric swell decreases. This is one of the greatest finding for use as liner material. There is no problem for alternate swelling and shrinkage for liner as it is always placed under wet tailings.

(vii) The age of specimens used for the measurements varied significantly in different methods and it is 5 weeks for FP techniques.

(viii) The measurement of total suction in compacted specimens provides values corresponding to a transient state. Redistribution of water is believed to occur as time elapses.

(ix) The results obtained from the FP technique show that the compaction technique plays no significant role in the magnitude of total suction for the compacted soil–bentonite mixture.

(x) Values of total suction for all the eight combinations show that suction characteristics for combination E (local soil + 20% bentonite) is very much improved. This fact justifies the high % of bentonite. Values for combination H (local soil + 20% fly ash + 20% bentonite) closely follows the values for combination E. 10 keeping in views of suction properties and economic combination H is most suitable for liner application.

References

Blatz JA, Graham J, Chander NA (2002) Influence of suction on the strength and stiffness of compacted sad-bentonite. Can Geotech J 39(5):1015. doi:10.1139/t02-056

Brackley JA (1980) Prediction of soil heave from suction measurements. In: Proceedings of the 7th regional conference for Africa on soil mechanics and foundation engineering. Accra, vol 1, pp 159–166

Bulut R, Lytton RL, Wray WK (2001) Soil suction measurements by filter paper. In: Vipulanandan C, Addison MB, Hasen M (eds) Expansive clay soils and vegetative influence on shallow foundation. American Society of Civil Engineering Geotechnical Special Publication No. 115, Reston, VA, pp 243–261

Chen FH (1988) Foundation on expensive soils. Elsevier, Amsterdam

Dineen K, Colmenares JE, Ridley AM, Burland JB (1999) Suction and volume changes of a bentonite-enriched sand. Proc Inst Civil Eng Geotech Eng 137:197–201

Fredlund DG (1979) Second Canadia Geotechnical Colloquim: appropriate concepts and technology for unsaturated soils. Can Geotech J 16(1):121–139. doi:10.1139/t79-011

Fredlund DG, Rahardjo H (1993) Soil mechanics for unsaturated soils. Wiley, New York

Lee HC, Wray WK (1995) Technique to evaluate soil suction—a vital unsaturated soil water variable. In: International conference on unsaturated soils, Paris, pp 615–622

Likos WJ, Lu N (2003) Automated humidity system for measuring suction characteristics of clay. Geotech Test J 26(2):179–190

Lucian C (2009) Spatial variability of expensive soil properties at different scale within Kibaha, Tanzania. Global J Agric Sci 8(1):95–100

McKeen RG (1981) Characterization of expansive soils. Transportation Research Borad Washington, D.C., Transportation research record 790, pp 73–78

Mckeen RG (1992) A model for predicting expansive soil behavior. Proc 7th Int Conf Expansive Soils 1:1–6

Rao SM, Shivananda P (2005) Role of osmotic in swelling of salt-amended clays. Can Geotech J 42:307–315

Ridley AM, Dineen K, Burland JB, Vaughan PR (2003) Soil matrix suction: some examples of its measurement and application in geotechnical engineering. Geotech 53(2):241–253

Sreedeep S, Singh DN (2006) Methodology for determination of osmotic suction of soils. J Geotech Geol Eng 24(5):1469–1479

Stewart DI, Cousens TW, Students PG, Tay YY (1999) Design parameters for bentonite-enhance sands as a landfill liner. Proc Inst Civil Eng Geotech Eng 137:189–195

Studds PG, Stewart DJ, Cousens TW (1998) The effects of salt solutions on the properties of bentonite-sand mixtures. Clay Miner 33:651–660

Yong RN (1999) Soil suction and soil water potential in swelling clays in engineering clay barriers. Eng Geol 54(1–2):3–13. doi:10.1016/s0013-7952(99)00056-3

Long-Term Water Planning: A Review of Kenya National Water Master Plan 2030

J. Kibiiy and J. R. Kosgei

Abstract The National Water Master Plan 2030 (NWMP), Kenya's water resources development and management blue print provides estimates of the country's water resources and its planning for the period 2010–2030. The water demands were estimated for the year 2010 and projected for the years 2030 and 2050. The projection for 2030 was intended to formulate the NWMP while the projection for 2050 was intended to assess future vulnerability of water resources considering the effects of climate change. However, the plan is constrained by challenges latent in its preparation and assumptions made on key issues, among them, the promulgation of the Constitution of Kenya 2010 which impacted the country's water sector in a very critical manner. This study evaluated the merits and weaknesses of the plan. An analysis of the NWMP was made considering the process, content and implementation. The findings suggest that NWMP was developed using inadequate data, projections for key sectors of the economy have large uncertainties, and that the management arrangements stipulated in the new constitution were not entirely taken into account. It was concluded that these weaknesses undermine its feasibility as an operational document. A review of the plan to address the identified gaps was recommended.

Introduction

General Background

The logo of the Ministry of Water in Kenya has the proclamation '*water is life*'. This phrase expresses a basic fact that water is at the base of all life. It sustains plant

J. Kibiiy (✉) · J. R. Kosgei
Department of Civil and Structural Engineering, Moi University, 3900-30100, Eldoret, Kenya
e-mail: Kibiiy@yahoo.com

J. R. Kosgei
e-mail: masultiony@gmail.com

V. P. Singh et al. (eds.), *Water Resources Management*, Water Science and Technology Library 78, https://doi.org/10.1007/978-981-10-5711-3_14

and animal life and is a foundation for agriculture, industry and social development in modern society. Water is a medium of transport for sediments, wastes and pollutants. Water also supports navigation, energy production and is necessary for sustenance of ecological balance.

There is plenty of water on earth. Its abundance on the planet makes it, 'the water world', (Zeman et al. 2006) distinct from other planets. Water covers over 70% of the surface area of the globe. However, most of this water is saline, located in the seas and oceans of the world and unfit for human use. It is estimated that only about 2.5% of all water on earth is freshwater. The bulk of this fresh water is held up in the polar ice caps. Groundwater constitutes a major proportion of the remainder. Only a miniscule 0.01% of all water on earth is available as fresh water in a form that is accessible and usable by man (Zeman et al. 2006).

Fresh water is unevenly distributed on the surface of the earth and its availability also varies with time making it both scarce and excess at different places at different times. Both scarcity and excess of water cause suffering and misery in the form of droughts and floods respectively. The impacts of the scarcity and excess of water together with the competition between its different uses makes water resources planning a vital necessity.

Historical Background

In 1974, the government of Kenya embarked on the National Water Master Plan Initiative with the slogan '*water for all by the year 2000*'. In order to support the initiative, the Department of Water Development in the Ministry of Agriculture was elevated to become the Ministry of Water Development. In the course of time, it became apparent that the target set for the year 2000 would not be realized. In 1992, the 'National Water Master Plan (NWMP-1992)' was published to address the needs of the water sector in Kenya. Later the same year, the International Conference on Water and the Environment, held in Dublin, laid out the principles that form the basis for integrated water resources management (IWRM). Only some of the projects proposed in NWMP-1992 had been implemented by 1999 when the water sector reforms to adopt IWRM were started. The National Water Policy was developed to guide water resources management and development. A law was passed in 2002 to implement the policy (GoK 2002). The law entrenched IWRM and completely transformed operations in the water sector in the country. In 2007, the Kenya Vision 2030 (GoK 2007) was launched as the country's new socio-economic development blueprint. The 20-year span of the NWMP-1992 was coming to an end and the Vision 2030 only served to accentuate the need to revise it, a process which started in 2008.

Vision 2030 was aimed at transforming Kenya into a newly industrialising, middle-income country providing a high quality of life to all its citizens by the year 2030. The development targets on the water sector are fourfold: to ensure that improved water and sanitation are available and accessible to all by 2030; to increase the area under irrigation to 1.2 million ha by 2030 for increased agricultural production; to be a nation that has a clean, secure and sustainable environment by 2030; and, to generate more energy and increase efficiency in the energy sector.

Study Area

Kenya is located on the equator on the east on the African continent washing the Indian Ocean (Fig. 1a). It lies between latitudes 5.5° N and 5° S and longitudes 33.8° E and 42.0° E. It borders Somali Republic, Ethiopia, Sudan, Uganda and Tanzania. Kenya has a territorial area of 592,262 km^2 of which 11,250 km^2 is inland waters consisting of territorial waters on Lake Victoria, Lake Turkana and a number of small lakes most of which are in the Rift Valley.

Kenya has a varied relief with altitude varying from zero at the Indian Ocean coast to about 5,200 m at the peak of snow-capped Mt. Kenya. The Great Rift Valley cuts through the country as a 50 km wide depression about 1,000 m deep forming an internal drainage area. On either side of the Rift Valley are highland plateaux ranging in altitude from 1,200 to 2,700 m. Volcanic mountains like Mt. Elgon, Mt. Kenya and the Aberdares rise from these plateaux forming drainage divides. A drainage basin or catchment is described as an area of land draining into a river at a given point or into a water body (Chow et al. 1988).

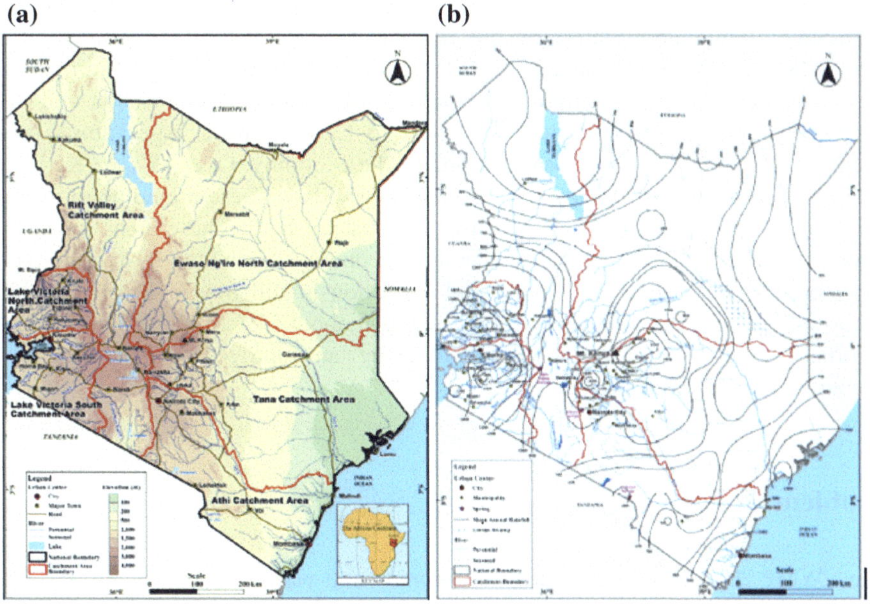

Fig. 1 **a** Location map of Kenya and **b** distribution of mean annual rainfall in Kenya (Muhindi et al. 2001)

Kenya is divided into five drainage basins or catchment areas (Fig. 1a): Lake Victoria Catchment (administratively divided into L. Victoria North and L. Victoria South Catchments); Rift Valley Catchment; Athi River Catchment; Tana River catchment; and Ewaso Ng'iro North Catchment. Mean annual rainfall over the country is 680 mm (Republic of Kenya 2013a, b). It varies from 200 mm in the extremely dry areas to about 1800 mm in the humid zones (Fig. 1b). Rainfall in Kenya is extremely seasonal with two rainy seasons separated by relatively dry periods: the long rains start towards the end of March peaking in May and decreasing through to August; the short rains come in October and November. The spatial distribution of rainfall over the country is very uneven. Over two-thirds of the country receives less than 500 mm of rain per year and only 11% of the country receives more than 1,000 mm per year. This is the rainfall that effectively forms Kenya's water resource base since the water resources of a country are constituted by the sum of the annual surface runoff and sustainable groundwater yield. It is the available water resources for development in a year.

Socio-economic Development

The national population census of 2009, estimated Kenya's population at 38.6 million with an annual growth rate of about 3% (KNBS 2010). About 67.7% were living in the rural areas and 32.3% were living in urban centres. The World Bank has estimated that 42% of income in Kenya is controlled by less than 10% of the population, while about 46% of the population lives below the poverty line (Fitzgibbon 2012). Economic activity is largely centred in the cities of Nairobi and Mombasa, and the fertile densely populated area around Mount Kenya and the highlands across the Rift Valley westwards to Lake Victoria. Agricultural production in these areas forms the backbone of the economy. Over 80% of Kenya's land mass constitutes arid and semi-arid lands (ASALs). These are home to nearly one-third of the population and 70% of the livestock herd. These areas are characterized by low and erratic rainfall with an economy dominated by mobile pastoralism. The level of water supply and sanitation is still low in Kenya. People using 'improved water sources' were estimated at about 55% of the population and a corresponding 67% of the population as using 'an improved sanitation facility' (Republic of Kenya 2013a, b).

Problem Statement

The NWMP 2030 was finalized in 2013, more than 6 years since its launch in 2008. It constitutes Kenya's water sector strategic plan for the period 2010 up to 2030 with climate change considerations in 2050. It was drawn by the Ministry of Water and Irrigation with the technical and financial support of Japan International

Cooperation Agency (JICA). It was developed to be complementary to '*Kenya Vision 2030*', the country's development blue print—hence its name.

Integrated water resources planning takes into consideration all sources of water, covering the entire country and encompassing all water use sectors. A national master plan serves to guide an entire economy and should be based on comprehensive data and sound information. Accurate information is required to answer questions commonly associated with water planning such as: what is the quantity of water available? Where is it located? What is the future demand? What will be the deficit? What are the management options? However, it is apparent that the preparation of NWMP 2030 was constrained by limited data (JICA 2013). In addition the NWMP came at a time when the country was undergoing critical socio-political change. In 2010, Kenya adopted a new constitution, Constitution of Kenya 2010, which provides for two levels of government, the national government and 47 county governments. This drastically changed the legal and institutional environment in the water sector in Kenya. The new constitution gives the county governments jurisdiction over water resources in a manner that seems to reverse the gains of the so-called water sector reforms that started in 1999.

The National Policy on Water Resources Management and Development (NWP 1999) was developed with an aim of achieving sustainable development and management in the water sector by entrenching the framework of integrated water resources management (IWRM) set out in the Dublin principles. Based on this policy, the Water Act 2002 was introduced to replace the Water Act Cap 372. The water sector underwent a complete transformation with new institutions being established and the roles of existing ones being redefined in a very basic manner. Water Resources Management Authority (WRMA) was established in 2003 as the lead agency for national water resources management while the role of government was redefined and limited to policy formulation and regulation. The water resources management was changed from administrative units to catchment basis. The NWMP 2030 was formulated based on this set up. However, the changes envisaged in the reforms were yet to be fully realized by the time the Constitution of Kenya was adopted in 2010 heralding a change in the basic law affecting the water sector.

Methodology

The NWMP 2030 was evaluated taking into account expert judgment contained in the plan on the data used, comparison with the previous plan produced in 1992 and 1999 (NWP 1999) in addition to other literature. In many places, the authors have indicated their opinion on the data used for projections, which are mostly reservations on their inadequacy.

The NWMP1992 was prepared by JICA in collaboration with the Ministry of Water Development. The plan was intended to form a basis for an orderly and sustainable development of the country's water resources for the period up to the year 2010. NWMP 1992 is perhaps the most comprehensive and authoritative document on the water resources of Kenya. The results of the assessment of available water resources in the NWMP 1992 were compared to the results of the NWMP 2030.

There is an abundance of literature on water resources management and related topics including reports prepared by United Nations agencies comprising assessments of the status of water supply and sanitation in many countries as well as scenario analyses of future water demands. The approach used in the NWMP for assessing the potential future effects of climate change on water resources situation in Kenya was evaluated.

Results and Discussion

The implications on the legal and institutional setup of the water sector in the country are presented and discussed. The estimated quantities of available water resources of both surface water and groundwater are presented. The national water demands are presented on a sectoral basis as at 1990, the current demands (2010) and the projections for 2030 based on Kenya's development blue print, dubbed Vision 2030 (GoK 2007) and 2050. Since the plan also considers the categories of consumptive and non-consumptive use, allocation of water for environmental conservation, international obligations and inter-basin water transfer is considered along with impacts of climate change on the resource base. Population projections play a vital role in the estimation of water demand as well as per capita water quantities of resources.

The Legal and Institutional Setup

There are several agencies active in the water sector in Kenya having narrow subsector interests in water resources management and development. In addition, there are other ministries involved in the water include those whose mandate cover Agriculture, Livestock and Fisheries, Energy, Industry, Roads, Transport and Infrastructure. Their operational overlap is not always smooth.

The NWMP 2030 was formulated based on NWP 1999 and Water Act 2002. They adopt the Dublin principles whose thrust is that water should be managed on a catchment basis and treated as an economic good in order to attain sustainability in water resources management and development. The law sought to transform the water sector in Kenya accordingly and all state water service entities were transformed into public-owned companies of water supply and sanitation. The role of government was reduced to policy formulation and regulation.

More than a decade later, there is no evidence that the companies of water and sanitation have performed better than the state entities they replaced, neither is there any indication that the water sector in Kenya driven by profit motivation can attract private investments to meet the growing water demands. This is partly due to the fact that the new companies and institutions spent a lot of time and resources to establish new offices in different parts of the country. The old government offices remained, albeit with reduced status. The multiplicity of water offices across the country has remained to date stretching limited finances and diverting it from the implementation of infrastructure.

The advent of the new constitution in 2010 constituted a great setback to the water sector reforms. The constitution created 47 county governments and a central government. Each county was given the mandate to develop the water resources within its jurisdiction, i.e. a return to management in administrative units. Efforts to amend the law to align it with the new constitution have taken considerable time. Although the planning in NWMP 2030 used the administrative boundaries of the 47 counties, no effort is shown in the master plan to suggest management structures for effective engagement with the current system under Water Resources Management Authority (WRMA). WRMA management structure is based on regions that do not conform to the county boundaries. The water towers, which are also trans-county, have faced serious invasion from human settlement and agricultural activities. Co-ordination of catchment management and conservation activities is already facing challenges.

Climate Change

Global warming and climate change is now a reality that is bringing with it numerous environmental and developmental challenges in many parts of the world. According to the report of the Intergovernmental Panel on Climate Change (IPCC 2014), concerted global effort is required to curtail emissions of CO_2 to levels consistent with a 2 °C temperature limit. Continued temperature rise is likely to cause changes in the hydrological cycle by affecting precipitation and evaporation (Huntington 2006).

The emission scenario selected for the climate change analysis in the NWMP 2030 was A1B scenario, because it is physically plausible and consistent and has a sufficient number of variables on a spatial and temporal scale that allows for impact assessment. According to the GCMs, an increase in surface air temperature seems to be unavoidable in the future. The surface air temperature will increase around 1 °C by 2030 and 2 °C by 2050 uniformly for the current climate.

Further, NWMP 2030 estimated potential evapotranspiration using the adjusted by FAO Penman–Monteith method based on monthly statistical meteorological data such as atmospheric pressure, temperature, relative humidity, daylight hours and radiation. The estimated mean annual rainfall and for the whole country in 2030 and 2050 are 750 and 801 mm, respectively, while the corresponding mean annual evapotranspiration are 675 and 723 mm. Both are predicted to increase toward 2050 but the temporal and spatial distributions of the mean annual rainfalls in 2030 and 2050 are similar to the current distribution (Fig. 1b).

Population Projections

The population figures used in the water demand calculations was based on the population census of 2009 whose results were released in 2010. The base value used in the projections is thus quite accurate. However, the projected populations are subject to uncertainty since the growth rate is itself uncertain. In fact, the population estimate for 2030 was obtained from Vision 2030 while that for 2050 depended on some projections made by United Nations. The ratios of rural to urban populations were assumed to be the same as those of 2030 (Table 1). It is noted that according to the projections, by 2050 two-thirds of the population will be living in urban areas. This is an inversion of the current situation (2010), where two-thirds of the population lives in rural areas. This population distribution pattern will demand redirection of effort to meeting high urban water demands as well as water for irrigation to meet their food requirement.

Water Availability

The renewable water resources of a country are equal to the difference between precipitation and evapotranspiration. This water is partitioned into groundwater recharge and surface runoff. Theoretically, this is the amount of water available for human exploitation. However, not all groundwater recharge is available for use.

The amount of water going into groundwater recharge can be estimated from rainfall–runoff analysis. The available annual fresh water resources of a country are the sum of the surface runoff and sustainable groundwater yield. Available water resources (million cubic metres per year (MCM/year)) for Kenya are given in Table 2. Evapotranspiration was estimated using the FAO Penman–Monteith method. Climate change has been factored in the evapotranspiration to get estimates of available fresh water resources for 2030 and 2050. The 1990 values are those from the NWMP 1992.

Groundwater recharge depends on the land use/land cover, soils and geology. The soils and geology of Kenya vary greatly across the country. An assumption of a constant groundwater recharge rate across the whole country would introduce large uncertainties in the estimates of sustainable groundwater potential calculated for

Table 1 Population projections up to 2050 (million people)

Year	2010		2030		2050	
Area	Population	%	Population	%	Population	%
Urban	13.08	33.9	46.02	67.8	65.69	67.8
Rural	25.45	66.1	21.82	32.2	31.20	32.2
Total	38.53	100	67.84	100	96.89	100

different catchments and the whole country. Due to limited data, the NWMP 2030 used 10% of annual recharge as sustainable yield of aquifers in Kenya.

The per capita fresh water amount of a country is found by taking the net difference between precipitation and evapotranspiration and adding it to any other surface inflows and groundwater recharges coming into a country in a year and dividing by the total population of the country (Table 3). This amount is an index of the relative freshwater endowment of a country: a country is said to be water scarce if it has a fresh water availability of <1,700 m^3 per person, per year.

Water Uses

The NWMP 2030 identifies and makes planning and management projections for six water use categories, namely, domestic, industrial, irrigation, livestock, wildlife and inland fisheries. However, it recognizes that some of these are broad categories that may be further subdivided. For example, domestic water demand comprises residential, institutional and commercial.

A summary of water demands for the various sectors considered in the plan is given in Table 4.

Table 2 Available water resources of Kenya (MCM/year)

Item	1990	2010	2030	2050
Available water resources	20,209	22,564	26,634	28,437
Surface water runoff	19,590	20,637	24,894	26,709
Groundwater recharge	619	1,927	1,740	1,728

Table 3 Per capita renewable water resources of Kenya

Year	2010	2030	2050
Population (million)	38.53	67.84	96.89
Per capita (m^3/c/year)	586	393	294

Table 4 Summary of water demands by sub-sector (MCM/year)

Subsector	2010 (a)	2030 (b)	(b)/(a)%	2050 (c)	(c)/(a)%
Domestic	1,186	2,561	216	3,657	308
Industrial	125	280	224	613	490
Irrigation	1,602	18,048	1,127	18,048	1,127
Livestock	255	497	195	710	278
Wildlife	8	8	100	8	100
Fisheries	42	74	176	105	250
Total	3,218	21,468	667	23,141	719

Hydropower power generation constitutes non-consumptive use of water. Therefore, current and future water use in hydropower production would be easily estimated. However, because it is non-consumptive use, these water use amounts were not incorporated in the water demand for the water balance calculations. It is thus observed that, this position may not be quite accurate because large hydropower projects require storage reservoirs which lead to considerable loss of water through evaporation, and in some cases, also through seepage.

Industrial development requires a commensurate development of power sources. The current and proposed hydropower projects will not be sufficient to meet the power demand for the newly industrialized country status projected for Kenya in *Vision 2030*. The absence of water allocation for cooling in thermoelectric power stations in the NWMP 2030 is a glaring omission.

The water demands for all the sectors were calculated based on a meticulous accounting for all the relevant factors. However, data on current water use amounts and demand rates were not available or were inadequate. These figures are the starting points of the projections. They were, therefore, estimated based on certain assumptions or expert opinion. The projections were based on the annual renewable water resources in Kenya. These figures are however misleading since some of the water resources have been polluted by waste water in degraded catchments. It is observed that NWMP 2030 does not seem to appreciate the impact of pollution and has not taken it into account.

Water for irrigation is projected to be constant between 2030 and 2050. No explanation has been provided for this. The change projected between 2010 and 2030 is an enormous increase of over 1,000% (see Table 4). There is no convincing justification for this apparent assumption that the irrigation demand will have peaked terminally by 2030 especially when the population over the 2030–2050 period is projected to rise by about 43% from 67.84 to 96.89 million. In the sub-sector projections, environmental flow requirements seem not to have been factored in. This is likely to affect the allocations once this is factored because there is an increasing awareness on water to sustain biodiversity.

Balance of Demand and Available Water Resources

The assessment of total water demand by catchment showed that the demand will keep rising. Demand for the current period, i.e. 2010, already shows a high water demand–water resources ratio of 76% in the Athi catchment area (Table 5). By 2030, the Athi catchment area will have high water deficit and by 2050, all the catchments will have demand–resource ratios higher than 40%. It is interesting to note that the 2050 ratio between water resources and demands is almost the same as that for 2030. This is attributed to climate change which is expected to bring in greater rainfall. The result will be an increase in water resources that almost matches the increase in demand over the period.

Table 5 Comparison of water demand and available water resources by catchment (MCM/year)

Catchment	2010			2030			2050		
	Resources (a)	Demand (b)	(b/a)%	Resources (c)	Demand (d)	(d/c)%	Resources (e)	Demand (f)	(f/e)%
Lake Victoria North	4,742	228	5	5,077	1,337	26	5,595	1,573	28
Lake Victoria South	4,976	385	8	5,937	2,953	50	7,195	3,251	45
Rift Valley	2,559	357	14	3,147	1,494	47	3,903	1,689	43
Athi	1,503	1,145	76	1,634	4,586	281	2,043	5,202	255
Tana	6,533	891	14	7,828	8,241	105	7,891	8,476	107
Ewaso Ng'iro North	2,251	212	9	3,011	2,857	95	1,810	2,950	163
Total	22,564	3,218	14	26,634	21,468	81	28,437	23,141	81

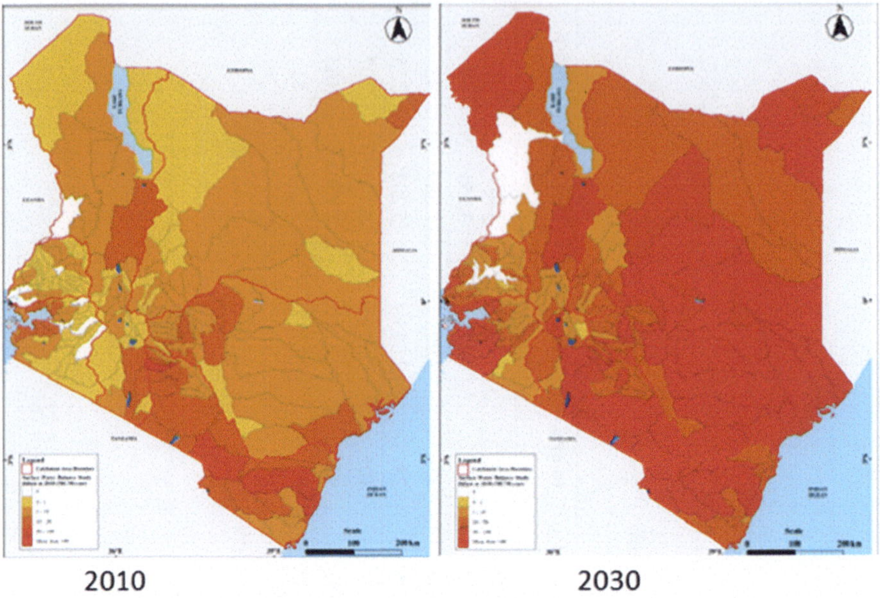

Fig. 2 Annual water deficit for a 10-year probable drought under current and future demands

However, the water deficit situations reflected by the ratios above are not to be relied on because of the uncertainty of projection of resources and demands. The projected water deficits in the different catchments for the current period (2010) and in future (2030) calculated on the 10-year probable drought are shown in Fig. 2.

Challenges in Long-Term Water Planning

Any planning presupposes the availability of ample data of good quality for analysis and synthesis. Deficiency in relevant data is one of the basic challenges in the study and planning for water resources in Kenya. This is mentioned repeatedly in NWMP 2030. The shortcoming is in two forms: either the data is of a relatively short duration; or, the data has significant gaps. The former is the result of abandoned stations or the proliferation of new ones. In many instances, new stations are established to generate data for specific projects in research or in infrastructure development. The latter, however, is a rather common problem resulting from institutional weaknesses. A study of rainfall data from stations in the River Nzoia basin in western Kenya indicated continually declining numbers of stations with complete data between 1970 and 2008 (Fig. 3).

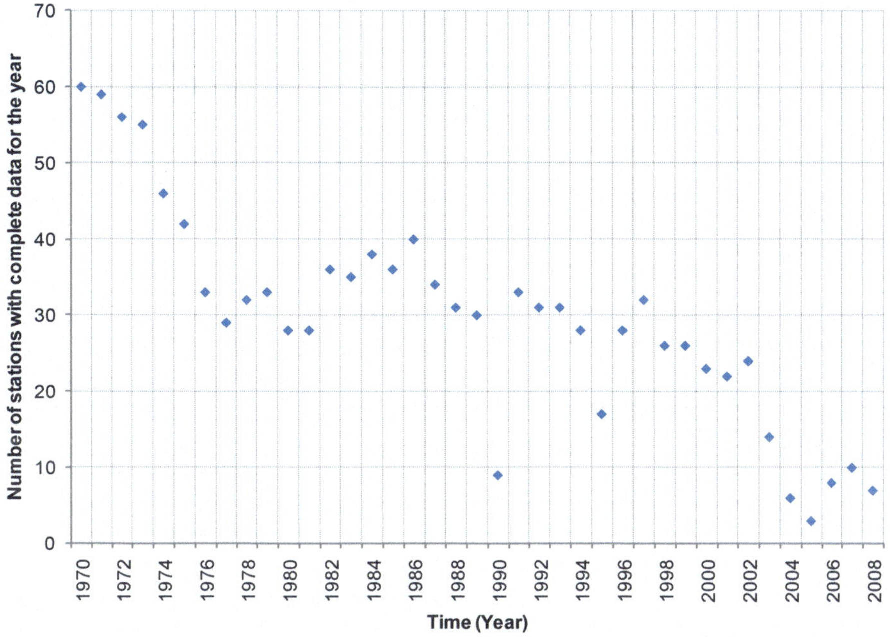

Fig. 3 Number of stations with complete rainfall data in R. Nzoia basin, Kenya (*Source* Kibiiy et al. 2015)

This trend of data availability fits well into observations that the tropics are generally poorly monitored (Ogden and Harmon 2012) and also that hydrologic data collection in the tropics is in rapid decline (Wohl et al. 2012).

The length of design life and that of plan period in water resources are based practical considerations and is determined from calculations and projections from relevant data. The practice in Kenya has been to adopt a 20-year period (GoK 1986). This period is reasonable in circumstances where the population has not stabilized and is undergoing rapid growth—as in Kenya.

It is also prudent to adopt a short plan period in cases where the data available is of a relatively short duration or is characterized by high uncertainty. The case of groundwater in Turkana County in northern Kenya is very illustrative of the difficulties of planning based scant data: in May 2015, UNESCO and the Ministry of Water entered into an agreement to start work on the Kenya Groundwater Mapping Programme aimed at building a comprehensive database of a country's groundwater resources (Sunday Standard2015). This programme was prompted by the reported discovery of two enormous groundwater reservoirs estimated to hold about 250 billion cubic metres in the dryland region Turkana County in northern Kenya.

Conclusions and Recommendations

The following conclusions and recommendations were arrived at from the above review and analysis of the NWMP-2030.

Conclusions

The failure to conclude the revision of the national water policy and law more than 4 years after the promulgation of the Constitution of Kenya 2010 is indicative of the difficulties in the water sector. A prolonged gap in legislation negatively impacts on operations in any sector, not least, the water sector.

The NWMP-2030 estimates the sustainable yield of groundwater resources at 1,947 MCM/year while the NWMP-1992 estimates it at 619 MCM/year. These estimates differ by a factor of three. The reason for this is unclear. The authors of the NWMP-2030 have indicated the inadequacy of groundwater data at their disposal but a change of this magnitude would require a sound justification.

The reported discovery of two gigantic groundwater reservoirs in the arid/semi-arid Turkana County with an estimated yield of about 250 billion cubic metres radically alters the water balance estimates of the country. If confirmed, this would render the NWMP-2030 invalid because these reserves are much larger than the annual available resources of the country estimated at 22,564 MCM/year (Table 2).

Kenya Vision 2030 (GoK 2007) envisions an industrialized country by 2030. This status presupposes adequate energy supplies to power industry, agriculture and society. Industrial production depends on water not only for processing but also for cooling in thermoelectric power plants. However, NWMP-2030 has only considered the hydropower potential and has not allocated water for use in thermoelectric power stations. This is a glaring omission because the hydropower potential of the country is not sufficient to meet the power requirement of the country in 2030. NWMP-2030 postulates that the impact of climate change on Kenya's water resources is to increase the country's water resources through increased precipitation. The available water resources are projected to rise by 26% from 22,564 to 28,437 MCM/year in 2050. This was estimated from the results of general climate models (GCMs) that are difficult to downscale and is thus subject to high uncertainty. Basing the plan on this uncertain increase in resources constitutes an unjustified risk.

The projection of the plan over the 40-year period 2010 to 2050 using inadequate data and unreliable information also constitutes an unjustified risk.

Recommendations

The long time taken to review the law and policy to align them with the Constitution of Kenya 2010 to provide a proper legal and institutional framework for operations in the water sector is indicative of the difficulties in the sector. The law should subsequently be revisited to ensure that it is streamlined for sustainable water resources development.

Reliable standardized data form the foundation for planning and management. Hydrological and meteorological data collection and storage should be strengthened in order to obtain high quality reliable data necessary for future water resources planning. The government should commission the re-assessment of the country's water resources with emphasis on groundwater in order to get a better estimate of the available resources.

Capacity should be developed locally for downscaling of results of general climate models (GCMs) for better prediction of the long-term impacts of global warming and climate change.

Acknowledgements The authors wish to acknowledge the financial support of the VLIR-UOS Programme at Moi University.

References

Chow VT, Maidment DR, Mays LW (1988) Applied hydrology (1st edn). McGraw-Hill, Singapore

Fitzgibbon C (2012) Economics of resilience study—Kenya country report

Government of Kenya (GoK) (2002) Water Act No. 8 of 2002. Government Printers. Nairobi, Kenya

Government of Kenya (GoK) (1986) Design manual for water supply in Kenya

Government of Kenya (GoK) (2007) Kenya vision 2030

Huntington TG (2006) Evidence for intensification of the global water cycle: review and synthesis. J Hydrol 319(1–4):83–95

IPCC (2014) Impacts, adaptation, and vulnerability. Part B: regional aspects. Contribution of working group II to the fifth assessment report of the Intergovernmental Panel on Climate Change

JICA (2013) National water master plan 2030. Draft 1. JICA. Nairobi, Kenya

Kenya National Bureau of Statistics (KNBS) (2010) Kenya demographic health survey 2008–2009. Kenya National Bureau of Statistics, Nairobi, Kenya

Kibiiy J, Kosgei J, Muku L, Willems P (2015) Why the sum is less than the parts: some recent and ongoing research on Nzoia river basin. In: Proceedings of the CREATE conference on integrated ecosystem and water management in the Lake Victoria Basin of Kenya

Muhindi JK, Ndichu RW, Oloo PN (2001) Rainfall atlas for Kenya. Drought Monitoring Centre-Nairobi

NWP (1999) The national policy on water resources management and development

Ogden FL, Harmon RS (2012) Editorial: Special issue of J. Hydrology on Tropical Hydrology. J Hydrol 462–463:1–3

Republic of Kenya (2013) The project on the development of the national water master plan 2030

Republic of Kenya (2013) Millennium development goals status report 2013

Sunday Standard, November 1 (2015) State maps groundwater to stem drought. Reuters. Nairobi, Kenya

Wohl E, Barros A, Brunsell N, Chappell NA, Coe M, Giambelluca T, Goldsmith S, Harmon RS, Hendrickx LMH, Juvik J, McDonnell J, Ogden FL (2012) A research vision for hydrology in the humid tropics: balancing water, energy, and land use. Nat Climate Change 2:655–662

Zeman C, Rich M, Rose J (2006) World water resources: trends, challenges, and solutions. Rev Environ Sci BioTechnol 2006(5):333–346

Integrated Water Management at Local Government Level for Southern Africa

J. A. Du Plessis

Abstract Southern Africa is a land rich in mineral resources, but developments are restricted in the largest part of the sub-continent due to a lack of adequate access to available water resources. To address this shortfall an understanding is required of the management structures responsible for the delivery of water to the user. To understand the drivers responsible for effective water supply, a study of the full water supply chain in South Africa, Namibia, Botswana, Zambia and Zimbabwe was done, comparing the different management models and the problems associated with the implementation of these different management models. A scorecard was developed to identify the readiness of selected local authorities to ensure effective WDM, taking water resources, purification, distribution, consumer demands, return flows and institutional aspects into consideration. The study concluded that, while significant differences do exist among water authorities, the biggest challenge rests with the management of return flows, while purification management seems to be the best addressed step in the water chain.

Introduction

As the pressure on the available water resources all over the world increases, due to the increase in population growth, increased levels of development and climate change issues; it will be required from the institutions responsible for the management of water resources to fully understand the challenges facing them. In many cases, it is a specific level of governance that has the responsibility to manage these water resources and their responsibility needs to be evaluated within the framework of local governance in general.

In an ever-changing environment, the availability of appropriately trained engineers to accept and deal with these challenges remains a problem. It is perhaps

J. A. Du Plessis (✉)
Department of Civil Engineering, Faculty of Engineering,
Stellenbosch University, P/Bag X1, Matieland 7602, South Africa
e-mail: jadup@sun.ac.za

V. P. Singh et al. (eds.), *Water Resources Management*, Water Science
and Technology Library 78, https://doi.org/10.1007/978-981-10-5711-3_15

best illustrated with the level of engineering staff available at local authorities in South Africa. Lawless (2005) provides statistics that summarise the extent of the problem: 16.5% of all municipalities in South Africa have only one civil technician on their staff, while 16% of all municipalities only have technical staff in service that are younger than 35 years in age.

South Africa is a country with an extremely rich biodiversity. This biodiversity is, however, increasingly under threat with the demand for land for agricultural and urban use, increasing as poverty in the country increases. This immense pressure on the biodiversity has a great impact on the water resources of the country (Turpie et al. 2008). Mathipa and Le Roux (2009) relate to this and define South Africa as a water stressed country; therefore legislation and strict managerial structures are required to ensure the protection and development of sustainable water resources. These challenges are not unique to South Africa, but also play an important role in Southern Africa.

In this context, research was, therefore, aimed at five Southern Africa counties, i.e. South Africa, Namibia, Botswana, Zambia and Zimbabwe. These countries are all interlinked with each other due to economic developments and support being rendered among these neighbours. They are all rich in minerals and other natural resources, but the availability of water does hamper their full development potential.

Main Objectives and Methodology

The Development Bank of Southern Africa (DBSA) and various other countries, such as China, are keen and willing to contribute towards these African countries, but the focus of their support to ensure sustainability remains a difficult issue. While it is recognised that many factors do play a role in a decision to provide assistance, this research focused on only the water cycle. The main objective of this research is to determine the main problem areas in the water cycle and in doing so, aims to provide some guidance as to where support needs to be provided within the water cycle.

For this to be fully understood, it is important to first have a better understanding of what is meant with the water cycle and what constitutes the different steps in an appropriate management approach towards integrated WDM. This needs to be seen within the different governance structures responsible for the management of the water cycle, which, as will be discussed, can be quite complex. For this reason, the different governance structures and models need to be discussed.

Having set the management environment, the main objective will be to provide an overview as to how well the different municipalities in the different countries are equipped to deal with the water management issues. A model, consisting of various decision support systems (DSS), was developed for this purpose (Du Plessis 2013) for each of the water cycle steps. To enable a quantitative comparison of how well the different municipalities managed these different water cycle steps as highlighted through the different DSS's, which are required to ensure effective WDM, these DSS will be evaluated using a scorecard system. The approach used in this research was to visit ten municipalities/water authorities and to meet with their representatives; mostly the managers of the water/technical departments. They then assisted the author in providing an overview of these institutions' readiness to manage their water, at that time, using the criteria identified during the development of the scorecard system.

The development of the model will be discussed briefly in this paper, but the focus will be on the result of the study based on the scorecard evaluation, which involves ten different water supply institutions, mostly municipalities.

Water Cycle

The water cycle, as applicable to a municipality, will be discussed in the context of water demand management (WDM) and water conservation (WC). WDM and WC are generally considered to be the same concepts by ordinary water users. There are, however, well-defined differences and, it is important to take note of these differences. WDM has been described (Johnson et al. 2002; Vairavamoorthy and Mansoon 2006) as "The adaptation and implementation of a strategy by a water institution or consumer to influence the water demand and usage of water in order to meet any of the following objectives:

- Economic efficiency
- Social development
- Social equity
- Environmental protection
- Sustainability
- Political acceptability".

Brooks (2006: 524) provided a more pragmatic approach, and defines WDM as being any action, whether of a technical, economic or social nature, which will achieve at least one of the following targets:

- Reduce the quantity or quality of water required;
- Adjust a specific task to ensure that it requires less water for the same outcome;
- Reduce losses between resource and outcome;
- Shift the timing of use from peak to off-peak periods and
- Increase the ability of water resources to maintain society during times of short supply.

Water conservation, in contrast, is described (Johnson et al. 2002; Vairavamoorthy and Mansoon 2006) as 'The minimisation of losses or waste, care and protection of water resources and the efficient and effective use of water'.

For most water service providers and municipalities, WDM means taking necessary steps to reduce the volume of water purchased or allocated to them by authorities. Their focus will be on reducing losses in their distribution systems and treating raw water with the least possible water losses, convincing their end-users to reduce the water use for garden purposes and general use, and eradicating ineffective water-using devices such as automatic flushing urinals.

WDM can be carried out at various levels within government structures, as suggested by Johnson et al. (2002: 5). Figure 1 illustrates six different stages representing the full water management cycle. WDM can be implemented with different levels of success and impacts at each of these stages.

To reduce the complexity of a single model for the full water management cycle, a decision support system (DSS) was developed for each of the steps in the water management cycle. These DSSs will be discussed further in more detail.

The DSSs were used to evaluate the different municipalities' or water institutions' readiness to do effective WDM for the full water cycle, from the resource back to the resource. It is, however, important to first understand the different management structure and the legal framework in which these municipalities need to function. For this purpose, a general overview of the institutional framework for municipalities in the Southern African countries included in this research will be discussed in the next section, with a specific focus on the South African municipalities.

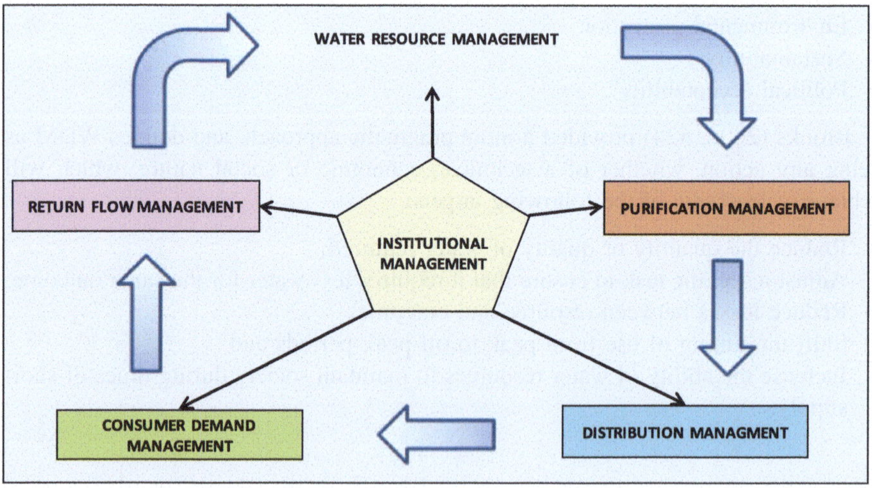

Fig. 1 Water management cycle

Institutional Models

The different institutional models in the countries visited during the research presented in this paper, will highlight the complexity of the relevant legislation and will serve to demonstrate the need for a structured approach towards the evaluation of the effectiveness of these institutions in managing their full water cycle.

South Africa

South Africa is a land with unevenly distributed water resources and huge climatic differences. According to the United Nations (UNPF 2001), a country can be considered as water stressed if the available resources are less than 1,700 m^3 per person per year. Southern Africa's estimated available water is only 1,289 m^3 per person per year. In general terms, South Africa can be classified as an arid or semi-arid region, with 21% of the country receiving less than 200 mm/year, while 44% receives between 200 and 500 mm/year. The rainfall distribution, as illustrated in Fig. 2 (DWAF 2004), highlights the diversity of the rainfall in South Africa.

Within South Africa, approximately 62% of the water is used for agricultural purposes. The other main users are urban (23%), rural (4%), bulk industries and

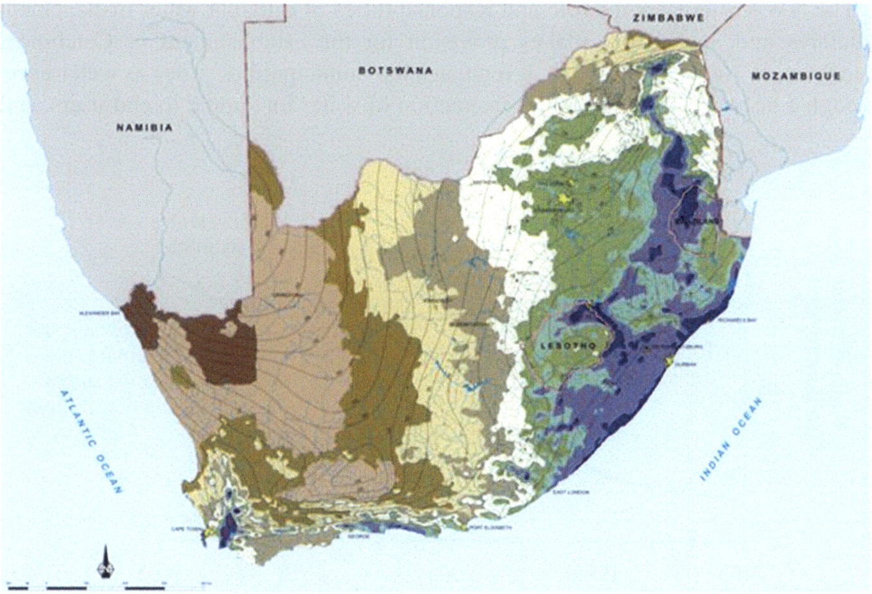

Fig. 2 Rainfall distribution in South Africa

mining (12%), cooling water for power generation (2%) and forestation (3%) (DWAF 2004).

Based on a number of possible water demand scenarios, the critical situation with the availability of water resources in South Africa is illustrated in Fig. 3 (Turton and Henwood 2002). Turton et al. (2006) further stated that, in the SADC region, it is not only South Africa that faces serious water problems; Botswana, Namibia and Zimbabwe are all reaching a point where economic growth can be put under pressure due to water scarcity.

According to legislation in South Africa and specifically the National Water Act (NWA) (Act 36 of 1998), ownership of water resides with the government, with a system of licensing introduced to authorise the water use rights on a 5-year renewable term to individuals or organisations (RSA 1998).

A number of Acts and Regulations play a role in the effective management of water resources by municipalities. Municipalities form the so-called third sphere of government in South Africa, with the national and provincial departments the first and second spheres respectively (DWAF, [s.a.]). In the case of water-related issues, the second sphere, or provincial government, plays a limited direct role. A municipality's main influence in the water sector is through the integrated development plans (IDP), of which the WSDP forms a key chapter. Legislation governing the water-related issues is primarily seated in the national DWA.

The division of South Africa's national water resources, and the institutions responsible for the management of each portion, are illustrated in Fig. 4 (Redrawn from DWAF, [s.a.]).

The NWA stipulates the role and responsibilities of different water management structures and specifically makes provision for the establishment of Catchment Management Agencies (CMAs). It regulates the municipalities' roles as water users through a licensing process for the abstraction of water for supply to end-users, and

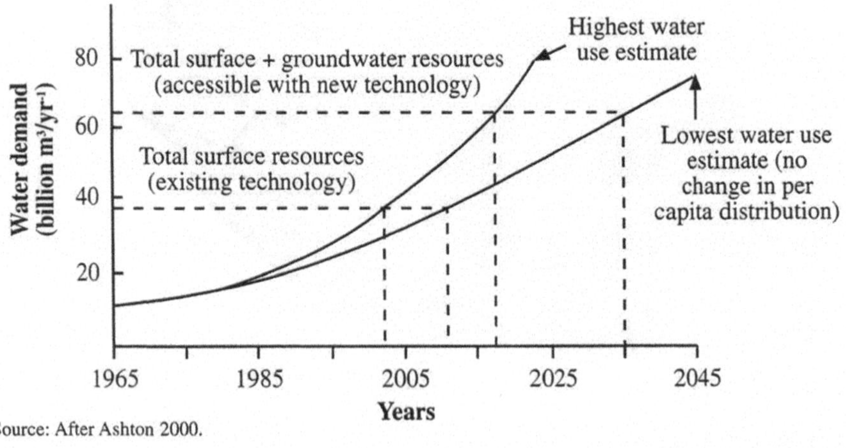

Fig. 3 Projected water demand for South Africa

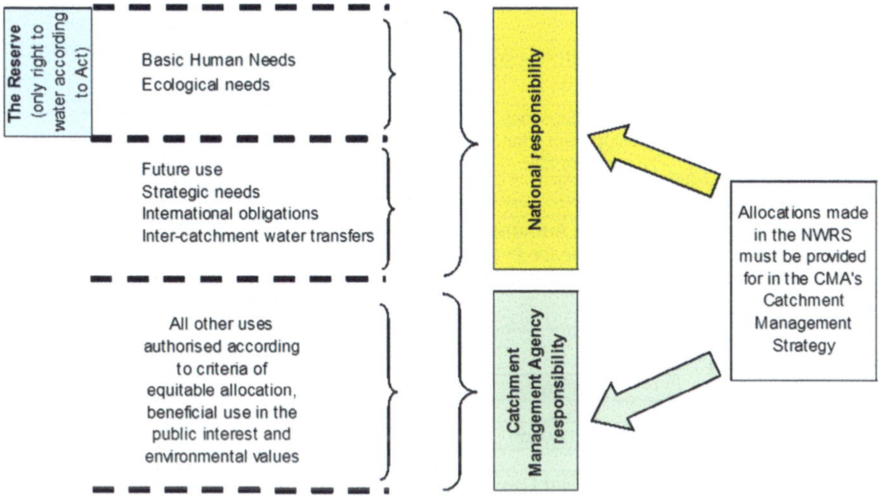

Fig. 4 Water users and the responsible authorities

it also governs the way in which a municipality can return effluent and other waste water back to the resource.

The main purpose (DWAF 2001) of the Water Services Act (Act 108 of 1997) is to assist municipalities in undertaking their roles in the water sector. The Act provides for the establishment of a number of water institutions, such as a water authority (e.g. municipalities), a water services provider, a water board and a water services committee. The Act ensures that the provision of water services is done in a fair and equitable manner, and that it promotes effective resource management and conservation.

Decision-making in the water sector does not always rest specifically and fully at the municipal level and it is therefore necessary to put the role and responsibility of water institutions regarding the decision-making process into context within the operating framework. Figure 5 illustrates the basic space of decision-making for the different institutions for each of the stages of the full water management cycles, as discussed before. It also highlights the Acts, as well as the main documentation that contains the information on which decision-making is based.

It is clear that decision-making and governance structure for municipal water management in South Africa are complex and not possible without appropriate management structures and cooperation.

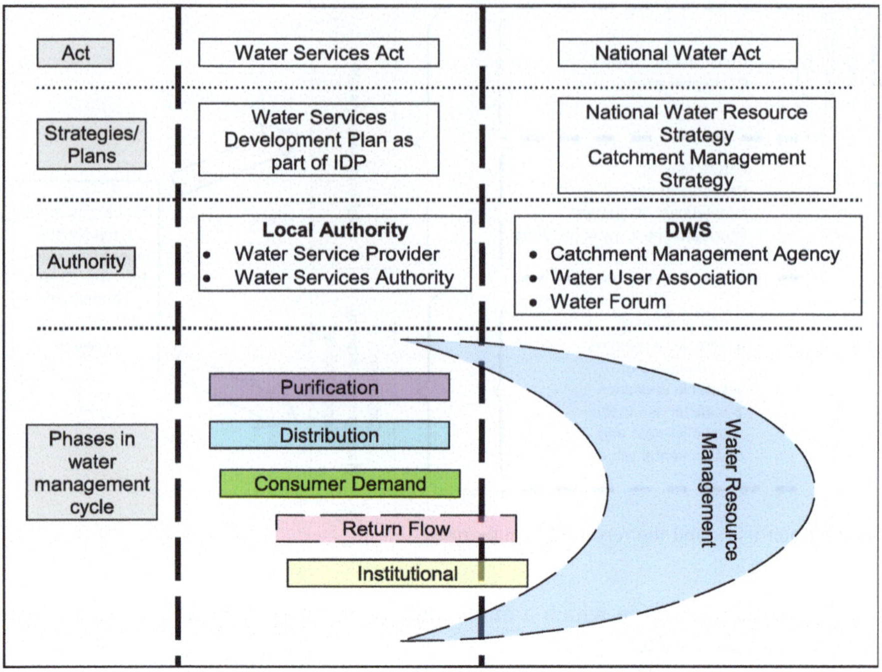

Fig. 5 Decision-making responsibility

Namibia

Namibia is seen as one of the most arid countries in the world as more than 80% of the country consists of semi-desert or desert (Lahnsteiner and Lempert 2007) with a mean annual rainfall less than 50 mm at the coast with a high of more than 600 mm in the North-east (Bethune et al. 2005).The country experiences frequent droughts while, according to Lange (1998), water use has continually grown over the past 25 years with an increased concern that the available water resources are not being used sustainably.

Up until independence in 1990, Namibia's water was managed according the Water Act (Act 45 of 1956) governed by South Africa. However, with independence an environmental clause was introduced in the Namibian Constitution, in the early 1990s. Article 95(1) of the Namibian Constitution provides for the sustainable use of all natural resources and states that: 'the state shall actively promote and maintain the welfare of people by adopting policies aimed at the maintenance of ecosystems, essential ecological processes and biological diversity of Namibia and utilisation of living natural resources on a sustainable basis for the benefit of all Namibians, both present and future' (Bethune et al. 2005).

The Namibia Water Corporation Act No. 12 (1997) makes provision for Namibia Water Corporation (NamWater) to be the bulk water supplier for Namibia and it requires NamWater to perform in an environmentally sustainable manner, with proper regards to the conservation and protection of water resources (Bethune et al. 2005). In 1998, the draft Environmental Management Bill followed and required that: ‘equitable access to sufficient water of ecological systems should be fulfilled to ensure the sustainability of such systems’ (MET 1998). This was followed by the National Water Policy approved in 2000, which states that water is an essential resource to life and that a sufficient supply of potable water is a basic human need. The policy also makes provision for natural ecosystems and states that all social and economic activity and development depend on healthy aquatic ecosystems.

This was followed by the Water Resources Management Act 2004 (Republic of Namibia 2004), which classifies water resources as national assets, and provides a modern legal framework for managing water resources based on the principles of integrated water resources management. However, the Local Authorities Act 23 of 1992 stipulates that each local authority must supply water to all the residents. In 2013, a new Water Resources Management Act was promulgated. This act makes provision for the establishment of Basin Management Committees, the management of rural water through Water Point Committees, Integrated Water Resource Management Plans and the issue of water licenses, just to name a few. This Act is at present still in the implementation phase.

Zimbabwe

Zimbabwe is situated in the southern part of Africa just North of South Africa, between Botswana and Mozambique. According to Chenje et al. (1998) water demand in Zimbabwe is increasing rapidly due to growing demands for agriculture, domestic and industrial water needs.

The lack of clean water is an increasing problem with the rapid growth in population. The safe water coverage levels, which were more than 95% in 1990, have dropped to less than 40% today in most cities. This fast deterioration of Zimbabwe’s water supply services is mainly due to aged infrastructure and a lack of maintenance and repair services (Ndedzu et al. 2012) and the need for appropriate legislation is therefore obvious.

In 1999, the Zimbabwean government introduced new legislation, namely the Zimbabwe Water Act No. 31, 1998 and the Zimbabwe National Water Authority Act, 1998 (Republic of Zimbabwe 1998a; b) in order to replace the Water Act of 1976. The cornerstone of the Water Act of 1976, which was the private ownership of water, was dismantled and all water is now vested in the President (public trust) (Ndedzu et al. 2012).

This new legislation is, however, not worth much without the proper management thereof and therefore the new Water Act also introduced the principle of managing water on hydrological boundaries. The Act gives the Minister (of Rural Resources and Water Development) the authority and obligation to establish catchment councils (seven) in conjunction with a National Water Authority (Ndedzu et al. 2012) and these Councils are responsible for the management of water which includes the function to issue water permits within their area of jurisdiction.

The tasks of the new catchment councils are mainly to assist in preparation of development plans, grant permits for water use, regulate and supervise the water use by permit holders and assuring compliance with the Act. Sub-catchment councils on their turn were created through the merging of the existing private and public water user organisations. Sub-catchment councils possess the power to levy rates upon permit holders within their jurisdiction in order to cover management costs. The objective of these sub-catchment councils is to monitor and manage the day-to-day water use within their jurisdiction (Ndedzu et al. 2012; Mtisi and Chaumba 2001).

Botswana

Botswana is a semi-arid country situated between Zimbabwe and Namibia, just north of South Africa. Rainfall throughout the country varies vastly from 250 mm/year in the south-western district to over 600 mm/year in the north-east (Swatuk and Rahm 2004). According to Moyo et al. (1993) the surface water resources in Botswana are limited as most rivers only flow for a few days per year. These riverbeds do, however, have a significant supply of groundwater storage and it is estimated that about 80% of all humans and animals in the country are dependent of this groundwater supply.

Water demand in Botswana is increasing rapidly mainly due to mining, an increased level of services and an increase in tourism with the associated economic benefits thereof. Appropriate legislation also provides a key element to ensure sustainability in Botswana (Swatuk and Rahm 2004).

After independence in 1966, the Republic of Botswana established its own Department of Water Affairs who implemented regulations under the 1967 Water Act (Sillery 1974). All water resources in Botswana belong to the state and are controlled by government. The state, however, delegated power to the Department of Water Affairs and the Water Apportion Board to issue water rights. The Department of Water Affairs was placed in the Ministry of Minerals, Energy and Water Affairs (MMEWA) who possesses the overall responsibility of the water policy. The Ministry is further assisted by the Department of Geological Surveys (DGS), Department of Water Affairs (DWA) as well as the Water Utilities Corporation (WUC) (Swatuk and Rahm 2004) to manage water appropriately.

The DGS, along with the DWA, are responsible for protection of surface water, supply of water to all villages, and the administration of the water legislation. The District Council, through the Ministry of Local Government, Lands and Housing (MLGLH), on their turn are responsible for the operation and maintenance of these village schemes (Republic of Botswana, no date).

Zambia

Zambia, a country situated just to the North of Botswana and Zimbabwe, is considered as one of the most urbanised countries in Africa with 5.19 million people out of a total of 13.1 million living in urban areas. Water governance in the country is complicated as the country is not ruled only by one government as in most countries. The Western Province of Zambia, better known as Sefula, is classified as traditional land and is ruled by a traditional ruler, also known as an Induna or local chief. Approximately 70% of Zambia consists of traditionally owned land and, therefore, the traditional water laws and regulations are frequently more significant than the water law established by the government (Chileshe et al. 2005).

All the water in Zambia is owned by the President and not by the state as in most countries. Water management, which falls within the jurisdiction of the government, is controlled by the Water Act of 1948, which makes provision for the ownership, use and control of surface water (Republic of Zambia 1949).

There are a number of government agencies and ministries which are involved in or have interest in water resource management, development or administration. The Department of Water Affairs along with the Water Development Board in the Ministry of Mines, Energy and Water Development are the two major institutions responsible for water resource development.

The Ministry of Environment and Natural Resources along with the Environmental Council of Zambia (ECZ) on their turn are responsible for water resource conservation. All functions related to provision of water supply and sanitation services are the responsibility of the Local Authorities (LAs) under the overall supervision and support of the Ministry of Local Government and Housing (MLGH) according to the Local Government Act, the National Water Policy and the Water Supply and Sanitation Act No. 28 of 1997.

The National Water Supply and Sanitation Council (NWASCO) is responsible for regulation of water supply and sanitation service providers while the Ministry of Health is responsible for water quality regulation and the Zambia Environmental Management Agency (ZEMA) is responsible for environmental protection (Phiri et al. 2000).

The management of traditional water rights, as applicable in the Sefula Province, is even more complicated with the involvement of the Indunas, Water Boards, water Committees, Wildlife authority and many more, depending of the nature of use (Meinzen-Dick and Bakker 2001).

The WDM Model

A water demand management model that needs to provide some guidance at a municipal level, needs to be able to involve all role-players and needs to make provision for a structured implementation program. From Fig. 5, it is also clear that the bulk of the responsibility of water-related service delivery rests with local government and that the model needs to take the local governance issues into consideration. Such a model was developed and discussed by Du Plessis (2013) and is illustrated in Fig. 6. The model consists of two processes and five DSSs, one for each step of the full water cycle.

While the detail of the two processes (Implementation and public participation) will not be discussed any further in this paper, the five decision support systems will be discussed briefly. These DSSs consist of flow diagrams indicating the logical sequence or steps to be considered to ensure that all aspects of WDM are attended to at a municipal level. These logical steps are guided with conditional questions that need to be answered, i.e. either positive (Y = Yes) or negative (N = No). The DSS for the water resource management phase of the full water cycle is demonstrated in Fig. 7.

The water resource management DSS highlights the importance of issues such as representation on formal management structures, the need to have knowledge of the safe yield of the various systems, the legal status of allocations from various sources, the need for an operational plan and the management of the catchments to deal for example invasive plant species.

The purification management refers to the treatment processes and includes aspects such as back-wash cycles, the management of the chemical dosing process, the effective metering of water during each stage of treatment and the re-use of settling tank water.

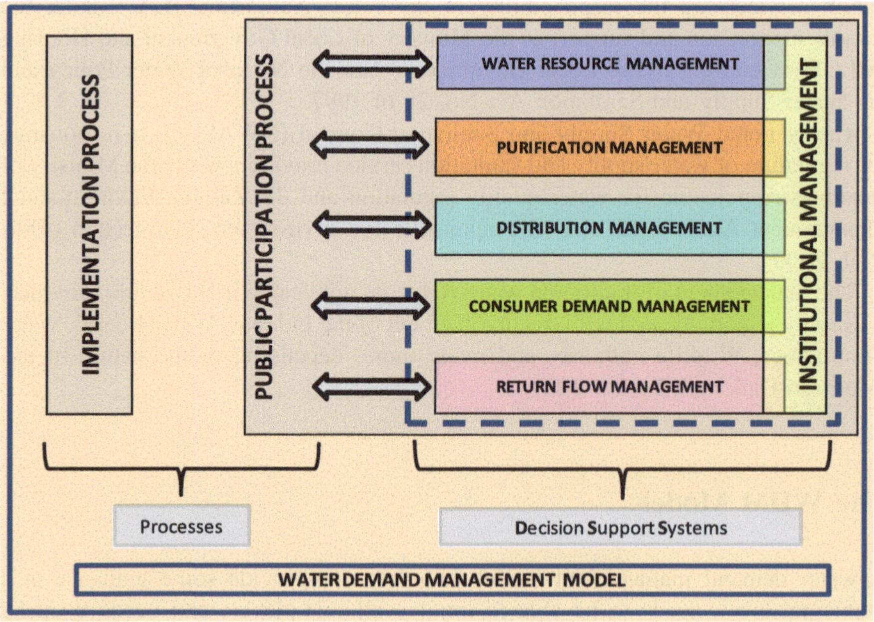

Fig. 6 WDM model

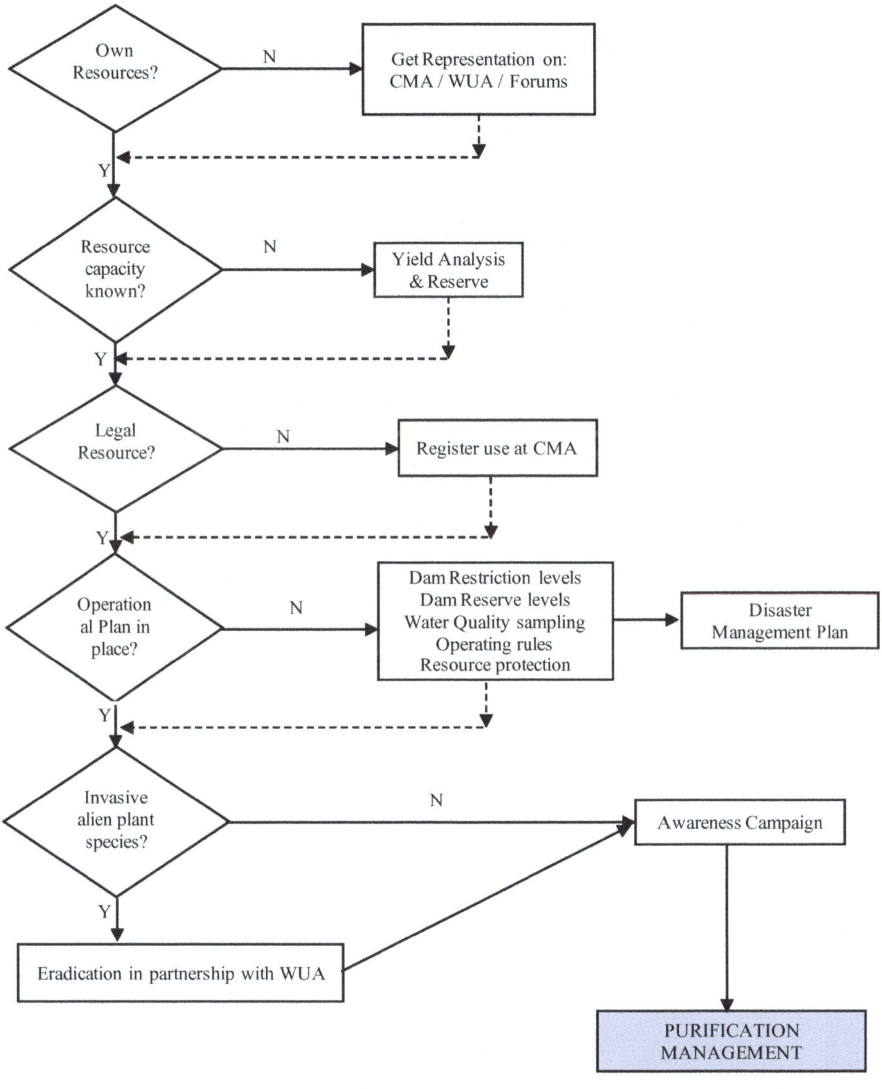

Fig. 7 Water resource management DSS

Distribution management refers to the steps to ensure effective metering, billing, the comparison of water consumption figures with acceptable benchmark values, water audits and law enforcement, just to name a few.

Consumer demand management includes consumer awareness, retrofit, re-use, water wise gardening actions and the repair of in-house leaks. The return flow management steps focus on the polluter pay principal, the re-use of treated affluent and the management of the infiltration of storm water into the sewer system, while the institutional management lastly deals with capacity building, training, appropriate bylaws, complaint centres and awareness among the public.

Evaluation Criteria

With the DSS available, a scoring system was developed (Du Plessis 2013) which was used in the research to quantify the readiness of an institution to ensure effective WDM. In these score cards (one for each DSS), the management steps or focus areas provided in each DSS are listed, and guidelines have been provided to facilitate the allocation of points to reflect compliance. The score card for the Return Flow DSS is shown in Fig. 8 as an example.

The score card process involves the allocation of one point (1) for each management focus area in the column that reflects the level of compliance, based on a value between 0 and 3. The score below the score card is then calculated by adding all the ones (1) in each column. The final scores for each water management cycle step are then calculated by firstly multiplying the number of ones with the category/ column value (0–3) and then adding all of the values to provide a score per stage, shown as a single value to the right of the array of score values.

The performance ranges provided below each of the score cards give an indication of the maximum number of points (number of ones (1) multiplied by the column value) that can be achieved for a specific step in the water management cycle, which are used to express progress achieved as a percentage of the maximum expected value for full compliance.

The final scores can then be calculated and used for comparison between different participating institutions or to measure progress over time for the same institution.

Water Cycle Step	Management Focus		Evaluation Criteria			None	Bad	Acceptable	Good	
			Bad	Acceptable	Good	0	1	2	3	
RETURN FLOW MANAGEMENT	Sewer system	Maintenance on sewage system	Removal of blockages on request	According to set program	Program in place and well funded					
	Point source pollution	Sampling	Once every 6 months	Every Month	Every month and during suspect conditions					
		Analysis	Sent samples to central lab (waiting time, weeks)	Sent samples to local lab (waiting time, days)	Analysis at own lab (Results immediately)					
	WWTW	Inflow measured	Irregular or Monthly	Daily and complete records	Continues electronic records					
		Quality of inflow	Samples analysed monthly	Samples analysed weekly	Samples analysed daily					
		Benchmark values	Available for some of aspects, but at least ratio between water sold against water received at WWTW	Available for most aspects, but at least ratios between water sold, water received and water returned to source.	Available for all aspects to be compared nationally and own benchmarks.					
		Capacity known	Only hydraulic capacity known	Hydraulic capacity and design loads known	Hydraulic capacity and design load measured against inflow data					
		Sampling	Irregular or Monthly	Weekly	Daily					
		Final Effluent Quality	Comply with licence conditions less than 70% of time	Comply with licence conditions more than 70% of time	Comply always with licence conditions					
		Re-use of treated effluent	Between 0 and 15% of annual final effluent	Between 15 and 50% of annual final effluent	More than 50% of annual final effluent					
		Maintenance program	Only during emergencies	On request	According to set program, including preventative maintenance					
	Treatment cost		Treatment cost estimated, seen as part of full water budget	Separate costing done and budget based on these figures	Separate costing done and sewage tariffs based on actual costs					
					Score	0	0	0	0	0
					Performance range	0	12	24	36	

Fig. 8 Scorecard for return flow management

Evaluation

While it is recognised that the data and results are only representative of a specific time period, it is also well known that changes in decision-making structures typically take a long time and it is therefore reasonable that the general tendencies resulting from this research do provide a fair overview of the situation in general. It also needs to be acknowledged that the municipalities who participated varied significantly in size, which was done purposefully to ensure a fair understanding of the general situation. In this regard the populations served by the municipalities varied between approximately 1 million to as little as 10,000 people.

Of the ten evaluations conducted, two were done in South Africa, four in Namibia (two from the same town/city, but one from the bulk supplier, while the other one was for the municipality), one for Zambia, two for Zimbabwe and one for Botswana.

Results

Based on the analysis of the data collected during the research period it was clear that all Southern African countries participating in this research experience typically the same problems. A summary of the results in terms of the readiness to manage the full water cycle effectively is presented in Fig. 9. These results indicated that the most challenging part of the water cycle is the treatment of the wastewater and the return thereof back to the resource, with scores less than 50%. This was only marginally less effective than the consumer demand management step.

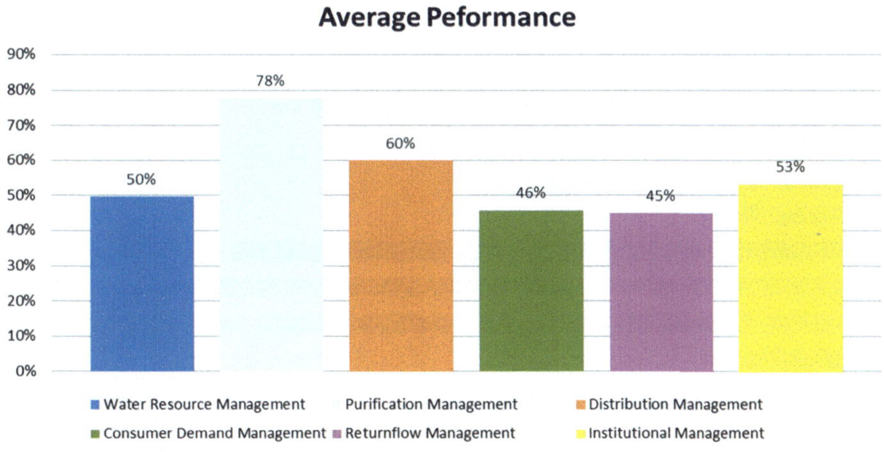

Fig. 9 Performance per water cycle step

The best well managed part of the water cycle was the water purification management step, with 78% compliance, followed by the distribution management step with 60% compliance.

When considering the water resource management cycle, Fig. 10 provides an overview of how the different participants performed in comparison with each other. Most of the participants indicated a clear understanding of the available yield of their systems and do have appropriate operational plans in place with reference to restriction levels. All the participants indicated that their water resources have been registered at the appropriate authorities. Lacking, however, was the appropriate participation at a catchment management level.

It was interesting to note that the performances of the top four municipalities were all from four different countries. They all did very well with scores varying between 67 and 80%. Similarly, the three poorest performances came from three different countries that all have done extremely poor with a score of only 17%. These are, however, municipalities who do not have their own resources and therefore very little control, but they also have very few steps in place to ensure awareness and to participate in whatever structures are available to provide some input to ensure an effective management of the catchments. In general, five participants reported fairly good scores, well above the average of 50%, with three reporting scores significantly below the average.

When comparing the respondent's readiness to treat their raw water to potable standard it was clear that almost all do take regular samples of their incoming raw water, most do measure their final treated water leaving the water treatment facilities and most do have a set of benchmark values against which they are measuring their compliance regarding final water quality. Only five of the participants do actually operate water treatment facilities and their scores are reflected in Fig. 11, with an average of 78%.

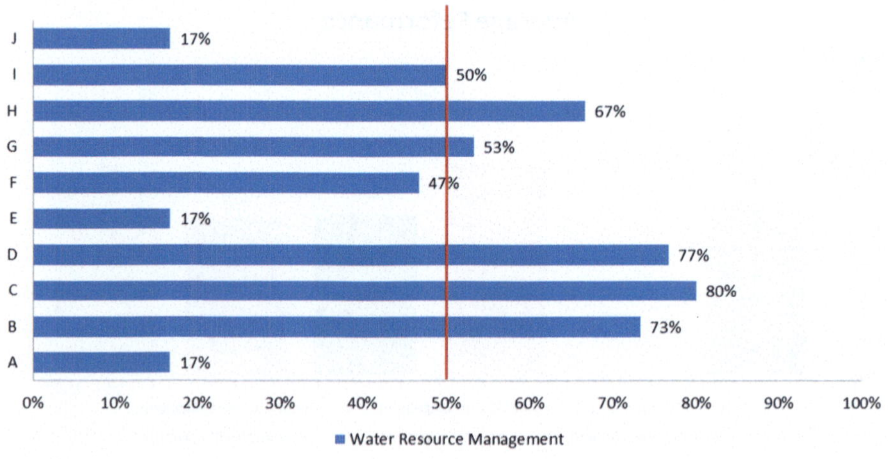

Fig. 10 Performance at water resource management cycle step

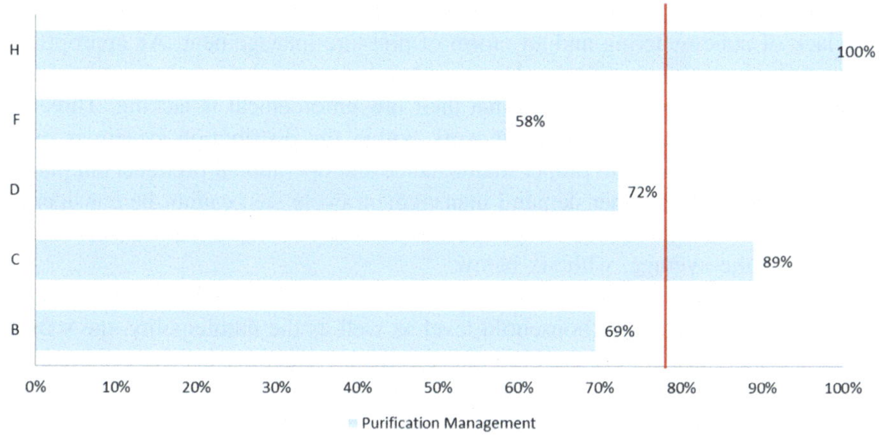

Fig. 11 Performance at purification management cycle step

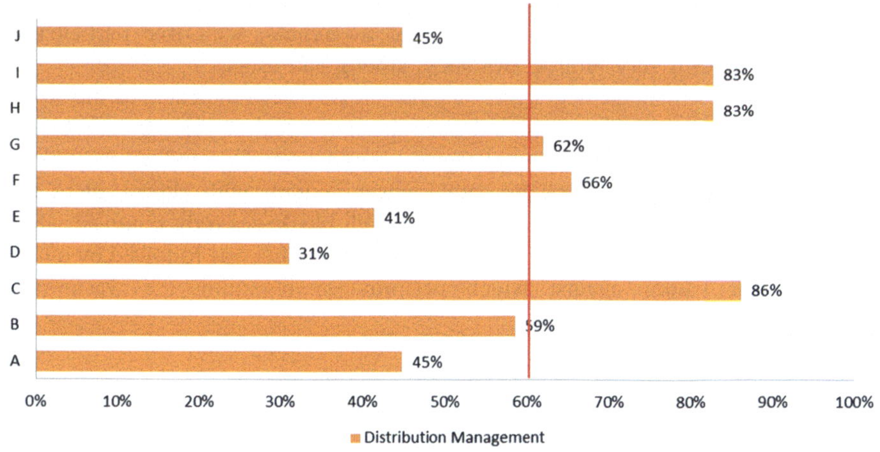

Fig. 12 Performance at the distribution management cycle step

The worst score in the purification management cycle step was attributed to the lack of appropriate steps to re-use some of the wastewater stream generated during the treatment process such as the back-wash water. Measuring of the back-wash water was also rarely achieved.

The performances of the distribution management cycle step are illustrated in Fig. 12. Eight of the participants indicated that they do measure all water entering their system on a regular base and four of them reported that they do have this information available electronically on a continuous basis. The average score amounts to 60%, with 3 participants reporting a score of more than 80%. The lowest score was reported to be 31%.

The most challenging aspect of the distribution management cycle step proves to be a lack of zone metering and any form of pressure management. An appropriate meter replacement program was also considered to be inadequate by seven of the participants, while seven indicated that their law enforcement is lacking. Three of the participants indicated that repair work within the distribution system is only done when needed, with no proper maintenance and operational producers in place.

In general, the consumer demand management cycle step cannot be considered as being effective, given an average score 46%. Only four participants reported a score above the average, with six below.

Although it is accepted that the responsibility towards demand management rests both with the end user at a household level as well as the municipality, the scores reflected in Fig. 13 only represent the municipal readiness to ensure effective WDM. While five of the participants indicated that a block tariff with more than three steps is in use, three participants did not have a rising block tariff in place. Of particular concern is that none of the participants uses their billing system to provide information that can be used by the end user to improve WDM at a household level.

On a request to raise an opinion regarding the end-users awareness, only three participants indicated an acceptable level of awareness, with five participants indicating none or a very limited awareness among end users.

The return flow management cycle step has already been indicated as the worst step in the water cycle. Figure 14 indicates the individual scores for each of the participants. Further analysis showed that the worst part of the return flow management is the management of the actual sewer systems with almost all of the participants indicating a lack in performance, with blockages only removed when they occur with no set maintenance program available. This was followed by a lack of benchmark values against which performance regarding the wastewater

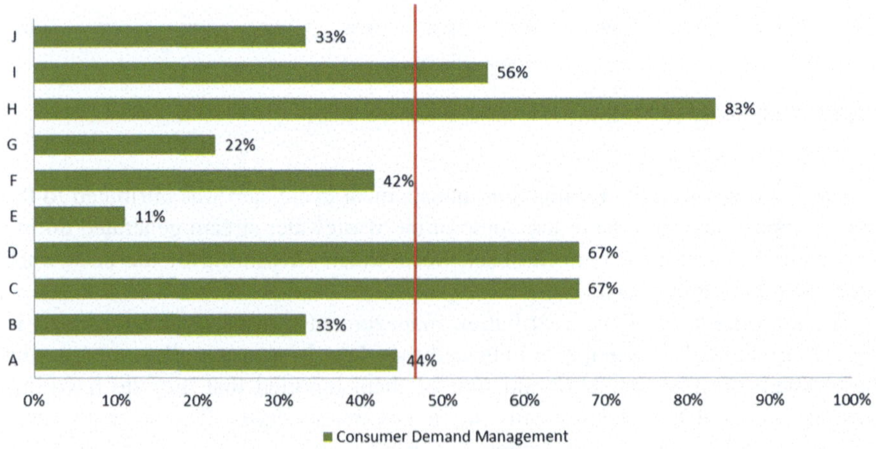

Fig. 13 Performance at consumer demand management cycle step

Fig. 14 Performance at return flow management cycle step

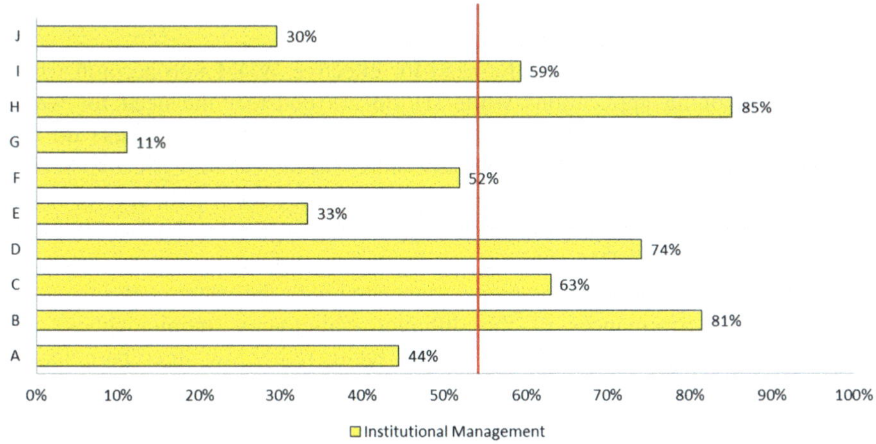

Fig. 15 Institutional management cycle step

treatment works operations can be measured. The lack of appropriate steps to ensure that proper sampling does take place and the associated lack to adhere to quality standards set by regulators are also evident.

There was, however, a reasonable agreement among all that they do have a clear understanding of the costs involved with the treatment of the wastewater stream, while it was also evident that, in terms of the availability of laboratory facilities to do the appropriate tests on water samples, these were seldom available, typically either at the bigger institutions or completely lacking.

The level of institutional arrangements in place to support and to enable effective WDM is reflected in Fig. 15. While three participants scored values of 74% or

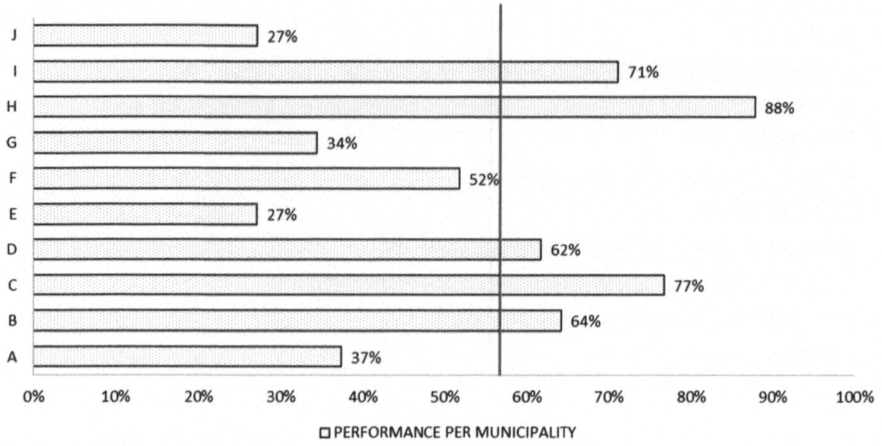

Fig. 16 Performance per municipality

higher, three also scored 33% or less with an average of 54%, which clearly indicates some reason for concern.

Seven participants do not take the need for effective WDM into account when considering rezoning applications while six do not consider the need for water saving devices before any building plans are approved. Five participants do not make provision for a designated WDM team or person to deal with effective WDM within the organisation, while a further two lack any structure to do so.

Nine of the participants indicated that they do have appropriate by-laws in place to assist with effective WDM, while eight do have a complaint centre to deal with consumers who need to report problems associated with the water cycle.

The overall performances of the different participants (municipalities) are reflected in Fig. 16, while the average of 54% seems rather low, three participants scored a value of 71% or higher. Four participants however scored values of 37% or less.

If these performances were arranged according to population served (or size of institution), it was also clear that there is a significant difference in the capacity to manage the full water cycle, with the institution with a population size towards 100,000 or more people significantly better off than the smaller institutions (10,000–60,000).

Conclusions

The development of an appropriate model to guide water managers through a process to ensure effective WDM can contribute significantly towards sustainable water use, specifically since it is clear that municipalities increasingly running a risk of finding appropriately trained technical staff to deal with these issues. The score card system developed to facilitate the DSSs developed as part of the model was used to provide a general overview of the WDM situation in five Southern Africa counties.

From the results, it was clear that all countries experience problems and that shortfalls are not limited to any specific country. It was also clear that the smaller municipalities, as perhaps could be expected, have significantly more challenges to ensure effective WDM than the larger municipalities.

While the treatment of raw water to potable water seems to be the best-managed step in the full water cycle, the treatment of the wastewater seems to be the most neglected part of the water cycle, followed closely by the inability of municipalities to involve the end-users as part of the consumer demand management.

Limited attention is given to the protection of the catchments, which provide the bulk of the water resource available, and the institutions' capacity to manage is also severely limited.

Acknowledgement The author wants to acknowledge the contribution of all the representatives of the different water institutions who participate in the evaluation, as well as the National Research Foundation of South Africa for making funds available to disseminate the results.

References

Bethune S, Amakali M, Roberts K (2005) Review of Namibian legislation and policies pertinent to environmental flows. Phys Chem Earth A/B/C, pp 894–902

Brooks DB (2006) An operational definition of water demand management. Water Resour Develop, 521–528

Chenje M, Sola L, Paleczny D (1998) The state of Zimbabwe's environment, 1998. Ministry of Mines, Environment and Tourism

Chileshe P, Trottier J, Wilson L (2005) Translation of water rights and water management in Zambia. International workshop on African water laws: plural legislative frameworks for rural water management in Africa, pp 26–28

Department of Water Affairs and Forestry (DWAF) [s.a.] Water Management Institutions—overview. Chief Director: Communication Services. Pretoria. pp 1–39

Department of Water Affairs and Forestry (DWAF) (2001) Water and sanitation business: the role and responsibilities of local government and related institutions. March 2001. pp 1–36

Department of Water Affairs and Forestry (DWAF) (2004) National water resource strategy. Our blue print for survival. National Department of Water Affairs and Forestry. First Edition. September 2004

Du Plessis JA (2013) Integrated water management for municipalities in SouthAfrica. Municipal Engineer 167 (ME2):77–88

Johnson EH, Stephens A, Fakir S (2002) Development of a methodology to determine the effectiveness of water conservation and water demand management measures. Water Research Commission Report WRC 1273/1/02

Lahnsteiner J, Lempert G (2007) Water management in Windhoek, Namibia.Water Sci Technol 55 (1):441–445

Lange G (1998) An approach to sustainable water management in Southern Africa using natural resource accounts: the experience in Namibia. Ecol Econ 26(3):299–311

Lawless A (2005) Numbers and needs—addressing imbalances in the civil engineering profession. s.l., SAICE
Mathipa KS, Le Roux CS (2009) Determining water management training needs through stakeholder consultation: building users' capacity to manage their water demand. Water SA 35 (3):253–260
Meinzen-Dick R, Bakker M (2001) Water rights and multiple water uses—framework and application to Kirindi Oya irrigation system, Sri Lanka. Irrig Drainage Syst 15(2):129–148
MET (1998) Environmental management bill, 1998, Draft
Moyo S, O'Keefe P, Sill M, et al (1993) The southern African environment: profiles of the SADC countries. Cambridge University Press
Mtisi SMS, Chaumba J (2001) Sustainable lively hoods in Southern Africa
Ndedzu D, Muhaji E, Kunguma IC (2012) Household demand for improved water supply services in high density urban areas: the case of Mabvuku in Harare. Journal of Strategic Studies: A Journal of the Southern Bureau of Strategic Studies Trust: Joint Research Programme of the Ministry of Finance, the Zimbabwe Economic Policy and Research Unit and the University of Zimbabwe 3.1, pp 126–140
Phiri Z et al (2000) Water law, water rights and water supply in Zambia–issues and perspectives. In: Proceedings of the 1st WARFSA/WaterNet symposium: sustainable use of water resources, Maputo, 9
Republic of Botswana (no date) National development plan 8. Government Printer
Republic of Namibia (2004) Water Resource Management Act (Act No. 24 of 2004). Government Gazette 3357
Republic of South Africa (RSA) (1998) National Water Act (Act No. 36 of 1998). Pretoria. Government Printer
Republic of Zambia (1949) The Water Act. Government Printer
Republic of Zimbabwe (1998a) Zimbabwe National Water Authority Act. GovernmentPrinter
Republic of Zimbabwe (1998b) Zimbabwe Water Act (Act No. 31 of 1998). Government Printer Chapter 20:24
Sillery A (1974) Botswana: a short political history. vol 10. Methuen
Swatuk LA, Rahm D (2004) Integrating policy, disintegrating practice: water resources management in Botswana. Phys Chem Earth A/B/C 29(15):1357–1364
Turpie JK, Marais C, Blignaut JN (2008) The working for water programme: evolution of a payments for ecosystem services mechanism that addresses both poverty and ecosystem service delivery in South Africa. Ecol Econ 65(4):788–798
Turton A, Henwood R (2002) Hydropolitics in the developing world. A southern African perspective. African Water Issues Research Unit. Centre for International Political Studies. University of Pretoria. Pretoria
Turton A, Schultz C, Buckle H, Kgomongoe M, Malungani T, Drackner M (2006) Gold, scorched earth and water: the hydropolitics of Johannesburg. Water Resour Develop 22(2):313–335
United Nations Population Fund (UNPF) (2001) The state of the world population 2001. United Nations population fund report. http://www.unfpa.org/swp/2001/english/index.html
Vairavamoorthy K, Mansoon MA (2006) Demand management in developing countries. In: Butler D, Memon FA (eds) Water demand management. Department of Civil and Environmental Engineering. Imperial College London. IWA Publishing. London. pp 180–214

Integrated Water Resources Management in Malaysia: Some Initiatives at the Basin Level

Rahmah Elfithri and Mazlin Bin Mokhtar

Abstract The concept of Integrated Water Resources Management (IWRM) was first introduced in Malaysia since 1990s in various forms. However, the key issue remained how to implement it? IWRM can be best implemented at the basin level by using an Integrated River Basin Management (IRBM) approach. IRBM as a subset of IWRM is an approach to water resource management that takes into account all the factors linked to the resources, including social and economic activities. The implementation of IWRM at basin level in Malaysia has been initiated in managing water resources and river basin in Langat River Basin, Malaysia. The Langat River Basin is a transboundary river basin and is one of the Evolving UNESCO-IHP HELP Basins since 2004. The basin is small but has inherited many problems of a large river basin. This is because the river plays an important role in conservation, agriculture and potable water supply, but is facing threat from rapid development in the industry sectors and urbanization in the basin. This is why the components of IWRM and IRBM is needed to be implemented as practical approaches toward sustainable water resources and river basin management in Langat. This paper highlights some initiatives that have been made in managing water resources in this basin, as well as describes the current development and existing commitments towards achieving sustainability of water resources and river basin. The implementation of IWRM and IRBM need to be enhanced and further strengthened in order to ensure the sustainability of the water resources and river basin in Langat.

Keywords IWRM · IRBM · Water resources · River basin · Integrated Management

R. Elfithri (✉) · M. B. Mokhtar
Institute for Environment and Development (LESTARI), Universiti Kebangsaan Malaysia, 43600 UKM Bangi, Selangor, Malaysia
e-mail: elfith_ukm@yahoo.com

M. B. Mokhtar
Office of the Deputy Vice Chancellor (Research and Innovation Affairs), Universiti Kebangsaan Malaysia, 43600 UKM Bangi, Selangor, Malaysia

V. P. Singh et al. (eds.), *Water Resources Management*, Water Science and Technology Library 78, https://doi.org/10.1007/978-981-10-5711-3_16

Introduction

Water is vital for human survival and health, and it is a fundamental resource for sustaining human economic development. While water is a renewable resource, its availability in space (at a specific location) and time (at different periods of the year) is limited, being determined by climatic, geographical and political conditions, by affordable technological solutions that permit its exploitation, and by the efficiency with which water is conserved and used (Nigam et al. 1998).

Today, the world's freshwater resources have become subjected to increasing pressure. Among the reasons are (1) water resources are increasingly under pressure from population growth, economic activity and intensifying competition for the water among users; (2) water withdrawals have increased more than twice as fast as population growth and currently one third of the world's population live in countries that experience medium to high water stress; (3) pollution is further enhancing water scarcity by reducing water usability downstream; (4) shortcomings in the management of water, a focus on developing new sources rather than managing existing ones better, and top-down sector approaches to water management result in uncoordinated development and management of the resource; (5) more and more development means greater impacts on the environment; and (6) current concerns about climate variability and climate change demand improved management of water resources to cope with more intense floods and droughts (Mokhtar et al. 2009c, d).

More and more countries are faced with increasing challenges related to water in their struggle for economic and social development, and Malaysia is no exception. With greater recognition of the IWRM process and its benefits in the context of sustainable development, IWRM has emerged as an accepted approach that can assist countries in their endeavour to deal with water issues in a cost-effective and sustainable way.

Integrated Water Resources Management

Integrated Water Resources Management (IWRM) is defined as a process which promotes the coordinated development and management of water, land and related resources in order to maximize the resultant economic and social welfare in an equitable manner without compromising the sustainability of vital ecosystem (GWP 2000). IWRM is necessary to combat increasing water scarcity and pollution. Methods include water conservation and reuse, water harvesting and waste management. An appropriate mix of legislation, pricing policies and enforcement measures is essential to optimize water conservation and protection (UNDP 1991).

IWRM includes social, economic and environmental factors in the planning, development, monitoring and protection of land and water resources. It is a comprehensive approach to the development and management of water, addressing its management both as a resources and the framework for provision of water services.

It is very broad in scope to not only cover water, but also environmental management aspects such as land use issues, pollution control, development pressures and biodiversity conservation. The concepts of IWRM and IRBM treat water as a finite and vulnerable resource, water as an economic good and water governance should be based on a participatory approach involving all levels of stakeholders.

The concept of Integrated Water Resources Management (IWRM) was already recognized in Agenda 21 of the United Nations Earth Summit on Environment and Development that was held in Rio de Janeiro in 1992. In Malaysia, the IWRM concept has already been accepted as an innovative approach in managing its water resources. It was first introduced in Malaysia in various forms in the early 1990s by water technical agencies such as the Department of Irrigation and Drainage (DID). It was also supported by the formation of institutions such as the Malaysian Water Partnership (MyWP) in 1997 and followed by the establishment of Malaysian Capacity Building Network for IWRM (MyCapNet) in 2001 to promote IWRM and capacity building in Malaysia (Mokhtar et al. 2009a, b).

These are among key drivers spearheading the initial IWRM movement in Malaysia. In term of policy, IWRM has also recognized by the Malaysian Government since 2001 in different policy documents such as the Third Outline Perspective Plan, Malaysia (OPP3) 2001–2010, the 8th Malaysia Plan documents (MP8) 2001–2005, the 9th Malaysia Plan (MP9) 2006–2010 and the 10th Malaysia Plan (MP10) 2011–2015, which emphasizes the need for IWRM and to build capacity for its implementation.

At the basin scale such as, Langat River Basin, a certain degree of activities pertaining to realizing goals of an IWRM approach have been implemented, especially focusing on the Integrated River Basin Management (IRBM), i.e. another integrated approach in water management as a subset of IWRM based on river basin as a geographical unit or area, with the objective of balancing man's needs with necessity of conserving resources to ensure their sustainability (Elfithri et al. 2011). It requires broad stakeholder participation in water planning and operating decisions, and is another strong tool for encouraging such new civil orientation.

However, the problem of integrated management in this river basin lies in developing cooperation and collaboration among the agencies to ensure the well being of the basin itself. This includes issues of conflict resolution, information management and transboundary organization (Elfithri 2008).

Langat River Basin in Malaysia

The Langat River Basin is a unique and special basin in Malaysia because it lies within three different administrations of Selangor State, Negeri Sembilan State and the Putrajaya Federal Territory. It is situated approximately 27 km south of Kuala Lumpur (the capital city of Malaysia). The catchment of the Langat River is about 2350 km^2 and the length 200 km and its flows out of Klang Valley and ends to the Straits of Melaka.

The Langat River Basin located adjacent to the highly developed Klang Valley metropolitan in Peninsular Malaysia is currently the fastest developing area in the country. A number of large scale social-economic projects are either currently taking shape or already completed in the Basin. These include; the new township of Putrajaya (new Federal Government Administration Center), the Multimedia Super Corridor (MSC) for the information technology industry, Cyberjaya (the paperless electronic village and township), the Malaysian BioValley project for biotechnology research/industry, the Kuala Lumpur International Airport (KLIA), the Sepang Formula One Grand Prix Circuit and several other institutions of higher learning including universities.

The Langat River is located at latitude 02° 50′ 48″ N and longitude 101° 40′ 48″ E with the highest peak at 820.8 m (2691 ft) above mean sea level (amsl) and almost 75% of the catchment is hilly terrain with the average slope of 6–9″ and another 25% of the areas are less than 6″ with some swamps along the river. The River has few tributaries, out of them, the main tributaries are: Sg Semenyih River, Lui River and Beranang River. Langat is the main river in the catchment. The Langat catchment has two reservoirs located at the upstream of the Langat and Semenyih Rivers. The purposes of these reservoirs are to supply water to the treatment plants situated downstream of the catchment. Figure 1 shows a map of Langat River Basin and its tributaries.

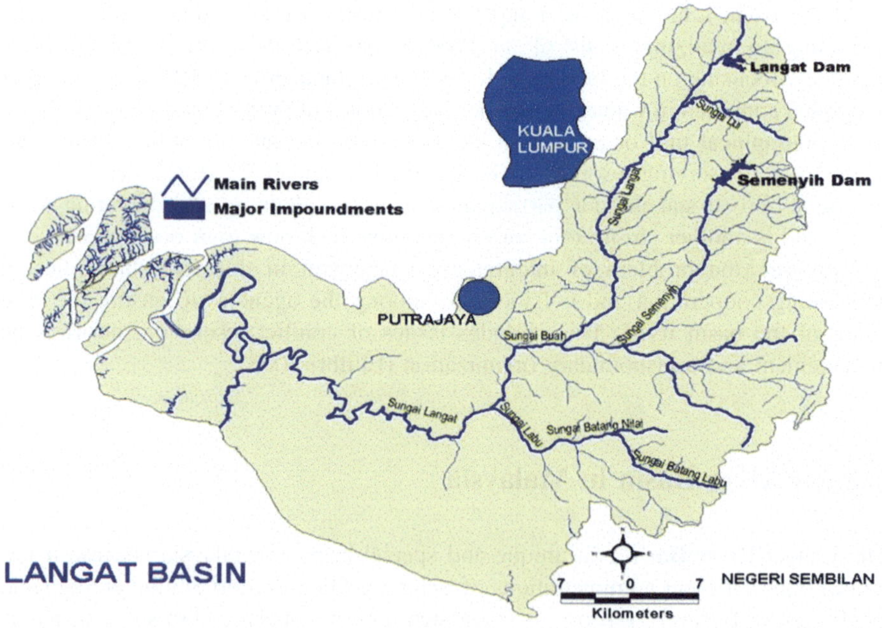

Fig. 1 Map of Langat River Basin, Malaysia. *Source* Elfithri (2006)

In 2004, Langat River Basin was declared as an "Evolving" River Basin under the framework of UNESCO-IHP HELP Network. This recognition was initiated by the Institute for Environment and Development (LESTARI), Universiti Kebangsaan Malaysia (UKM). This basin is currently the only River Basin in Malaysia that has already connected to the HELP Programme of UNESCO-IHP.

HELP (Hydrology for Environment, Life and Policy) is a cross cutting and transdisciplinary initiative of the UNESCO led by the International Hydrological Programme (IHP) with objectives to deliver social, economic and environmental benefits to stakeholders through sustainable and appropriate use of water by directing hydrological science toward improved integrated catchment management basins and also implementation of research in collaboration between scientists, managers and stakeholders (UNESCO 2004). HELP is creating a new approach to integrated catchment management through the creation of a framework for water law and policy experts, water resource managers and water scientists to work together on water-related problems (UNESCO 2010).

Currently there are 91 HELP basins, including Langat, across the globe (in Australia, Asia, Africa, North America and Latin America) to demonstrate how HELP principles can be put in practice. The current HELP Network involves 67 Member States of UNESCO and is classified under the four categories as shown in the Table 1. Selection of the network has been made based on the international programme that is a catchment based activity which is interfacing scientific research with stakeholders needs. The HELP classification is there to measure the catchments progress toward implementing the HELP philosophy, which might help to secure support from national donors.

Table 1 HELP category

Category	Description
Group D: World demonstration HELP basin	This is seen as demonstrating best practices in HELP and IWRM, and serves as a model or demonstration basin for other basins—with something to offer other basins
Group O: Operational HELP basin	This is an established basin which may become a world demonstration basin in due course. It is implementing the HELP philosophy in an integrated manner and is involved with stakeholders in basin management
Group E: Evolving HELP basin	This is a basin which is not yet fully operational, but Well-developed plans conforming to the HELP philosophy which are beginning to be implemented
Group P: Proposed HELP basin	Further work is required to develop plans and activities in an integrated way that supports the HELP philosophy

Source UNESCO 2004

Water Resources Management Issues in Langat

The water in the basin is used mainly for agriculture/aquaculture and for water supply to major cities in the basin and to the adjacent Klang River basin where Kuala Lumpur is located. The demand of water for Year 2000 was 468 million liters/day that was projected to increase 2141 million liters/day by 2050. The rapid urbanization in the basin has led a large influx of people into the region.

Among the issues related to water resources management facing in Langat River Basin are mainly in term of water quality and environmental degradation due to fastest and lots of mega projects development in this basin, which are impacting ecosystem.

The Langat River pollution is also come from sewage and suspended solids resulted from land clearing and discharge of untreated or incompletely treated sewage. The problem of pollution is a concern of many, including government, NGOs and the public, as the Langat River is the main potable water resource for the whole Langat Basin and nearby area (the Klang Valley including Kuala Lumpur) where almost a million of people depend on the Langat for drinking water.

The Langat river basin is small but has inherited many problems of a large river basin. This is because, the river plays an important role in conservation, agriculture and potable water supply but it faces threat from rapid development in the industry sectors and urbanization in the basin. The contradict land use in the basin has led to a deterioration of the river water quality in the past few years and water use conflicts have emerged frequently. The complexity of the problems cannot be addressed by normal river basin management concept. There is a need to develop an ecosystem approach to find solutions to the problems. This ecosystem approach can be integrated with water, land use, people and management policy where the river can be viewed as an ecosystem that provides services to people. This integration will allow water resources in the Langat River Basin in managing the river ecosystem health approach sustainably. The outcome of such an approach may be beneficial to river basins in other parts of the world that are facing the similar problem as Langat.

Some Initiatives at Basin Level

Integrated water resources and river basin management concepts and approaches have been adopted and implemented at certain level in managing water resources in Langat River Basin, Malaysia. There are some issues and challenges faced in managing Langat River Basin and there are some initiatives and approaches undertaken to deal with these issues.

Through Establishment of IWRM Research Group

The National University of Malaysia (UKM) has established an IWRM Research Group in December 2007 under the framework of regional sustainable development research Niche of UKM. The group consists of lecturers and researchers from various faculties and institutes in UKM who are engaged in IWRM research and activities. The researches are mainly focused on four fields, namely, governance, science and technology, economy and health at the local, national and global levels.

Some of the actions taken by the research groups were building up consensus to formulate an IWRM research blueprint (2008–2020), activities related to IWRM and involvement of multidisciplines and multi-stakeholders to move forward toward the implementation of IWRM in Malaysia with the main focus on Langat River Basin. These are done through outreach programmes and multi-stakeholder participation with special consideration for special groups. Figure 2 shows IWRM research structure.

Various plans and key actions have been set up under this research in terms of short, medium and long-term goals with IWRM as the main agenda. The IWRM research group's strategies and plans on governance, capacity building and research are useful as guidelines and reference to implement IWRM, and to achieve the targets within a certain period. The agenda of this IWRM research group can help state governments to establish an authority such as Langat River Basin Authority (LRBA) by the year 2020. Guidelines and manuals that take into account different aspects and perspectives of different target groups are also being produced by this group as advocacy material on IWRBM/IWRM for its implementation.

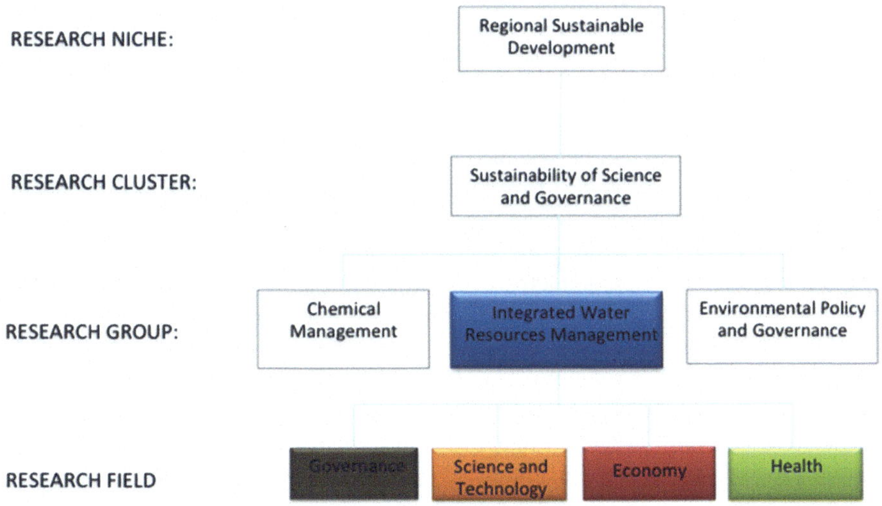

Fig. 2 Integrated water resources management research structure. *Source* IWRM Research Group of UKM (2008)

The multidisciplinary and multi-sector research on IWRM carried out by the research group members could provide a guideline for achieving the common goals from the different perspectives and overviews. Various inputs and expert opinions can also be used to produce the group's outputs towards the implementation of IWRM.

Through IWRM R&D Projects

A number of Research and Development (R&D) projects have been initiated for water resources and river basin management in Langat River Basin. These R&D projects aimed at implementing the IWRM through integrated and holistic approaches for sustainability science and governance focusing on categorized issues listed in Table 2.

In 1997, a study on the application of the concept of ecosystem approach in management and conservation of natural resources and environment of the Langat River Basin in Malaysia was conducted. The knowledge and databases gathered on issues of land use, demography, social environment, economic development, physical environment, industry, chemical environment, biotic environment and forestry can be used as indicator of baseline conditions in Langat Basin.

The study on Ecosystem Health of the Langat Basin has completed in 2004. Since then, the ecosystem health concept in environmental management is gaining popularity and there is no doubt that many other river basins have begun to adopt such a concept in the management of water resource. However, the application of ecosystem health is very much dependent on the nature and location of the river basin and due to high variability, no two basins can operate in an exact manner using ecosystem health approach. Therefore, the concept of river ecosystem health for the Langat River basin can be used as a good guide for other river basin of similar characteristics.

Through research, the identification and prioritization of issues and challenges for Langat River Basin can be obtained directly from the grassroot, these will lead to strategically and well planned management of water and river basin in Langat. There are also number of published research articles, books and proceedings on Langat River Basin. These are good sources of information on Langat River Basin.

More research will be conducted to identify more practical and simple approaches that can be used to help in promoting some noble principles, such as "user pays principle", "precautionary principle", and "prevention is better than cure". This may help in reducing water demand and optimizing water consumption. Some research fund allocations are needed for conducting more research studies on Langat River Basin in order to improve its current status of Evolving River Basin towards Operational or Demonstration level in the future.

Table 2 IWRM research focus

No.	Research focus	Example of research project
1.	Focus on strategic management plan	• Development of sustainable resource management and planning system towards sustainable ecosystem • Harmonizing environmental considerations with sustainable development potential of river basins • Integrated catchment management of Putrajaya lake and wetlands • Development of sustainable development direction for water resources • Strengthening IWRM research and collaboration network towards good water governance in Malaysia
2.	Focus on developing capacity and awareness	• MyIWRM, what can I do? Developing capacity for the practical implementation of integrated water resources management in Malaysia • Water and media: rising up to the challenge • Learning for change: education for sustainable development in Malaysia • Promoting green lifestyle to UKM community
3.	Focus on knowledge transfer	• Access to water information on Langat Basin: communicating environmental awareness • System of rice intensification (SRI) for sustainable rice security and heritage • Nurturing sustainable future: knowledge transfer to eco volunteers and community • Development of ecosystem discovery journey module for sustainable future
4.	Focus on new assessment methods and tools	• New approach on decision support system (DSS) for sustainable Langat River Basin management • Preliminary study of sustainable mosque design with the efficient use of water and energy on the modern mosque design in Malaysia • Analysis of watershed sustainability index (WSI) at catchment level for Langat River Basin • Water footprint for producing crude palm oil industry in Malaysia • Water quality analysis using solar powered real-time monitoring system with GSM at Tasek Cempaka Bangi
5.	Focus on new development, application and technology	• Development of comprehensive model for erosion and sediment control towards good water governance: UKM Campus as a watershed model • Development of a model for water governance of selected urban areas in Langat Basin, Malaysia • Development of an innovative payment mechanism for environmental services: a case study for Malaysia • The application of multivariate analysis for catchment classification model in Langat river floodplain, Malaysia • Flood mitigation through integrated now-casting model for the Langat River Basin, Malaysia • Upscaling of MSMA eco-hydrology at catchment level (Sg. Langat)

Through IWRM Capacity Building Programmes

Capacity Building is a very important component in order to develop capacity of various stakeholders and to make sure certain level of improvement and enhancement of knowledge, experience and expertise in specific field were obtained, in this case is on water resources and river basin management.

For Langat River Basin, capacity building programmes and activities were done in form of practical implementation through conducting a series of training and education, as well as awareness programmes for different stakeholder groups. Table 3 shows some examples of Capacity Building programmes on IWRM and Sustainable Ecosystem Management that have been carried out.

In term of training, some IWRM capacity building modules and manuals, including the concept of "training of trainers" for different levels of stakeholders have been developed to schematically and systematically introduce IWRM and its subsets. An integrated module on Ecosystem Discovery Journey is also being

Table 3 IWRM capacity building programmes

No.	Capacity building	Example of programme
1.	Training programmes	• Developing capacity of NGOs for the practical implementation of integrated water resources management in Malaysia • Developing capacity of academia for the practical implementation of integrated water resources management in Malaysia • Developing capacity of government officers for the practical implementation of integrated water resources management in Malaysia • Training of trainers course: managing water pollution • Training of trainers course: ecosystem services and integrated water resources management
2.	Educational programmes	• Formal education course: master of environmental science (IWRM) at Open University of Malaysia • Non-formal education course: water and ecosystem related studies offered in various faculties and institutes in UKM, as well as in co-curriculum subjects • Water watch programme for young leaders • Water and media: rising up to the challenge • Learning for change workshop: education for sustainable development in Malaysia
3.	Awareness Programmes	• Round table dialogue with campus community on water resources management • Integrated water resources and river basin management: sharing experience at the global level • Environment, ecosystem and educational programme (3EP) and healthy community healthy ecosystem (HCHE) • Promoting green lifestyle to broader community • River cleaning campaign • Water saving campaign

developed with cover broader issues related to Sustainable Ecosystem Management.

In term of Education, even though IWRM is not specifically taught as a subject or offered in any faculty programme, some innovative approaches in introducing and implementing IWRM have been made. It includes promoting IWRM concept in any programmes, during lectures or classes, or encouraging Masters and Ph.D. students to do their thesis related to IWRM. It shows that education can be accomplished through formal, non-formal and informal approaches.

In term of awareness, several initiatives were taken especially in improving the level of awareness and understanding of various stakeholders in water resources and river basin management. These have been done through interactive dialogue with specific target groups of stakeholders, talkative campaigns as well as enthusiastic promotion of sustainable and integrated concepts and approaches. In some cases, we are also assisting the government and other stakeholders to develop educational and awareness packages for different levels of society.

Through Stakeholders and Community Engagements

All the people of the basin area has been considered as primary stakeholders while all the agencies, NGOs, private sector organizations and universities, which are responsible for providing services in water sector in the basin area, have been considered as secondary stakeholders. Both categories of stakeholders need to be involved in the social learning process.

Stakeholder participation issues are brought under formal higher level research. Elfithri (2006) has conducted a study on Stakeholders Participation through Collaborative Decision Making (CDM). She has conducted stakeholder's analysis of the Langat River Basin in 2004–2006, where a total of 26 Government agencies, 10 Private Sectors and 15 NGOs were identified as the major stakeholders of Langat River Basin. Their involvement and participation is crucial in Langat River Basin and need to be involved and participated through Collaborative Decision Making (CDM). Figure 3 shows complete cycle of IWRM Stakeholders for Langat River Basin.

A number of Government agencies, private companies and universities in Malaysia can provide base line information on Langat River Basin. They were also committed to work together under the Stakeholders Research Committee for Integrated Langat River Basin Management through Decision Support System Development (SRC DSS-IRBM Langat) formed in December 2010. The development of integrated information system for Langat River Basin that has been done through development of Decision Support System (DSS) provides data and information on various issues and characteristics related to water resources management in Langat River Basin. It also provides reasonable and best results to decision makers in managing river basin for sustainable management of Langat River Basin.

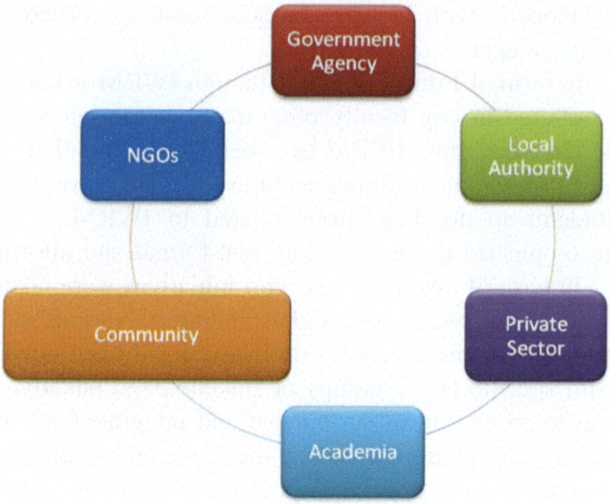

Fig. 3 IWRM stakeholders. *Source* IWRM Research Group of UKM (2008)

This DSS is designed to be used and applied in Langat River Basin and to be shared among various related stakeholders at various levels in this Basin in order to ensure the implementation of IWRM and IRBM for Langat River Basin.

Key Challenges

The main challenge of the basin is on how to improve the current status of Langat River Basin including strategy on how to improve and maintain water quality Class II of Sungai Langat to become better quality (Class I) by certain period of time. Aggregation of pollution from all the point sources is now the threat for the aquatic ecosystem of Langat River.

It also involves challenges in improving the current level/stage of Basin Acknowledgement from UNESCO-IHP HELP, i.e. from Evolving River Basin to become operational or demonstration basin in the future. It involves the "know how" to dialogue with stakeholders in LRB which involved various stakeholders in different area and jurisdiction, the "know how" to interface the water law and policy and governance, the "know how" to undertake the necessary scientific research where basin scientific infrastructure is lacking, the "know how" to develop criteria to better define "vulnerable" basins to global change (sensitivity to climatic variability and hydrological impacts of land use change), the "know how" to address upstream–downstream issues within IWRM from both a technical, management and policy perspective, and the most important thing is the "know how" to use the HELP approach to address national and transboundary basins policy issues connected with intra and inter basin conflicts connected with surface water and groundwater.

IWRM approach is needed to be implemented as a practical approach toward sustainable water resources management in this basin. The collaboration and cooperation from all related stakeholders in Langat River Basin are very important to achieve improvement of the current basin status. There is a need critical mass for organizational change at catchment level and also commitment from all levels to solve the problem at the Langat River Basin. All these contributions would be useful to the improvement of R&D work on Langat River Basin.

Development of Decision support system is also important process that can provide reasonable and best results to decision makers in managing Langat River Basin to sustainable for long term. Implementation of the concept of integrated river basin management and integrated water resources management can be achieved through proper system. At present, hydrologists working in urban areas are facing many new challenges imposed by the ever changing hydrological environment in cities. Emphasis should be put on managing the urban systems as good as possible by applying currently available information and technology, for example by implementation of real-time hydrological information systems—like the one outlined above. A part from managing urban hydrology in real time, many other challenges have to be addressed in the near future-integrated and holistic approached together with other disciplines, i.e. ecologists, social scientists, engineers and planners.

Conclusion

Langat River Basin needs a very careful management and development plans for its water resources to ensure its sustainability. The rapid urbanization in the Basin has led to a large influx of people into the region. Although many solutions and activities have been carried out to arrest the deterioration of the Langat River water quality but few have achieved success. The root to the problem is the absent of a truly integrated management system of water resources in the Basin. Therefore, there is a need for IWRM practices and implementation at all levels.

Some initiatives have been done to bring Langat at the national and international arena, such as improvement of multi-stakeholders collaboration and participation, multi disciplines research/in-depth studies on Langat River Basin, Best Management Practices (BMP's) to be adopted for Langat Basin, capacity building programmes, training and education that have been carried out at the basin scale, as well as the network enhancement and proactive funding mechanisms. Sustainable efforts and commitments are very important and need to be further strengthened in order to sustain the water resources management in this basin.

Overall, it is found that IWRM and IRBM are suitable approaches to be implemented in the management of water resources and river basin, especially for Langat River Basin. The implementation of IWRM and IRBM need to be continuously undertaken in order to sustain the water resources and ecosystem management in this basin.

References

Elfithri R (2006) Pembuatan Keputusan Kolaboratif dalam konteks Pengurusan Sumber Air Bersepadu (English translation: Collaborative decision making within the context of integrated water resources management). Ph.D. thesis, Universiti Kebangsaan Malaysia (UKM), Malaysia

Elfithri R, Mokhtar MB, Saad NMNFB (2008) Data and information for integrated water resources management (IWRM): needs and challenges. Asian J Water Environ Pollut 5(4):49–57. ISSN 0972-9860

Elfithri R, Toriman ME, Mokhtar MB, Juahir H (2011) Perspectives and initiatives on integrated river basin management in malaysia: a review. Soc Sci 6(2):169–176 (Medwell Journals. Scopus)

GWP (2000) Integrated water resources management. Global Water Partnership Technical Advisory Committee (TAC) Background Paper No. 4. Global Water Partnership, Denmark

IWRM Research Group of UKM (2008) Integrated water resources management research group, Universiti Kebangsaan Malaysia

Mokhtar MB, Elfithri R, Mathew D, Wan Nur'ashiqin WM (eds) (2009a) Training module on developing the capacity of NGOs for the practical implementation of IWRM in Malaysia. Institute for Environment and Development (LESTARI), Malaysian Capacity Building Network for IWRM (MyCapNet) and International Capacity Building Network for IWRM (CapNet). ISBN 978-983-9444-91-9

Mokhtar MB, Elfithri R, Wan Nur'ashiqin WM, Mathew D (eds) (2009b) Training module on developing the capacity of academia for the practical implementation of IWRM in Malaysia. Institute for Environment and Development (LESTARI), Malaysian Capacity Building Network for IWRM (MyCapNet) and International Capacity Building Network for IWRM (CapNet). ISBN 978-983-9444-98-8

Mokhtar MB, Elfithri R, Mathew D, Wan Nur'ashiqin WM (eds) (2009c) MyIWRM: what can I do? manual for NGOs. Institute for Environment and Development (LESTARI), Malaysian Capacity Building Network for IWRM (MyCapNet) and International Capacity Building Network for IWRM (CapNet). ISBN 978-983-9444-99-5

Mokhtar MB, Elfithri R, Wan Nur'ashiqin WM, Mathew D (eds) (2009d) MyIWRM: what can I do? manual for academia. Institute for Environment and Development (LESTARI), Malaysian Capacity Building Network for IWRM (MyCapNet) and International Capacity Building Network for IWRM (CapNet). ISBN 978-983-5227-00-4

Nigam A, Gujja B, Bandyopadhyay J, Talbort R (1998) Fresh water for India's children and nature: learning from local-level approaches. World Wide Fund for Nature (WWF) & United Nations Children's Fund (UNICEF)

UNDP (1991) Report of the global consultation on safe water and sanitation for the 1990s. United Nations Development Programme (UNDP). New York

UNESCO (2004) Hydrology for the Environment, Life and Policy (HELP) - United Nations Educational, Scientific and Cultural Organization (UNESCO). http://www.unesco.org/water/ihp/help

UNESCO (2010) Hydrology for the Environment, Life and Policy (HELP). HELP Brochure. UNESCO, Paris, France. http://unesdoc.unesco.org/images/0021/002145/214516E.pdf

A Mechanical System for Lake Environment Management

S. M. Mathur

Abstract This invention relates to a mechanical system for chopping and crushing of water hyacinth in a single machine comprises a frame accommodating a hopper disposed with a cutting cylinder and a shear plate at its bottom for cutting the plants, a crushing cylinder and a pressing roller provided under the hopper for crushing of the plants and a delivery conveyor for delivery of the chopped and crushed plants to a transporting unit. The system is capable of chopping and crushing of water hyacinth in a single pass and reduces its volume up to 65–70%, which results in reduction of cost for transportation.

Introduction

Water hyacinth (Eichhornia crassipes (Mart.) Solms) is the most predominant persistent and troublesome aquatic weed in the world and has posed ecological and economical problems in several countries. Utilization of water hyacinth as animal feed, as a fuel, handicrafts, furniture, biogas, compost, pollution abatement and paper pulp has been done. However, one of the major problems faced was bulk and the cost of transportation of freshly harvested water hyacinth from water bodies to the factories. The capacity of water hyacinth to grow and reproduce is phenomenal and in one growing season 25 plants can give rise to two million plants, covering 1 ha and weight as much as a fully loaded Jumbo Jet (Lindsey and Hirt 1999). The natural loss of water from the water surface by evaporation is thought to increase through transpiration from the leaves of water hyacinth by at least 40–50% (Lindsey and Hirt 1999). Due to this, sometimes it is called "Shokh Samunder" in India. The oxygen depleting load of 1 ha of water hyacinth mat is equivalent to the sewage created by 80 persons (Gopal and Sharma 1981).

S. M. Mathur (✉)
Department of FM & PE, College of Technology and Engineering,
MPUAT, Udaipur, India
e-mail: Shiloo2009@gmail.com

V. P. Singh et al. (eds.), *Water Resources Management*, Water Science and Technology Library 78, https://doi.org/10.1007/978-981-10-5711-3_17

Water Hyacinth Management in Lakes

Over the years, various control methods have been studied and tried including chemical, biological and mechanical means but with no lasting success (Julien et al. 2001). It is effective and relatively cheaper than the mechanical and manual control. However, their long-term effects on the environment are unknown and sprayed plants were left to rot in the water causing pollution and aggravated eutrophication. The plants cannot be completely controlled by biological method. Control by mechanical and manual harvesting has been widely practiced in many countries. The simplest mechanical method of control would be to chop the water hyacinth in situ and leave it to rot in the water. This, however, has two main drawbacks:

The plant is wasted (not utilized).

In the process of decay, the chopped plant extracts oxygen from the water causing suffocation to animal life in water body.

A major contributory factor to the failure of water hyacinth machinery is the large volume and moisture content of water hyacinth, which slows down the harvesting operation by increasing requirement for handling, transport and cuts down harvester productivity with increased harvesting cost. It also limits effective utilization due to low dry matter content and high drying equipment costs. Water hyacinth removed in large quantities from water body can be transported easily when chopped and crushed to higher densities. The review indicated that the most attractive alternative for solving the water hyacinth problem is integrated harvesting and utilization, which offers potential benefits in terms of generation of useful products (Mathur et al. 2003). The mechanical mean is the only safe and rapid method to control the growth of water hyacinth but the disposal costs are high because of its bulk. In order to solve this problem, water hyacinth has to be chopped and crushed. The chopped and crushed plants will not allow water hyacinth to re-grow and can be transported easily (Mathur 2005a, b).

The Invention

The present intention was to propose a system for chopping and crushing of water hyacinth which can be installed at the shore or on the harvester from which crushed plants can be taken to site for reuse.

To propose a system for chopping and crushing of water hyacinth which will reduces volume of the plants by 65–70% resulting in reduction of cost for transportation and does not cause environmental pollution.

Water Hyacinth Chopper Cum Crusher

To solve the problem of lake environment management in the present invention there is provided a mechanical system for chopping and crushing of water hyacinth comprises a frame accommodating a hopper disposed with a cutting cylinder and a shear plate at its bottom for cutting the plants, a crushing cylinder and a pressing roller provided under the hopper for crushing of the plants and a delivery conveyor for delivery of the chopped and crushed plants to a transporting unit (Mathur, 2006; Mathur and Singh, 2004). Figure 1 shows the front elevation of system of the present invention.

The system as shown in Fig. 1 comprises a conveyor system to transport the Hyacinth plants to a main hopper (1). The conveyor system comprises a hopper (22) to admit plants therein. A conveyor belt (10) is provided between two rollers (8 and 9) to deliver the plants into the main hopper (1). Further, the conveyor system comprises a chain and sprocket system (19) as shown in figure connected to the roller (9) to drive the same. The sprocket (19) is connected to a DC motor (20) as the speed of the conveyor system is required to be maintained at low level.

The main hopper (1) is provided at its bottom with a cutting roller (2) and a sheer plate (3) maintained at a clearance from each other. A pressing and crushing roller (6 and 5) are provided below the cutting roller (2) for crushing the chopped plants falling from above. Thereafter, chopped and crushed material is delivered to a transporting unit by means of a delivery conveyor (8–10, 15 and 16) disposed below the crushing and pressing rollers. The rollers (9) are provided with an adjusting screw (7) to control the tension in the conveyor belt (10) if required. The driving to the rollers is provided by prime mover (11) provided at the bottom. All the above rollers, hopper, etc., are mounted on a frame, to which the conveyor system is detachably fixed. The details of different components of the system are as follows:

Hopper: The hopper is trapezoidal in shape provided with slot in both the side walls and rear wall to facilitate fitting of shear plate (3) against the cutting cylinder (2).

Cutting Cylinder with Blades: The cutting cylinder (2) is provided with at least a set of rings, one at each end of the cylinder detachably mounted with a plurality of

Fig. 1 Water hyacinth plants

blades by means of connecting pieces such as square pieces fixedly held to the ring. Further, a circular plate is provided at each end of the cylinder so as to cover the same. The shear plate (3) is fixedly provided against the rotating cylinder to act as counter cutting edge. The blades and shear plate are preferably made of spring steel for longer life wherein the number of blades can be varied from 12 to 36 preferably 36.

Crushing Rollers: The crushing roller is provided with a plurality of uniformly distributed corrugations on its periphery for providing positive thrust to flow the chopped material and welded with a plate at each end.

Pressing Rollers: The pressing roller is disposed with a rubber lining through out its outer surface rotating against the crushing roller in opposite direction for crushing of chopped material. Provision of rubber lining is made to avoid noise pollution. This roller is also covered by a plate at each end like others. The prime mover (11) drives all the rollers and delivery conveyor through suitable pulleys and *v*-belts. The DC motor is used to drive the feeding conveyor through chain and sprockets to get a speed ratio of for example 1:6.

The electric power required for chopping and crushing increases with the increase in flow rate and speeds of cutting cylinder and crushing roller. The system of the present invention is capable of chopping and crushing the plants in a single pass. The system is covered with safety covers to avoid any accident. The chopped and crushed water hyacinth can be used as animal feed, mulch, compost, a source of pulp for making paper and board, rope, crafts, furnitures, fuel, a medicinal plant and pollution abatement. The system also allows removal of surface and free cell water.

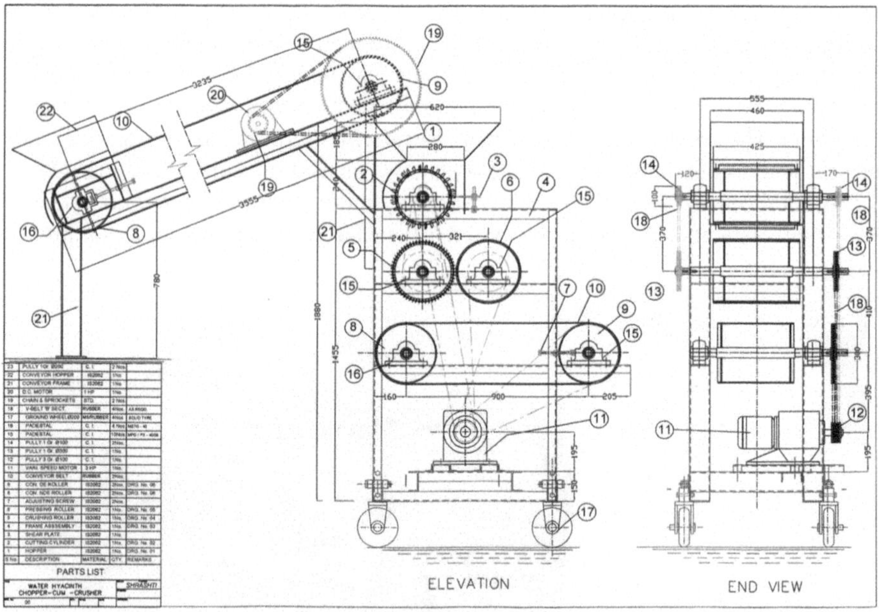

Significant Achievements and Advantages of the Developed Machine

- The machine can be easily fabricated locally.
- It can be easily transported to the harvesting site.
- Ease in handling and saving of labour in transportation.
- Environmental friendly and no need of application of any chemical in fresh water lakes.
- The average no load power to run the machine was 1.1 kW. The power required in chopping and crushing increased with the increase in feed rate and speeds of cutting cylinder and crushing roller.
- Specific chopping and crushing energy and total energy requirement did not change with change in no. of blades.
- The developed chopper cum crusher reduced the specific volume of water hyacinth by 73% and weight by 45% at recommended 36 no. of blades on cutting cylinder, cutting cylinder speed of 13.33 m/s (800 rpm), crushing roller speed of 6.66 m/s (400 rpm) and feed rate of 2t/h.
- The cost of chopping and crushing of freshly harvested water hyacinth was 16 Rs./t with highest capacity (2t/h) of system at recommended performance parameters.
- 65.7% reduction in cost of transportation can be achieved with the introduction of mechanical system and only 5 truck trips per day are required against 16 truck trips in conventional system.

References

Gopal B, Sharma KP (1981) Water hyacinth. Hindustan Publishers, D-35, C.C. Colony, Delhi, India

Julien MH, Griffiths MW, Stanely JN (2001) Biological control of water hyacinth 2. Australian Centre for International Agricultural Research, GPO Box 1571, Canberra, ACT 2601

Lindsey K, Hirt HM (1999) Use water hyacinth! a practical handbook of uses for the water hyacinth from across the world. Published by Drukerei Bauer, Winnenden, Germany

Mathur SM (2005a) Development and performance evaluation of water hyacinth chopper cum crusher. Report Published by the Department of Farm Machinery and Power Engineering, College of Technology and Engineering, Udaipur-313 001, India

Mathur SM (2005b) Key note paper on mechanical management of water hyacinth in fresh water lakes to boost lake environment. In: Mahavidyapeetha JSS (ed) International proceedings on water and health, held at Maysore, Jan 22–23, pp 26–30.

Mathur SM (2006) Aquatic weeds and its impact on environment. Int J Ecology, Environ Conserv (3)

Mathur SM, Singh P (2004) Development and performance evaluation of water hyacinth chopper cum crusher. Int J Biosyst Eng, U.K., 88(4):411–418

Mathur SM, Singh P, Mehta AK (2003). Influence of system parameters on compressibility of chopped water hyacinth. J Agric Eng, New Delhi, 40(4):27–36

Wetland Characterization and Implications on Agriculture in L. Victoria Basin

J. R. Kosgei

Abstract Wetlands in flood plains can be characterised by significant variations in water transitions. The Ombeyi watershed in Nyando Basin hosts a number of wetlands that are characterised by competing water uses, especially between expanding irrigated agriculture and availability of wetland products. The seasonal contribution of water from various sources to the wetland water is a function of season of the year, geology, land use/cover, rainfall and upstream water-use activities. These influence the extent to which the wetlands function. An electrical resistivity tomography (ERT) mapping was done in June 2014 along four transects that coincided with the location of shallow wells whose water levels were being monitored. To further understand the relationship between surface and groundwater, water samples from rainfall, streams, wetlands and the shallow wells were collected at weekly intervals for a period of 6 weeks between May and June 2014 in the Ombeyi watershed. Measurement of stable isotopes oxygen-18 (^{18}O) and Deuterium (^{2}H) on these samples was carried out in July 2014 at UNESCO-IHE. The ERT imaging revealed permeable zones that surface water infiltrates into shallow groundwater. The profiles also showed subsurface features that could explain the surface hydrological behaviour. The results of stable isotope content showed that shallow groundwater is recharged from local rainfall but suffers evaporation during transit from the supplying rivers through wetlands to downstream rivers. The Deuterium excess of the local rainfall was 5.4. This surface water—groundwater linkage calls for conjunctive management of the water sources in the prevailing circumstances of expanding land under rice irrigation.

J. R. Kosgei (✉)
Department of Civil & Structural Engineering, Moi University,
P.O. Box 3900, Eldoret 30100, Kenya
e-mail: masultiony@gmail.com

V. P. Singh et al. (eds.), *Water Resources Management*, Water Science and Technology Library 78, https://doi.org/10.1007/978-981-10-5711-3_18

Introduction

The hydraulic connectivity that exists in most types of landscapes between surface water and groundwater make surface water bodies to be integral parts of groundwater flow systems. Even when a surface water body is separated from the groundwater system by an unsaturated zone, seepage from the surface water could recharge groundwater. Because of the interchange of water between these two components of the hydrologic system, development, utilization or contamination of one, commonly affects the other.

The movement of surface water and groundwater is controlled to a large extent by the physiographic form and geology of an area. Also, climate, through the seasonal effects of precipitation and evapo-transpiration, affects the distribution of water to, and removal from landscapes. Human factors also influence the paths that water take through landscapes. Population, the structure of human settlements, water use activities and other socio-economic factors determine the quantity of water required, its source, amount consumed and management of wastewater and return flows.

Freshwater wetlands are characterised by significant temporal and spatial variations in terms of pattern and magnitude of water inflows and outflows (Clay et al. 2004) leading to distinctive water table fluctuations that reflect the hydrometeorological conditions and changes in the predominant water source (Winter 1999 and Bradley 2002 in Clay et al. 2004). The Ombeyi watershed in Kano flood plains, Nyando Basin, hosts some wetlands that are characterised by competing water uses, especially between expanding irrigated agriculture and availability of other wetland products. Water within the wetlands may comprise varying quantities of water derived from rainfall, surface water and groundwater sources. The seasonal contribution of each source to the wetland water is a function of season of the year, geology, land use/cover, rainfall and upstream water-use activities.

The movement of water through the wetland is influenced by the quantity of surface water, the relative position of surface water bodies with respect to groundwater flow systems and the localized geologic features beneath wetlands. Wetlands are ecologically important particularly because they store and retain nutrients (and potential contaminants), transform compounds biologically and chemically, provide refuge to invertebrates, and are a base of the aquatic food web. These influence the extent to which the wetlands function. According to Talabi (2013) understanding the hydrochemical processes and recharge source are critical to the sustainability and management of water resources.

The expansion in rice cultivation in Ombeyi watershed is likely to induce surface water resource alterations. According to Khisa et al. (2013), drainage of land for cultivation is among the factors that have contributed to the shrinking area of Nyando wetland. The authors (ibid) cautioned that if the land use trend continued unabated, then the increase in papyrus losses could pose a big challenge to the ecological functioning of the wetland and its support to sustaining community livelihoods. Furthermore, the increase in land being irrigated under usual water management practices leads to abstraction of much more surface water to irrigate additional land which in most cases is farther away from the water source, hence added conveyance losses. Reduced surface water is likely to affect the quantity, quality and timing of water flowing into the wetland and also groundwater recharge. This happens if the water for irrigation is abstracted from sources that supply the wetland and/or groundwater. The surface–groundwater interaction in Ombeyi watershed has not been studied in detail and thus there is insufficient knowledge of the influence of the expansion of rice cultivation on especially groundwater.

The connectivity between surface and groundwater can be studied through some methods among them by investigating the geologic composition of an area. Different geophysical in situ tests for stratigraphic information have been developed and include Seismic Methods, Electrical Methods, Gravity and Magnetic methods, near subsurface nuclear methods and Borehole methods. According to Bhoi (2012), the electrical resistivity tomography (ERT) has been used widely to study groundwater availability, ground water contamination survey, subsurface cavity, karst and fault, archaeological survey, characterization of fibres distribution in steel fibre reinforced concrete, delineation of seepage zones, soil layer and their thickness. Even though ERT surveys have been made for decades on hydrology and geotechnical investigations, Koch et al. (2009) cautioned that such geo-electrical patterns provide only a descriptive perspective of subsurface structures and water occurrence, and they require supporting ground truth data.

A complementary approach that can be used to study relationships of hydrological components is using environmental isotopes. Tekleab et al. (2014) reported many attempts at various spatial scales in which workers have investigated the use of stable isotopes to identify and compare samples of water from different sources. These include studies to examine runoff generation mechanisms, hydrograph separations, characterize water cycle components, identify source areas and flow pathways under different flow conditions, and estimate the mean residence time of catchments. Clay et al. (2004) used stable isotopes to investigate the seasonal dynamics of wetland water storage in the United Kingdom. Bowen et al. (2005) performed isotopic characterization of bottled waters sourced and distributed from locations around the world. Tekleab et al. (2014) used stable isotopes to characterize and better understand the dominant runoff components in the Abay/Upper Blue Nile basin, Ethiopia.

The increasing water demand for irrigation and the diminishing wetland area are primary factors that could precipitate shifts in hydrological flow paths in Ombeyi Watershed, which could affect groundwater recharge if a distinct connectivity exists between surface water and groundwater. In this study, measurement of stable isotopes (oxygen-18 and Deuterium) and ERT imaging were used to investigate whether there was an interaction between surface water and groundwater in Ombeyi watershed.

Materials and Methods

Study Area

Ombeyi Watershed is situated in Kano plains, Nyando drainage basin in Kenya. The main rivers in the watershed are Great Oroba and Little Oroba that originate from the Nandi Escarpment. The Ombeyi wetland ecosystem (Fig. 1) cuts across the Nyando and Kisumu East Sub-Counties within the Kisumu County and forms a complex ecosystem comprising of several streams, rivers and swamps (GoK 2014).

According to GoK (2014), the wetland ecosystem covers an area of about 1037 ha consisting of small swamps, streams and rivers. The Plains stretches from Miwani, Nyando and Lower Nyakach divisions to Kisumu East district is a lowland flat area with a topography that gentlyslopes from the north-eastern and

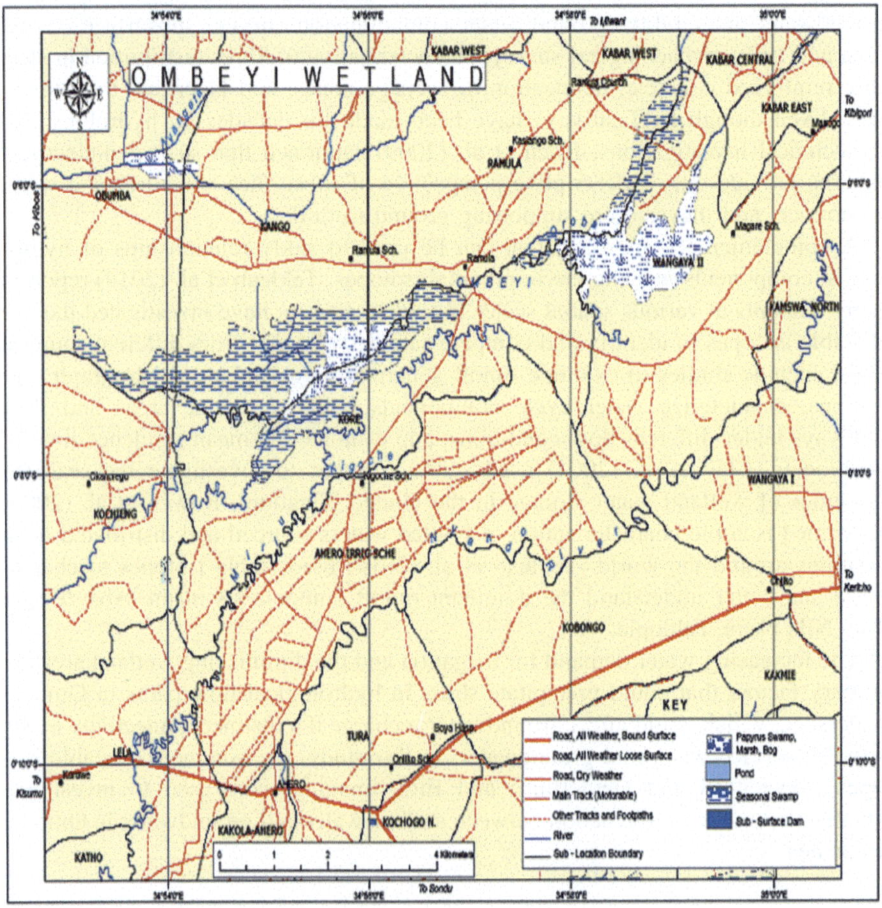

Fig. 1 Location and drainage of Ombeyi wetland ecosystem (GoK 2014)

south-western directions. Owing to impended soil drainage on account of prevalence of vertic clays, water logging occurs extensively in the plains during and immediately after the rains.

Eutric Vertisols with a sodic phase are dominant soil types on the flat to very gently undulating slopes of 0–1%. The soils are generally imperfectly drained to poorly drained, deep to very deep, dark greyish brown to black, firm to very firm, cracking, clay; in some places sodic deeper subsoil (ibid). Some sections of the West Kano area are dominated by sedimentary sandy loams which are too pervious for irrigation development. The main geological feature in the study area is the Kavirondo Rift. The faulting has allowed the accumulation of Pleistocene sediments and possibly tuffs of a vast thickness. These sediments and tuffs are now covered by alluvial material. Tertiary faulting was accompanied by sheet eruptions of phonolitic lavas which are found north and north-west of Kisumu Town (Binge 1962). Major factors that have influenced sedimentation are the supply of different types of sediments by rivers depositing gravel and sands during periods of high discharge, clay and silt during periods of low discharge, and the development of the Kavirondo Rift Valley which controlled the hydraulic gradient of the rivers and the deposition environment (land or lake).

Previously, about 75% of the plains were viewed unsuitable for economically viable small-scale farming due to the extremely heavy soils combined with a warm climate, relatively low rainfall and repeated flooding which made farming to be unattractive as an economic venture. However, rice irrigation and recent sugarcane plantations within the wetlands have completely introduced complex hydrological pathways especially as a result of canalization.

ERT Imaging

Four transects all running NW–SE were executed. Figure 2 shows the geological map and illustrates the profiles along which ERT imaging was done in June 2014.

A Terrameter complete with electrodes and a loop for electric sounding and 2D Imagery was used. Analysis of data and interpretation was done through Res_2D Software.

Transect 1

The transect stretched for 2000 m. It was the most upstream transect and coincided with three shallow wells which were being observed. The transect crossed two streams, Nyankoko and Great Oroba. The measurement went through various land use categories as shown in Plate 1.

The land use categories included (a) homesteads, subsistence maize and grazing land, (b) maize and sugarcane, (c) sorghum, grazing land and bare land and

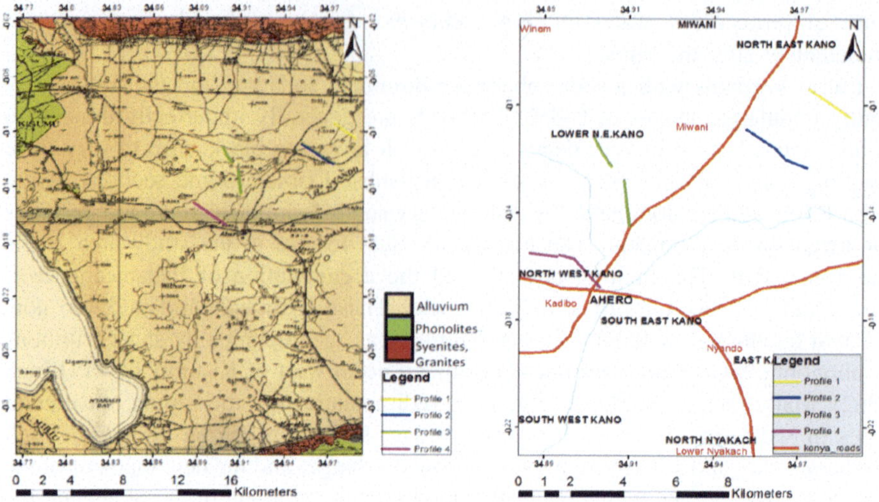

Fig. 2 Geological map and the four transects on which imaging was conducted

(d) maize, shrubs and stream banks. Others were harvested rice fields and sweet potato vines.

Transect 2

Transect 2 was 3000 m long. Among the dominant land use categories (Plate 2) through which this transect went through are (a) swamps, (b) arrow root, (c) papyrus on the banks of Great Oroba River and (d) harvested trice fields. Other categories include grazing land, subsistence maize, sugarcane and puddle rice fields ready for transplanting.

Transect 2 coincided with three shallow wells which were being monitored.

Transect 3

This transect was 3400 m long. Two wells existed in this transect. As opposed to the previous two transects that were executed within a day, measurements across this transect were done over a period of three days. This transect went through the main wetland, Great Oroba River, harvested rice fields, grazing land, homesteads and a school.

Plate 1 Measurements conducted in Transect 1 through various land use categories

Transect 4

This was the longest transect having a length of 3300 m. Measurements along this transect took three days. Three observation shallow wells were along this transect. Grazing land, subsistence maize, harvested rice fields and sugarcane were the dominant land use categories. The transect also crossed two streams, Ombeyi and Miriu as well as a wetland.

Stable Isotope Signatures

Water samples were collected between April 2014 and June 2014 from various sources in the Ombeyi Watershed. The sites (Fig. 2) were chosen strategically so that the first transect (Transect 1) was upstream of the wetlands while the last one

Plate 2 Land use categories through which Transect 2 traversed

(Transect 4) was relatively downstream of the irrigation fields and the wetlands. The two intermediate transects (Transect 2 and 3) cut across, among other land use categories, irrigation fields, swamps and the main wetland.

Shallow wells were requested from individuals whose wells were lined and raised off the ground to prevent runoff from getting into the wells. Along Transect 1, five samples were collected on a weekly basis, three of which were from shallow wells and the rest were from Great Oroba River and Nyankoko stream. However, there were a number of occasions when Nyankoko stream was dry. From Transect 2, three sample were collected from shallow wells and one sample obtained from the exit of a small wetland (Plate 2). Three samples were collected from Transect 3, two of which were from shallow wells and the remaining one obtained from the exit of the main wetland. Five samples were collected each week from Transect 4, three from shallow wells and the rest from rivers Miriu and Ombeyi. A rainfall sample was collected every week from a household in Transect 1 having an elaborate roof water harvesting system.

The water level from each shallow well (except one well in Transect 1 which had no access) was measured using an electric dipper (Plate 3) to establish weekly fluctuations.

All water samples were analyzed at UNESCO-IHE (Delft, the Netherlands) using an LGR liquid–water isotope analyser. The stable isotopic composition of

Plate 3 Measuring the water level in the shallow wells using an electric dipper

Oxygen-18 and Deuterium are reported using the δ notation, defined according to the Vienna Standard Mean Ocean Water (VSMOW) with $\delta^{18}O$ and $\delta^{2}H$. The accuracy of the LGR liquid–water isotope analyser measurements was 0.2‰ for $\delta^{18}O$ and 0.6‰ for $\delta^{2}H$, respectively.

Results and Discussion

ERT Imaging

A preliminary conceptual model of the geological setting was derived from a combination of previous geological work and geo-morphological field observations. The geophysical interpretation model used was largely based on resistivity values derived from relatively straightforward soundings from previous studies, and measurements correlated with expected geological formations in the absence of borehole geological logs. The model was used for the preliminary interpretation of tomography profiles and later refined to account for site-specific differences. Table 1 shows the resistivity range for the geological units of the study area (Kuria 2013). Differentiation between formations with overlapping resistivities was often possible using a combination of location and depth.

Profile 1

The profile stretches a total distance of 2000 m in a NW–SE direction. The model exhibits a range of resistivity values ranging from a low of 0.167 Ωm (deep blue) and a high of slightly over 1335 Ωm (purple).

Table 1 Common resistivity ranges around the study area (Kuria 2013)

Geophysical unit	Resistivity range (Ωm)
Sandy topsoil of the "high lands", generally iron-stained	100–2,000
Topsoil of clayey sands and silts (black cotton type) *When hard and cracked:*	20–100 *100–1,500*
Heavy clays (alluvial or intensive clay-producing weathering)	2–5
Clayey saprolite, dry	5–10
Very highly weathered rock (gneisses and schists)	20–100
Highly weathered and fractured rock (gneisses and schists)	100–500
Fractured rock (gneisses and schists)	500–1000
Massive and compact gneisses	1000–8000
Fresh Basement rocks, compact and dry	2500–10,000

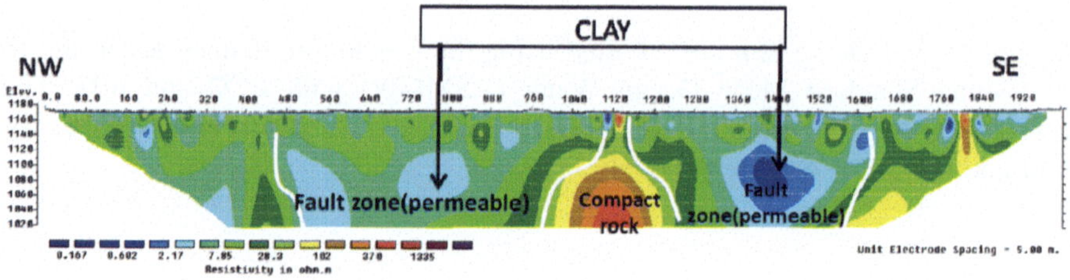

Fig. 3 Profile 1 showing the permeable zones within the faulted block where the rock is weathered

From Fig. 3, most of the profile records values in the range of 7.85–28.3 Ωm. These are the intermediate values. These values are representative of highly weathered and permeable material. This material fills the faulted and weak zones on the profile. These are mainly alluvial gravel sands and some clay content. The deep blue colours recording values lower than 2 Ωm represent the clays. The high value above 100 Ωm recorded at the central part of the profile are as a result of a compact fresh rock which in this area is probably the intrusive granodiorites, syenites or the contaminated Granites. The fault blocks have their fault lines at about 420, 1120, 1140 and 1640 m. These faults extend through the entire probed depth of about 150 m. This observation is also made in the other profiles.

Profile 2

The Profile 2 stretches over a distance of about 3 km in length in the NW–SE direction. The resistivity distribution within this profile is confined within a narrow range (0.5–367 m) that commensurate lesser number geological formations probed. The interpreted ERT model shows a layer of low resistivity values forming circular to elliptically shaped features that are randomly distributed within the shallow depth, i.e. below 50 m. However, at 960 and 2080 m marks on the horizontal distance the clays are encountered to a depth of 110 and 90 m, respectively.

The permeable zones (8.5–30 m) occur in preferentially in areas not occupied by clays at shallow depths but characterize the entire basal layer of the profile. The shallow permeable zones act as conduits that channel groundwater vertically to deeper levels. On the other hand the deeper permeable resembling a continuum, marks aquifers augmenting surface and deeper groundwater. In addition, this deeper aquifer could possibly be recharging levels below 150 m (maximum probed depth) (Figs. 4, 5 and 6).

The high values (140–367 m) are representative of fractured rock matrix of intrusive nature that is localized within the alluvial deposits. At 1320 m, the localized high resistivity values are recorded at the swampy area. These fractured zones provide additional recharge where fractures are interconnected.

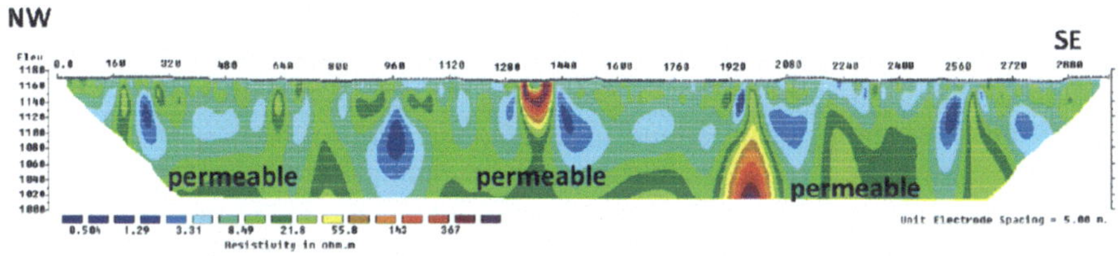

Fig. 4 Profile 2 showing the permeable zones within the entire profile

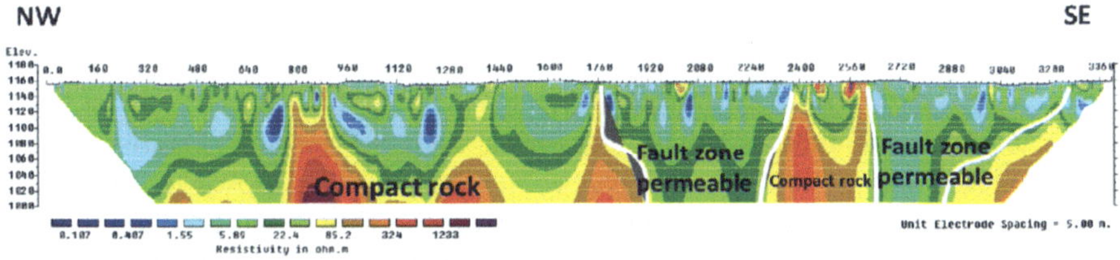

Fig. 5 Profile 3 showing two isolated permeable zones within the faulted zone extending below the probed depth

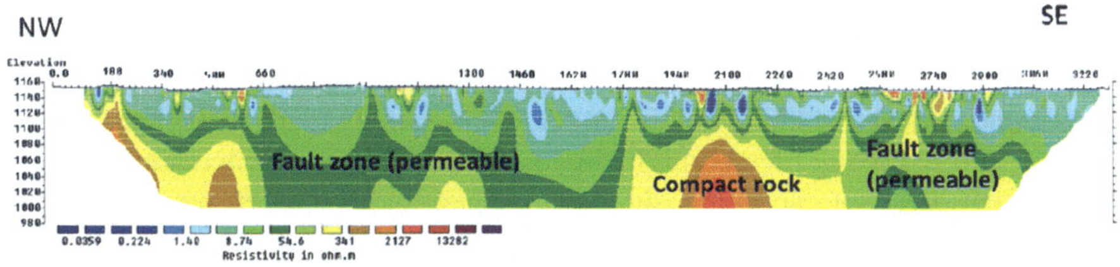

Fig. 6 Profile 4 showing an extensive permeable zone on the western part stretching over a distance of about 1900 m

Profile 3

The profile extends a total of 3400 m running in a NW–SE direction. Resistivity values range from 0.1 to 1247 vΩm. The intermediate values between 5.92 and 22.6 vΩm are representative of the permeable sediments.

This profile is dominated by permeable material intermittently broken by compact rock at 800, 1760 and 2400 m. This rock is more compact and more resistive than in profile 2 and appears as intrusives below the sediments. Faulted blocks are observed to the right end of the profile at 2080 and 2880 m. Towards the western part the faulting terminated at a depth of about 120 m below ground level.

Profile 4

This profile stretches 3300 m in a NW–SE direction. It records the biggest range of resistivity values from 0.05 to 14,166 Ωm the lowest values being those representing the clays while the highest representing the compact fresh rock.

The faulted zone to the west and to the farthest end to the east is filled with permeable sediments with values in the intermediate range of 11.5–67.9 Ωm. Clays occur in the sands mainly on the upper surface along the entire profile. Another small permeable zone is localized at the eastern terminal end. The permeable zone covers the entire probed depth, i.e. below a depth of 150 m.

Borehole Distribution in the Study Area

A large number of boreholes exist in the Plains. Figure 7 indicates the distribution of boreholes in and around the study area. The attributes are shown in Table 2.

The prevalence of these boreholes corroborates the ERT findings and attests to the presence of relatively deep aquifers recharged directly by rainwater and indirectly by the river waters.

Stable Isotope Signatures

In this study, ratios of the stable isotopes of oxygen (oxygen-18/oxygen-16, or $^{18}O/^{16}O$) and hydrogen (deuterium/hydrogen, or $^{2}H/H$) of the water samples were investigated. The isotopic composition of oxygen and hydrogen are reported in terms of differences of $^{18}O/^{16}O$ and $^{2}H/^{1}H$ ratios relative to a standard called the Vienna Standard Mean Ocean Water (VSMOW). Positive values of $\delta^{18}O$ and $\delta^{2}H$ indicate enrichment when compared to VSMOW, the negative values imply

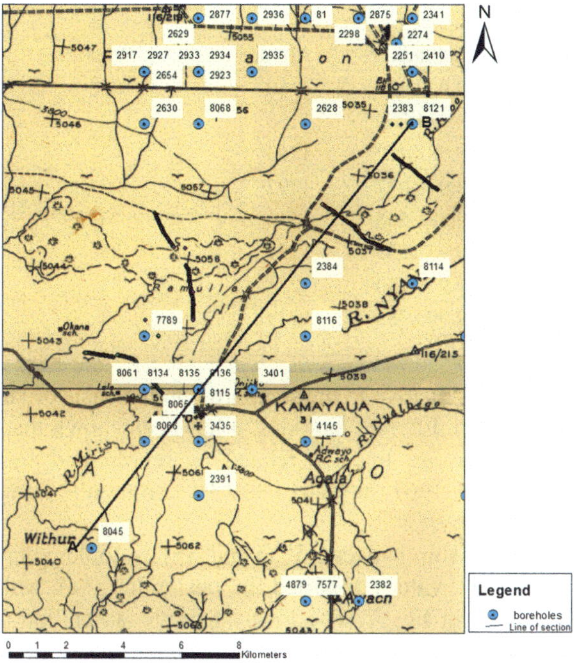

Fig. 7 Local geology and the distribution of boreholes in the study site

Table 2 Attributes of the boreholes in Fig. 7 (mwsl—mean water surface level)

ID	Longitude	Latitude	Altitude (m)	Depth (m)	mwsl (m)
2251	720,738	9,992,624	1219	107	72
2341	720,738	9,994,471	1220	46	18
2382	718,888	9,974,200	1219	76	9
2383	720,738	9,990,788	1219	39	7
2384	717,030	9,985,259	1219	46	5
2391	713,322	9,977,883	1219	91	50
2628	717,030	9,990,788	1220	64	
2650	264,456	9,970,964	1509	99	85
2654	711,464	9,992,624	1215	92	76
2875	718,890	9,994,471	1219	46	46
2877	713,323	9,994,471	1215	46	46
2923	713,323	9,992,624	1213	65	18
2935	715,171	9,992,624	1219	46	6
2936	715,171	9,994,471	1219	61	3
3401	715,170	9,981,565	1265	64	64
3435	713,322	9,979,730	351	109	103

(continued)

Table 2 (continued)

ID	Longitude	Latitude	Altitude (m)	Depth (m)	mwsl (m)
4145	717,030	9,979,729	1160	180	64
4879	717,029	9,974,200	1160	152	149
7789	711,463	9,983,412	1263	40	71
8045	709,603	9,976,037	1297	90	52
8061	711,463	9,981,566	1246	6	53
8065	713,322	9,981,565	1254	83	64
8066	711,463	9,979,730	1271	83	70
8068	713,323	9,990,788	1179	53	50

depletion of these samples relative to the standard. On the basis of large numbers of meteoric water collected at different latitudes, it has been shown that $\delta^{18}O$ and $\delta^{2}H$ values of meteoric water are linearly related as represented by the equation: $\delta^{2}H = 8\delta^{18}O + 10$ (Craig 1961 in Talabi 2013), commonly termed the Global Meteoric Water Line (GMWL).

The plot of the relationship between $\delta^{18}O$ and δ^{2} Hisotopic composition for the various sources, with all values reported in per mille (‰) with reference to VSMOW, are illustrated in Fig. 7. The local meteoric water line (LMWL) comprised isotope composition of the rainfall values from the study site. There is a clear indication that most of the samples irrespective of source conform more to LMWL rather than to GMWL. The equation of the LMWL is $\delta^{2}H = 3.7\delta^{18}O + 5.4$ with a coefficient determination (R^2) of 0.56.

Most of the shallow wells data are grouped fairly tightly to the lower left, indicating that there is not much variability in isotopic signatures in shallow groundwater among the wells sampled, even though they are from various areas in the study area, and that there could be little isotopic enrichment of deuterium and oxygen in these waters. The local rainfall, shallow wells and rivers in Transect 1 have clustered closely together. This suggests that stream flow is mainly as a result of direct surface runoff which is influenced by rainfall characteristics. Furthermore, the similarity in properties with that of most shallow groundwater wells is an indication that the source of recharge of shallow groundwater system in the entire watershed consists of a mixture of recharge from surface water (rivers) and local rainfall. Those samples that were plotted closer or on the line LMWL are likely to be recharged directly from local rainfall through preferential flow channels such as cracks or fractured zones with little evaporation (Fig. 8).

The wetland in Transect 2 also has fairly similar composition although much more enriched in terms of Oxygen and Deuterium. The mean and other descriptive statistics are provided in Table 3 and Fig. 9.

The range and means in Table 2 and Fig. 9 still suggest a closer relationship between rainfall and water samples from rivers in Transect 1, shallow wells in Transects 1, 2 and 3. The mean isotopic contents of the samples from the wetlands and rivers in Transect 4 are relatively more enriched. This can be attributed to effects of evaporation and detention time.

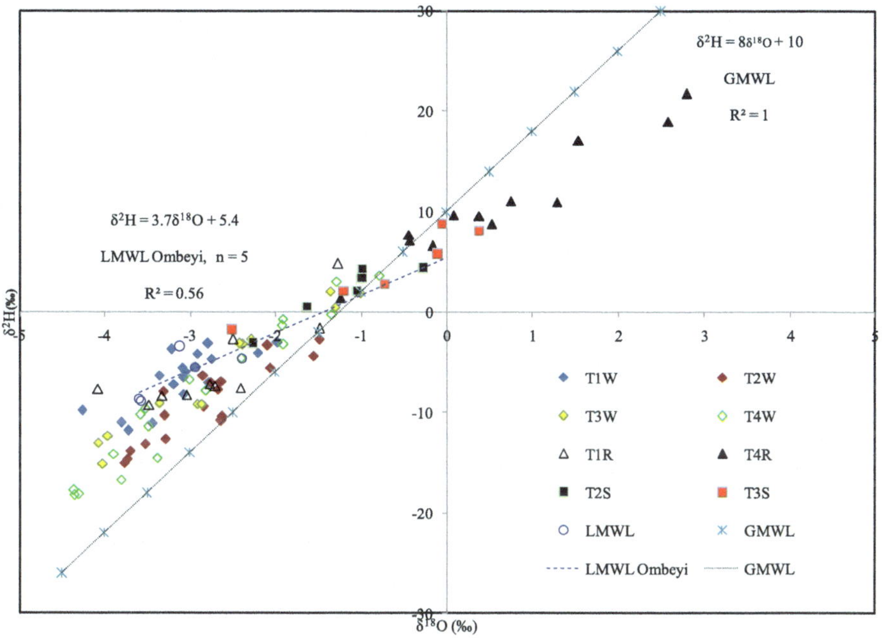

Fig. 8 Relationship between $\delta^{18}O$ and δ^2H isotopic composition for the various sources. The abbreviations in the legend are described in Table 3

Table 3 Descriptive statistics of δ^2H and $\delta^{18}O$ (‰, VSMOW) from various sources during different investigation periods

Description	Abbreviation	Mean (‰, VSMOW)		Minimum (‰, VSMOW)		Maximum (‰, VSMOW)		Standard dev. (‰, VSMOW)	
		δ^2H	$\delta^{18}O$	δ^2H	$\delta^{18}O$	δ^2H	$\delta^{18}O$	δ^2H	$\delta^{18}O$
Shallow wells, Transect 1	T1 W	−6.6	−3.1	−3.0	−2.0	−11.8	−4.3	2.8	0.5
Shallow wells, Transect 2	T2 W	−9.0	−2.8	−2.7	−1.5	−15.0	−3.8	3.9	0.7
Shallow wells, Transect 3	T3 W	−6.0	−2.7	2.1	−1.0	−15.1	−4.1	6.0	1.1
Shallow wells, Transect 4	T4 W	−8.2	−2.9	3.7	−0.8	−18.2	−4.4	7.4	1.1
Rivers, Transect 1	T1R	−5.5	−2.7	4.8	−1.3	−9.3	−4.1	4.4	0.9
Rivers, Transect 4	T4R	9.2	0.3	19	2.6	1.4	−1.2	4.4	1.1
Wetland, Transect 2	T2S	1.9	−1.2	4.4	−0.3	−3.1	−2.3	2.9	0.7
Wetland, Transect 3	T3S	4.3	−0.7	8.8	0.4	−1.7	−2.5	4.0	1.0

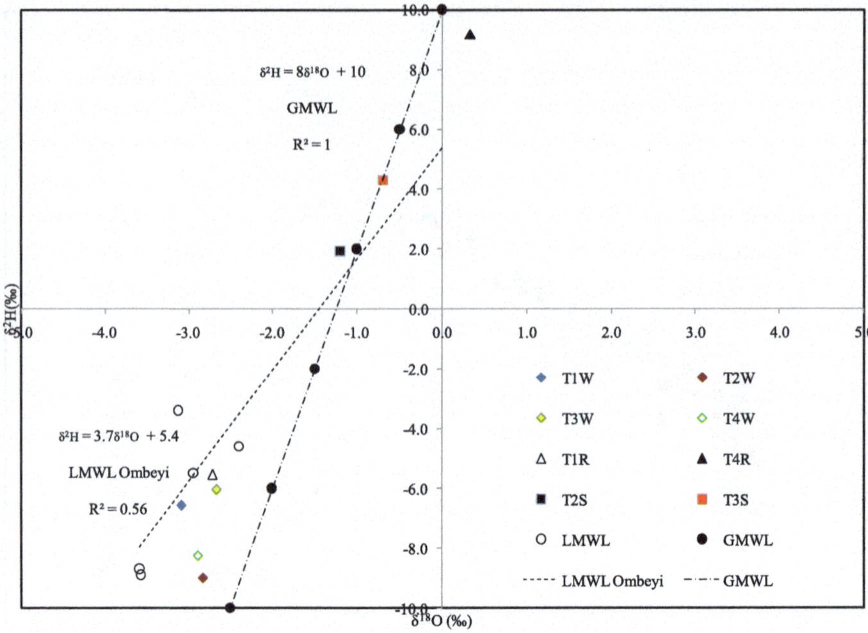

Fig. 9 Relationship between mean values of $\delta^{18}O$ and $\delta^{2}H$ isotopic composition for the various sources. The abbreviations in the legend are described in Table 3

Discussion

The ERT imaging conducted showed that there are numerous heterogeneities in the structure of the subsurface even over a fairly small area. The four profiles have unique features that may explain the local hydrological behaviour. For example, the swampy area in Plate 2a coincides with the shallow high resistivity zone of Profile 2 at chainage of about 1300 m (Fig. 4). Furthermore, wherever there is a solid compact rock even if it is quite deep, the area tends to be swampy. This is supported by the existence of the main wetland in Profile 3 (Fig. 5) and also a smaller one in Profile 4 (Fig. 6) between chainage 1800 and 2400 m.

Alluvial gravel sands, connected fractured rocks and fault lines, are the possible areas responsible for groundwater recharge. Profile 1 and Profile 3 have more visible fault lines as compared with Profiles 2 and 4. This could explain the slight difference in mean isotopic composition of water from shallow wells in Transects 2 and 4 (Fig. 9). It is possible that recharge through fault lines is much more rapid as compared to when through alluvial gravel sands or connected fractured rocks. According to EPA (2000), small-scale variations in sediment type can cause the locations and rates of seepage to vary substantially over small distance. In addition, anisotropy of the porous media affects the pattern of seepage. Thus, the mean isotopic composition of water from shallow wells that could be recharged through fault lines, have more similar properties to that of the contributing rainfall. Furthermore, in the case of Transects 2 and 4, (i) rainfall might have suffered some level of evaporation before reaching the groundwater table, (ii) groundwater might not have been recharged from the local precipitation but from distant sources and (iii) isotopic exchange between groundwater and geologic materials might have taken place (Pelig-Ba 2009).

The water samples from the two wetlands (T2S and T3S) and from the rivers downstream (T4R) are more enriched with δ^2H and $\delta^{18}O$. The spreading and detention time in the wetlands (Fig. 1) is considered to be the reasons behind this behaviour. Baldwin and Cook (2004) demonstrated evidence of isotopic enrichment of wastewater as a function of wastewater detention time. In their study, wastewater sample that had high residence times and thus exposed to significant evaporative effects, plotted far upper right of the plot field indicating substantial alteration of the wastewater isotopic signature. The samples from T4R were more enriched because the rivers Ombeyi and Miriu drain the wetlands and also due to the meanders they go through as the gradient reduces. This further exposes the water to more evaporation.

Conclusion

From the two investigative approaches, ERT imaging and stable isotope signatures, there is a local linkage between surface water and shallow groundwater. The existence of Alluvial gravel sands, connected fractured rocks and fault lines, all conduits of water at different degrees is an indication that where possible surface (rainfall/runoff, river and wetland) water infiltrate into the subsurface. The stable isotope composition of the shallow wells are close to that of rainfall and supplying rivers (T1R), further indicating that there is connectivity between surface water and groundwater in the study area.

The utilization of surface water is likely to have a direct bearing on the availability of shallow groundwater and could also affect deep groundwater. The expansion of rice cultivation increases the amount of water abstracted from the rivers and direct rainfall is also utilized by the crops. Cultivation modifies the soil structure so that in some cases, infiltration is inhibited which further diminishes recharge. With this understanding, the planning and use of surface water should be done in conjunction with that of groundwater.

Acknowledgements The author wishes to acknowledge the financial support received from UNESCO-IHE, Institute of Water Education through the project "Co-management of wetland ecosystems and irrigation for enhancing livelihoods in Lake Victoria Basin (Assignment No. 103864). The institutional support from UNESCO-IHE and personal input by Prof. Michael McClain, Prof. Charlotte de Fraiture, Dr. Jochen Wenninger and Prof. Kenneth Irvine all from UNESCO-IHE, Institute of Water Education, contributed immensely to the success of this work. Dr. Samantha Adey and Dr. Serena Adede Nasong'o who were part of other components of this work deserve recognition for the teamwork they offered.

Egerton University (Kenya) led by Prof. Nzula Kitaka and Prof. Julius Kipkemboi hosted the research and played the coordination role locally. Prof. J. B. Okeyo-Owuor of VIRED International played a noble role by hosting the research team in the field and also assisting in creating an appropriate working environment with the community and all stakeholders. Special gratitude is given to the Ombeyi Community for their land, labour and information that enabled the completion of this study.

References

Baldwin JA, Cook MJ (2004) Environmental isotope studies of wastewater and ground water at wastewater land treatment sites in Idaho. Ground Water Quality Technical Report No. 19

Bhoi AK (2012) An assessment of electrical resistivity soundings data by different interpretation techniques. Int J Biol Ecol Environ Sci (IJBEES) 1(3):108–112

Binge FW (1962) Geology of the Kericho area, geological survey of Kenya, ministry of natural resources, geological survey of Kenya. Report No. 50

Bowen GJ, Winter DA, Spero HJ, Zierenberg RA, Reeder MD, Cerling TE, Ehleringer JR (2005) Stable hydrogen and oxygen isotope ratios of bottled waters of the world. Rapid Commun Mass Spectrom 19:3442–3450

Clay A, Bradley C, Gerrard AJ, Leng MJ (2004) Using stable isotopes of water to infer wetland hydrological dynamics. Hydrol Earth Syst Sci 8(6):1164–1173

EPA (2000) Proceedings of the ground-water/surface-water interactions workshop. United States Environmental Protection Agency. EPA/542/R-00/007

GoK (2014) Ombeyi integrated wetland management plan

Khisa PS, Uhlenbrook S, van Dam AA, Wenninger J, van Griensven A, Abira M (2013) Ecohydrological characterization of Nyando Wetland, Lake Victoria, Kenya, A state of the system (SOS) analysis. Afr J Environ Sci Technol 7(6):417–434

Koch K, Wenninger J, Uhlenbrook S, Bonell M (2009) Joint interpretation of hydrological and geophysical data: electrical resistivity tomography results from a process hydrological research site in the Black Forest Mountains, Germany. Hydrol Process 23:1501–1513

Kuria ZN (2013) Groundwater distribution and aquifer characteristics in Kenya. Dev Earth surf processes 16:83–107

Pelig-Ba KB (2009) Analysis of stable isotope contents of surface and underground water in two main geological formations in the northern region of Ghana. West Afr J Appl Ecol 15

Tekleab S, Wenninger J, Uhlenbrook S (2014) Characterisation of stable isotopes to identify residence times and runoff components in two meso-scale catchments in the Abay/Upper Blue Nile Basin, Ethiopia. Hydrol Earth Syst Sci 18:2415–2431

Talabi AO (2013) Hydrogeochemistry and stable isotopes ($\delta^{18}O$ and $\delta^{2}H$) assessment of Ikogosi spring waters. Am J Water Resour 1(3):25–33

Coastal Land Use/Land Cover and Shoreline Studies for Dakshina Kannada Coast, Karnataka, India

Shabnum Suhura, B. Nithyapriya, L. Revanth Reddy, Noelle Philipose, Mary Manisha and G. S. Dwarakish

Abstract Coastal zone is highly dynamic in nature and undergoes rapid change. It plays an important role in the livelihood of 55 million people along Indian coast. The critical issue faced by the coastline includes coastal erosion, Sea Level Rise (SLR), urbanization, industrialization, and resource management. 23% of Indian coastline is facing coastal erosion and this problem is relatively severe in Dakshina Kannada, making it a critical stretch. As per NOAA report prediction, global SLR is of the order of 0.5–2 m over the next century which ultimately cause extensive loss of low lying coastal areas. The present study is carried out with a view to identify the coastal vulnerability sites along 40 km stretch of Mangalore coast, starting from Talapady in South to Mulky in North. Both conventional and remotely sensed data, coupled with field data were employed to identify critical erosion zones, over a period of 16 years. ERDAS Imagine (2014) software was used to analyze the satellite data to prepare Land Use/Land Cover map. This can be employed in identifying area of land submergence, in case of coastal inundation and to develop ICZMP for Mangalore Coast.

Introduction

Coastal zone plays an important role in the economic, social, and cultural sectors of all the countries. They are vital components of the local ecosystems and are generally densely populated. About 23% of the world's populations are estimated to live within 100 m of sea level and 100 km from a coast (Nicholls and Small 2002).

S. Suhura (✉) · B. Nithyapriya · L. Revanth Reddy · N. Philipose · M. Manisha · G. S. Dwarakish
Department of Applied Mechanics and Hydraulics, National Institute of Technology Karnataka, Surathkal, Mangalore 575025, Karnataka, India
e-mail: shabnumsuhura@gmail.com

V. P. Singh et al. (eds.), *Water Resources Management*, Water Science and Technology Library 78, https://doi.org/10.1007/978-981-10-5711-3_19

However, shorelines are highly dynamic in nature, which depend on a number of intricately linked complex processes like tectonic movements and climatic changes which are difficult to predict accurately. Impulsive change in shoreline and submergence of low lying areas due to sea level rise are the solemn issues that have to be addressed. Indian coastline of about 7516 km is under threat due to global warming and related human interventions. Remote sensing data products provide synoptic and repetitive view of the earth in various spatial, spectral, temporal and radiometric resolutions. Hence, it can be used in monitoring coastal areas on a temporal scale.

Remote sensing provides a platform for rapid delineation of the coastlines at relatively low cost. A detailed quantification of shoreline change can be obtained through repeated observations over the time. The earliest coastline can be obtained from maps dating back to the late 1800s. The long-term shoreline changes can be developed using topographic maps. Aerial photographs for developing shoreline change were popularized since 1920s (Anders and Byrnes 1991). The Geographic Information System (GIS) has been adopted worldwide to elude reliable shoreline change over the years. The digitized shorelines can be used as an input to Digital Shoreline Analysis System (DSAS), an extension to ArcGIS, for calculating shoreline change rate. DSAS computes rate of change statistics from multiple historic shoreline positions residing in a GIS. Karnataka coast has one major port and ten minor ports and is one of the most indented areas with a number of river mouths, bays, spits, sand dunes, lagoons, creeks, cliffs, and long beaches. This clearly unveils the problems associated with this coast. A numerous researchers have studied this coast either as a complete stretch or by considering a particular river mouth. Dakshina Kannada coast has two crucial river mouths namely Netravathi–Gurpur river mouth (with breakwater) and Mulky–Pavanje river mouth (without breakwater), which makes the stretch ideal for studying the effect of breakwater on coastal erosion. The studies on long- and short-term shoreline changes along Mangalore coast shows the impact of construction of seawalls in shifting erosion from one place to another and of breakwaters acting as barriers for littoral drift (Raghavan et al. 2001; Kumar and Jayappa 2009; Gumageri and Dwarakish 2011; Kankara et al. 2014; Hegde and Akshaya 2015).

Knowledge of the distribution of land use and land cover is essential for planning and management activities (Chauhan and Nayak 2005). The global environment change can reflect in LU/LC changes. SLR can occur as a result of storms, tsunami or global sea level change due to global warming. Inundation of coastal low lying areas is a direct outcome of sea level rise. This could lead to great economic and habitat losses, as both human constructions and wildlife habitats would be submerged. Hence, it is primary to quantify the perils associated with the SLR to guide decision maker in future coastal infrastructure development. Land Use/Land Cover (LU/LC) of the coastal area and its dynamics have to be considered in prior to contemplating the Sea Level Rise (SLR) owing to inundation losses. Murali (2014), Murali and Kumar (2015) have examined the implications of SLR on LU/LC classes of coastal zones of Orissa and Cochin respectively. Effect had been studied on the major LU/LC classes like Agriculture, Built-up, Water body, etc., by preparing inundation maps. They are useful in assessing the extent of SLR and potential impact on inundation. LU/LC map generated with conventional maximum likelihood classifier is merged with Digital Elevation Model for quantifying inundation. They considered the SLR of 1 and 2 m to calculate the habitat loss. Dwarakish et al. (2009) had done an extensive study along the Udupi coast of Karnataka, by calculating Coastal Vulnerability Index (CVI) along the coast and by preparing LU/LC map and inundation map of the study area.

This study is focused on understanding the shoreline changes of Dakshina Kannada coast over the years 1997–2013. Also, LU/LC classification has been done through various methods.

Study Area

For this study, a 40 km coastal stretch of Dakshina Kannada Coast starting from Thalapady in South to Mulki-Pavanje river mouth in North is considered. It lies between 12° 45′ 50″ and 13° 05′ 00″ North Latitude and 74° 46′ 30″ and 74° 52′ 48″ East Longitude. The length of the coastline is about 40 km and is oriented about the NNW–SSE direction. This stretch is characterized by the presence of two estuaries formed by Nethravathi and Gurupur rivers on southern side and Mulki and Pavanje rivers on Northern side. The New Mangalore Port is located along this coast with two breakwaters in the harbor. The establishment of this port has lead to rapid urbanization and industrialization along this coast. Fisheries, tourism and agriculture form other main economic activities of this coast. This implies the importance of studying coastal characteristics of this coast. The study area is depicted in Fig. 1.

Data Used

The data used in this study with their specific purposes are listed in Table 1.

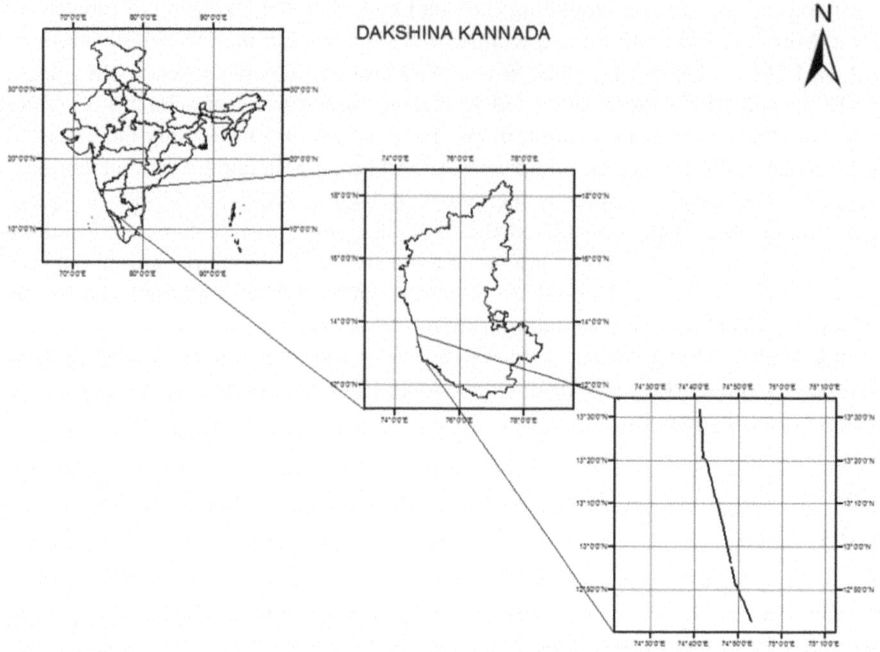

Fig. 1 Location Map of the Study Area

Table 1 Details of data used for the study

S. No.	Type of data	Source	Year	Purpose
1	Toposheet (1:50,000)	Survey of India (SOI)	1987	To prepare base map
2	IRS—1D LISS III (23.5 m resolution)	NRSC	1997	Shoreline extraction
3	IRS—1D LISS III (23.5 m resolution)	NRSC	2001	Shoreline extraction
4	IRS—1D LISS III (23.5 m resolution)	NRSC	2005	Shoreline extraction
5	IRS—P6 LISS III (23.5 m resolution)	NRSC	2013	Shoreline extraction and to prepare LU/LC map

Methodology

General Methodology

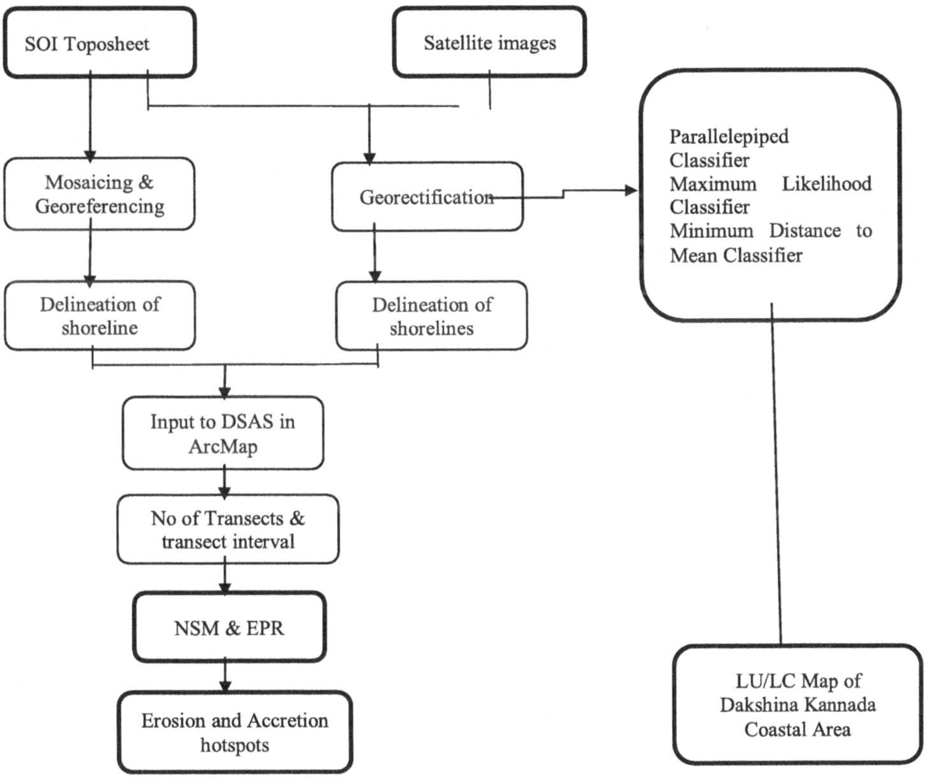

DSAS Digital Shoreline Analysis System
NSM Net Shoreline Movement
EPR End Point Rate

Shoreline Configuration

The mapping of shorelines was done using Arc GIS software. Initially the digital satellite images were geo-referenced to geographic coordinate system and then projected to UTM. Locating and delineating water bodies with remote sensing data are done most easily in Near Infra Red (NIR) wavelengths (0.7–1 μm), because of its absorption property (Lillesand and Kiefer 1994). Therefore, NIR band was considered for shoreline mapping. For the periods 1997–2013, High Tide Line (HTL) was used to represent shoreline. The tidal range and the distance between HTL and LTL (Low Tide Line) being low for Mangalore coast, the error of mapping HTL to represent shoreline for a LISS III image will be negligible. For the digitized shoreline, the rate of change analysis was carried out using Digital Shoreline Analysis System (DSAS) developed by United States Geological Survey. For this, after creating baselines, transects were developed at an interval of 50 m. For the interpretation of shoreline changes, Linear Regression Rate Method (LRR) and End Point Rate Method (EPR) were used. LRR was determined by fitting a least squares regression line to all the comparable shore points of different periods for a particular transect. EPR indicates basically the horizontal displacement rate of shoreline. This was calculated by dividing the distance of horizontal shoreline movement by the time elapsed between the earliest and latest measurements. The positive LRR and EPR represent the rate of accretion and a negative value for the rate of erosion.

LU/LC Mapping

The satellite image data for year 2013 had been used for developing land use/land cover maps of the area. After images were geo-referenced and geometrically rectified, Image classification and interpretation was performed using ERDAS Imagine 2014. For the classification with level I, seven LU/LC classes, i.e., barren land, built-up land, sandy area, forest land, sea water, river water and other vegetations were classified. For supervised classification, total of 210 training sites were made by demarcating a polygon or an area of interest for all the seven land use and land cover types. Many different strategies can be used to classify land use or land cover categories in remotely sensed images. The study used three different methods namely, the Maximum Likelihood algorithm, Minimum Distance to Mean algorithm, and Parallelepiped algorithm for classification Maximum Likelihood Classification (MLC) represents the most established approach, which assumes a normal gaussian data distribution of multivariate data with pixels allocated to the most likely output class, or a posterior probability of membership and dimensions equal to the number of bands in the original image (Richards and Jia 2006).

This method takes advantage of the probability of a pixel being a member of an information class in its decision making. The decision rule in the minimum distance to mean algorithm is based on the relativity among the spectral distances between the pixel in question and the center (mean) of all information classes that have been derived from the training samples. The parallelepiped algorithm assigns a pixel into one of the predefined information classes in terms of its value in relation to the DN range of each class in the same band. After supervised classification, accuracy assessment was carried out. It is the most important aspect to assess the reliability of map. No image classification is said to be complete unless its accuracy has been assessed. To determine the accuracy of the classification, a random sample of 50 pixels for each method is selected on the classified image and their class identity is compared with the ground reference data. Based on the results of accuracy assessment, best classification algorithm was identified and its results were utilized for the process of identifying sea level inundation.

Results and Discussion

Analysis of Shoreline Change

The Dakshina Kannada coast has undergone various changes in the last decade. The shoreline obtained for the various years are represented in the Fig. 4. From the analysis, EPR and LRR are obtained for the period of 2001–2013. The variation of EPR and LRR for each transects length along the shoreline is displayed in Figs. 2 and 3. The range of EPR and LRR for the coast obtained is shown in Table 2. These parameters can be considered to define the coastal management zones in future.

The entire beach shows both accretion and erosion along the coast. The open beach shows erosion during 1997–2001 and an accretion during 2001–2005 and further erosion in the period 2005–2013. Near the Nethravathi–Gurupur river

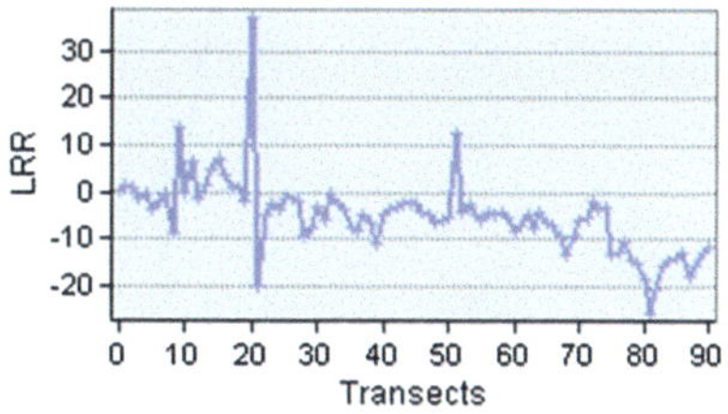

Fig. 2 LRR variation of shoreline change for 1997–2013

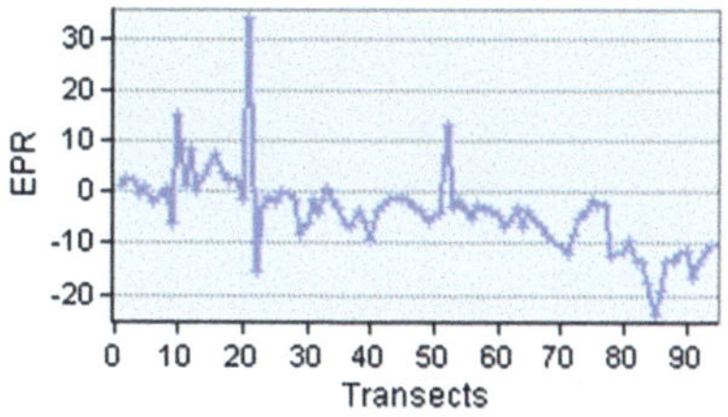

Fig. 3 EPR variation of shoreline change for 1997–2013

Table 2 LRR and EPR rates during the period 1997–2013

	LRR (m/year)	EPR (m/year)
Maximum	32.23	30.92
Average	−4.93	−3.79
Minimum	−25.63	−23.68

mouth, Bengre spit shows an accretion while Ullal has undergone erosion during the period. Similarly, Mulki spit showed an accretion during the period 1997–2013. On the other hand, Pavanje spit had been under erosion during this period. Slight deposition characteristics have been seen near the New Mangalore Port Breakwaters during this period (Fig. 4).

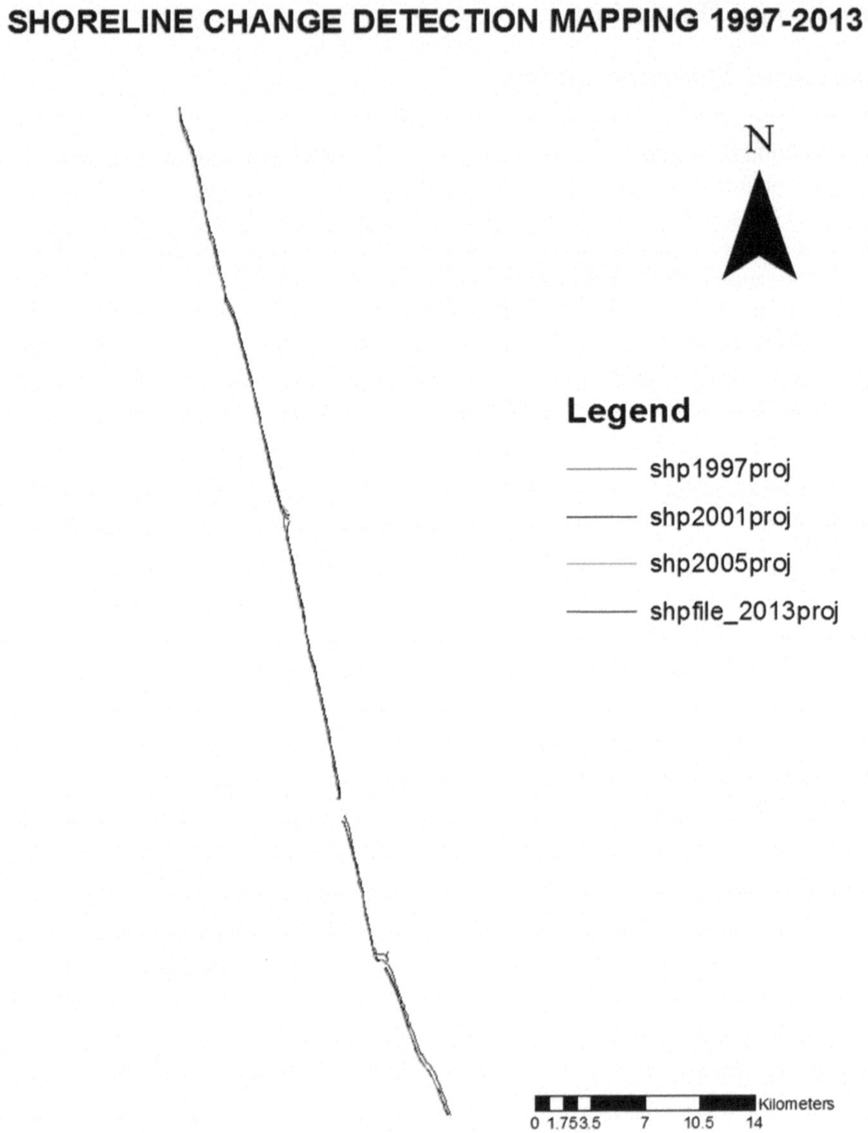

Fig. 4 Shoreline Dynamics During 1997-2013

Land Use Land Cover

The LU/LC classification was done for an area of 800 km^2. The classification was carried out by using three different methods: Maximum Likelihood Algorithm, Minimum Distance to Mean Algorithm, and Parallelepiped Algorithm and the classified area of different LU/LC classes is shown in Figs. 5, 6 and 7. The land use distribution of this area is shown in Table 3. The accuracy of the classified maps

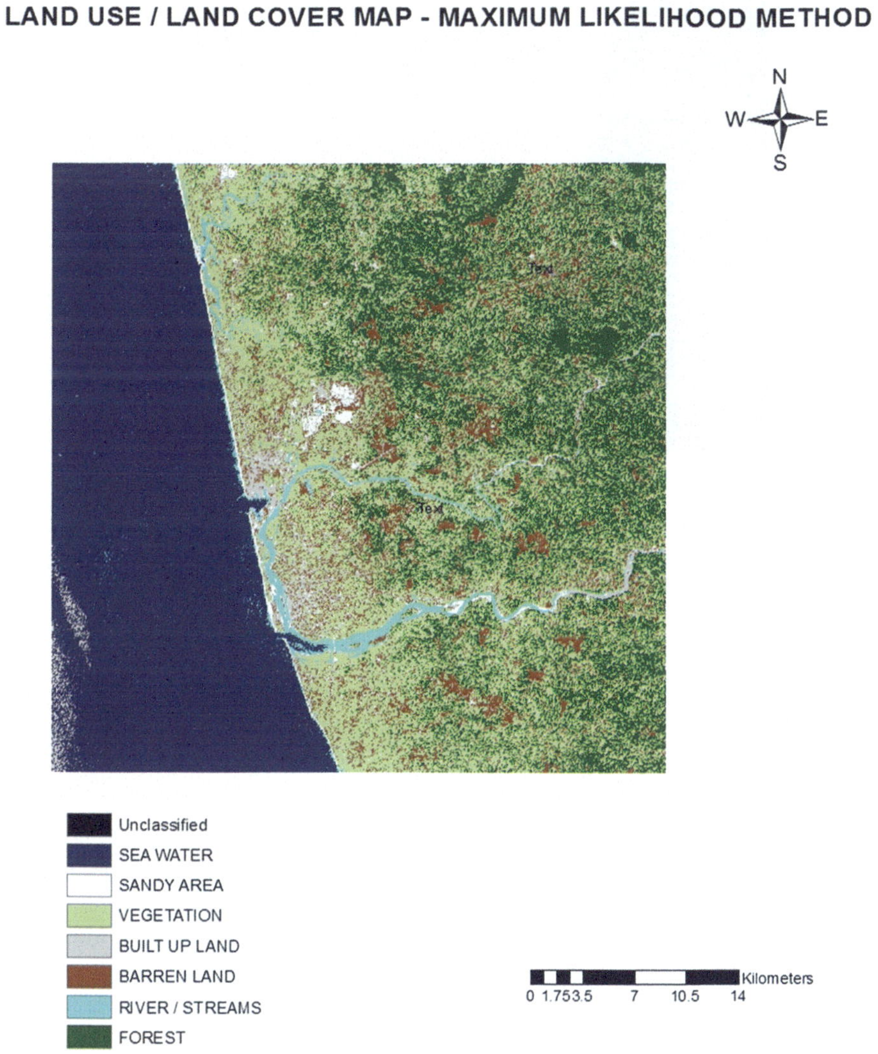

Fig. 5 LU/LC map—maximum likelihood method

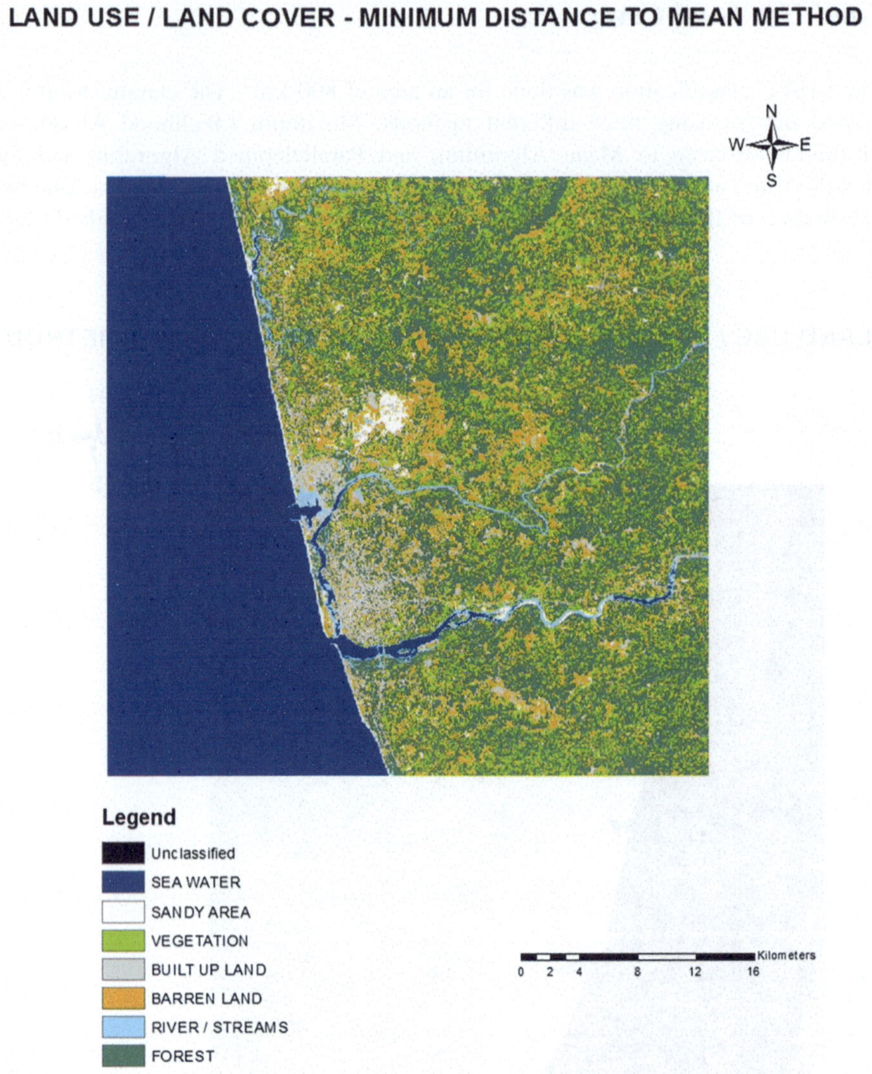

Fig. 6 LU/LC map—minimum distance to mean method

was analyzed by estimating overall accuracy. The result of accuracy assessment is as given in Table 4.

The study shows that more accuracy can be obtained through Maximum Likelihood Method than other two. The other two shows an overestimate of forest and barren land.

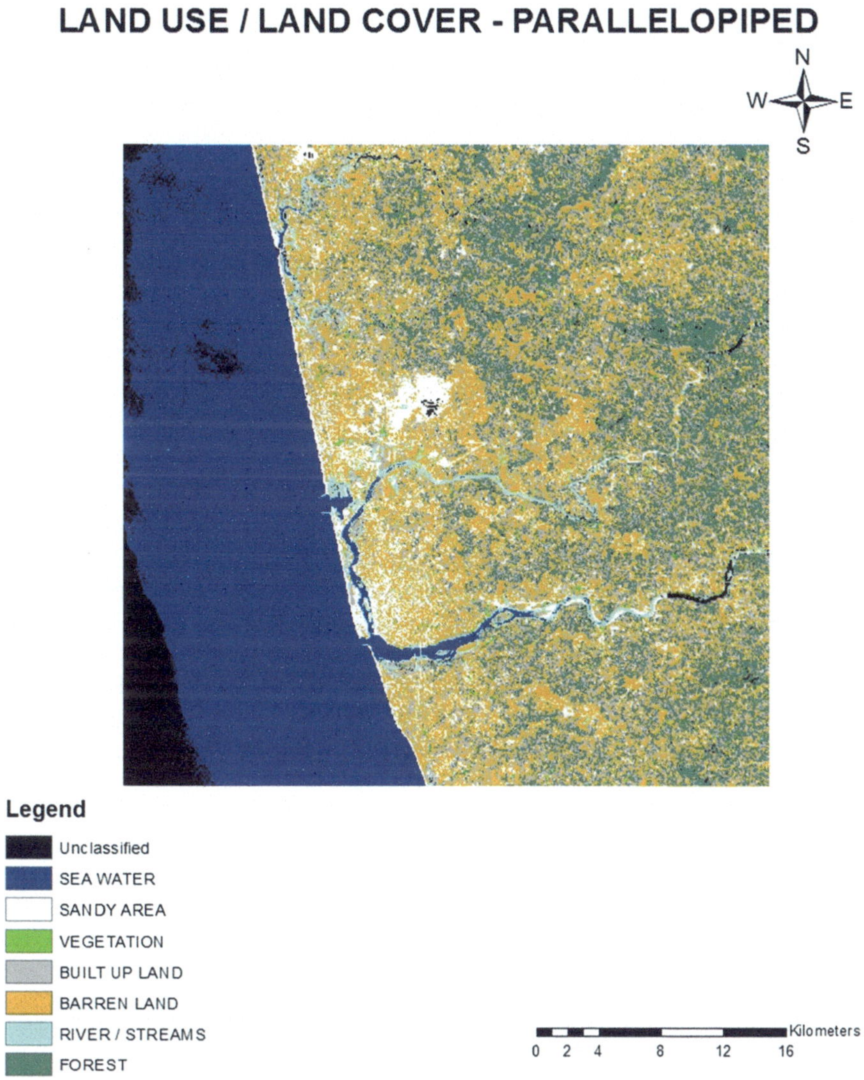

Fig. 7 LU/LC map—parallelopipped method

However, the result shows a lower accuracy for vegetation (80%) than other classes. This is due to the combination of omission and commission errors particularly mixing with forest class. Lower producer's accuracy was also observed for wetlands due to the omission and misclassification to agriculture than to water.

Table 3 Extent of land use/land cover

Class	Area (km^2)
Vegetation	538.77
Forest	354.3
Built-up land	464.9
Barren land	154.9

Table 4 Accuracy assessment of different types of classifier

Type	Accuracy (%)
Maximum likelihood	86
Minimum distance to mean	78
Parallelepiped	74

Conclusions

The study was focused on the coastline variation that had occurred over the years 1997–2013. The shoreline showed mostly an eroding trend throughout the period. The Ullal spit has undergone erosion during the period while Bengre spit has been growing. Similar characteristics have been shown in Mulki–Pavanje spit as well. The land use/land cover study through various methods show that maximum likelihood method gives a more accurate result with an accuracy of 83%. The reliability of these classification methods can be increased by considering more ground control points. Ultimately, these results can be used to prepare coastal inundation mapping which helps in developing coastal management plans.

References

Anders FJ, Byrnes MR (1991) Accuracy of shoreline change rates as determined from maps and aerial photographs. Shore Beach 59(1):17–26

Chauhan HB, Nayak S (2005) Land use/land cover changes near Hazira Region, Gujarat using remote sensing satellite data. J Indian Soc Remote Sens 33(3):413–420

Dwarakish GS, Vinay SA, Natesan U, Asano T, Kakinuma T, Venkataramana K, Jagadeesha Pai B, Babita MK (2009) Coastal vulnerability assessment of the future sea level rise in Udupi coastal zone of Karnataka state, west coast of India. Ocean Coast Manag 52(9):467–478

Gumageri N, Dwarakish GS (2011) Shoreline changes along the Mangalore cast, West coast of India. Int J Earth Sci Eng 04(03):63–70 (spl issue)

Hegde AV, Akshaya BJ (2015) Shoreline transformation study of Karnataka coast: geospatial approach. Aquatic Procedia 4:151–156

Kankara RS, Selvan SC, Rajan B, Arockiaraj S (2014) An adaptive approach to monitor the Shoreline changes in ICZM framework: a case study of Chennai coast. Indian J Mar Sci 43 (7):7

Kumar A, Jayappa KS (2009) Long and short term shoreline changes along Mangalore coast, India. Int J Environ Res 3(2):177–188

Lillesand TM, Kiefer RW (1994) Remote sensing and image interpretation. Willey, United States of America

Murali RM (2014) Application of geo-spatial technologies in coastal vulnerability studies due to sea level rise (SLR) along the Central Orissa Coast, India. In: Geospatial technologies and climate change, pp 187–199

Murali RM, Kumar PD (2015) Implications of sea level rise scenarios on land use/land cover classes of the coastal zones of Cochin, India. J Environ Manage 148:124–133

Nicholls RJ, Small C (2002) Improved estimates of coastal population and exposure to hazards released. Eos Trans Am Geophys Union 83(28):301–305

Raghavan BR, Vinod BT, Dimple KA, Prabhu HV, Udayashankar HN, Murthy TRS, Sreedhara TR (2001) Evaluation of the Nethravathi spit complex, west coast of India: integrated change detection study using topographic and remotely sensed data. Indian J Mar Sci 30(4):268–270

Richards JA, Jia X (2006) Remote sensing digital image analysis-hardback

Management of Coalbed Methane and Coal Mine Produced Water for Beneficial Use in Damodar Basin of India

Vinod Atmaram Mendhe, Subhashree Mishra, Awanindra Pratap Singh, Alka Damodhar Kamble, Mollika Bannerjee and Sunil Kumar Gupta

Abstract In India, coalbed methane (CBM) is rapidly emerging source of natural gas with current production levels of about 1.35 million metric standard cubic metres per day (mmscm/day) and expected to rise to 7 mmscm/day by 2020. CBM extraction associated with pumping of large amount of formation water to reduce hydrostatic pressure existing on gas-bearing coalbeds. At this time out of 32 awarded CBM blocks, only 6 blocks have commenced recovery of methane gas from about 200 wells, with water producing rate more than 20 m^3 per well per day. Out of 424 underground coal mines, only 10–20 places, mine water is being used for domestic, washing of coal and other industrial uses with or without treatment. The high cost of water disposal and lack of efficient technology for treatment are barriers to advance development of CBM reserves in the country. If large amount of CBM and mine-produced water handled economically and treated efficiently to make it acceptable for different uses or surface discharge, it may become a source of fresh water. Produced water samples were collected from CBM production wells and different coal mines water disposal heads in various locations of Damodar basin and analysed using ICPMS and water analysis kit for the assessment of water quality. In CBM water the physical parameters like pH, electrical conductivity, TDS and alkalinity observed in the range of 7.23–8.72, 1678–5436 μs/cm, 1124.26–3642.26 mg/L and 1650–2150 mg/L respectively, whereas in coal mine water, it varies from 6.78 to 8.58, 623–1513 μs/cm, 417.41–1013.71 mg/L and 100–800 mg/L

V. A. Mendhe (✉) · S. Mishra · A. P. Singh · M. Bannerjee
CSIR—Central Institute of Mining and Fuel Research, Barwa Road, Dhanbad 826015, Jharkhand, India
e-mail: vamendhe@gmail.com

A. D. Kamble
Department of Chemical Engineering, Indian Institute of Technology (ISM), Dhanbad 826004, Jharkhand, India

S. K. Gupta
Department of Environmental Science and Engineering, Indian Institute of Technology (ISM), Dhanbad 826004, Jharkhand, India

V. P. Singh et al. (eds.), *Water Resources Management*, Water Science and Technology Library 78, https://doi.org/10.1007/978-981-10-5711-3_20

respectively. CBM water is mainly of Na–HCO_3 type and coal mine water is Ca–Mg–SO_4 and HCO_3–Cl–SO_4 type. Much of the produced water has total dissolve solids (TDS) content <3000 mg/L and can potentially be put to beneficial use within and outside the CBM industry. Sodium adsorption ratio (SAR) was calculated for each sample using concentration of sodium to the sum of the concentrations of calcium and magnesium. The higher the SAR, the greater the potential for reduced permeability, which reduces infiltration, reduces hydraulic conductivity, and causes surface crusting. Trace metal concentrations have a very similar range of distribution in both CBM and coal mine water. With minimal processing, much of this water may be used for a variety of industrial and agricultural purposes controlling pH, electrical conductivity, alkalinity, bicarbonate, sodium, fluoride, metals and SAR values. Drinking water availability is the major issue in Damodar basin; however the large quantity of water generated from CBM production wells can be potential freshwater sources for various applications, including potable consumption after RO treatment. This investigation employs CBM and mine water management strategies considering the spectrum of geologic, hydrologic and geochemical parameters to ensure environmental protection, foster beneficial use, treatment options of produced waters and improving reservoir performance. The CBM-produced water is derived from virgin multiple deep aquifer system having higher concentrations of Na^+ and HCO_3^-, while mine water is of shallow aquifer continuously flushed by seasonal rain water percolation and water drainage system employed in underground coal mine. It also reviews specific water treatment options and associated economics for managing CBM-produced water in Damodar basin.

Keywords Coalbed methane · Coal mine · Produced water · Quality
Production · Management and beneficial use

Introduction

Coalbed methane constitutes an emerging source of natural gas currently under exploration and production development in India. CBM is mainly methane gas stored in underground coal seams saturated with groundwater. The methane gas is adsorbed within the micropore structure of the coals held in by hydrostatic pressure. CBM extraction involves initial pumping of formation water resulting in production of large amount of water from gas-bearing coal seams (Scharufnagel 1993; Moore 2012). To extract this gas, operators depressurize the coal seams by extracting large amounts of water, typically >20 m^3/day, depends on the type of aquifer and hydraulic conductivity. Therefore, water production is a critical measure for the success of CBM operations (Hamawand et al. 2013). The schematic of a coalbed methane production well is given in Fig. 1.

In India, since 2007 CBM industries are working in India and it is important to know how to manage large amount of produced water, considering local water deficiency issues without harming the surface water bodies, soil and drainage

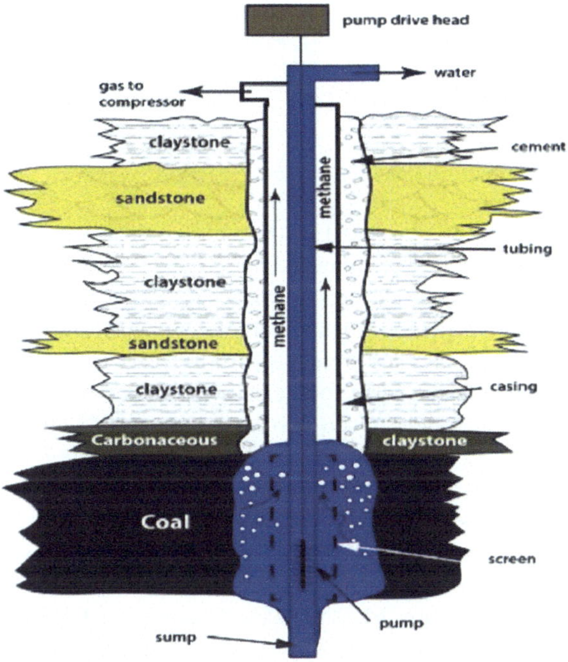

Fig. 1 Schematic of a typical CBM production well

system. CBM water chemistry is an important tool to assess potential environmental implications arising from disposal issues (Rice et al. 2002; Mendhe et al. 2010, 2015; Moore 2012; Taulis and Milke 2013). Water produced from CBM wells can vary from very high quality meeting statutory regulations to having very high total dissolved solids (TDS) concentration (up to 180,000 mg/L) which is not suitable for reuse (Rice et al. 2000; ALL Consulting 2003; Van Voast 2003). CBM industry, therefore, is both a boon for energy starved country like India and a bane as well in the form of huge amount of produced water that has to be dealt with at the beginning itself and managed properly for sustainable gas production (Nuccio 1997; Rice and Nuccio 2000).

It is regardless that the formation lithology, geological age and geochemical processes inherent with the conditions of methane occurrence and generation are understood to modify groundwater quality to a distinctive type that is easily recognized (Van Voast 2003). Therefore, reliable data on composition of water produced from CBM wells are needed so that operators in association with state and regulatory agencies can make informed decisions on handling, disposal and possible beneficial use of water. However, surface discharge or sub-surface injection, TDS is a governing parameter for mixing with any surface or groundwater for many decade and was invariably determined. This study may assist the CBM industry and provides a conceptual framework for the management of produced water from coal. This investigation employs CBM and mine water management strategies considering the spectrum of geologic, hydrologic and geochemical parameters to ensure environmental protection, foster beneficial use, treatment options of produced waters and improving reservoir performance.

Experimental

In this study, produced water samples were collected from CBM production wells and different coal mines water disposal heads. The standard methods for examination of water and waste water suggested by APHA.AWWA.WPCF (1992) were used for analysis of water samples drawn from CBM wells and coal mine heads. The water samples were kept in dry place under normal atmospheric temperature and then analysed for pH, electrical conductivity and turbidity. The water samples was filtered and divided into two halves. Half samples were acidified to pH 2.0 with concentric Nitric acid (HNO_3) and other half left as un-acidified. The un-acidified samples were analysed for anions such as SO_4^{2-}, Cl^-, F^-, and NO_3^- using Ion Chromatography (IC) and all cations (Ca^{2+}, Na^+, Mg^{2+}, and K^+) by Atomic Adsorption Spectrophotometry (AAS). Acidified water samples were analysed for metals like Fe, Al, Cr, Mn, Pb, Cu, As, Zn, Se, Mo, Cd, Ba and B by Inductively Coupled Plasma Mass Spectrophotometry (ICP-MS). Bicarbonate and total alkalinity on un-acidified samples was determined by acid titration method.

Results and Discussions

An investigation has been carried out for CBM and mine-produced water considering the water chemistry and its suitable beneficial use. There are wide variations of water production rates from coal beds in Damodar basin. Ease of dewatering any well depends on the coal's permeability, interference with other wells or mines and link to an aquifer or meteoric waters. Past mining in the area, even though presently inactive, may have depleted water in the seams. It is informative to study the average water production rate of wells in Damodar basin. In India, out of 32 awarded CBM blocks, only 6 blocks have commenced recovery of methane gas from about 200 wells, with water producing rate more than 20 m^3 per well per day. However, the production rate is dependent on the location, associated aquifer type and hydraulic conductivity, these initial rates decline to a much lower, steady value for most of the producing life of the well. The CBM production profiles containing water and gas are shown in Fig. 2.

Figure 2 shows the water and gas production behaviour of four different CBM wells, (a) the water production is very high at initial stage and gas recovery started after 2 months indicates potential aquifer acting over the CBM reservoir, (b) the rate of water production is very high even though gas production initiated at very beginning stage point towards high saturation of CBM reservoir with gas, (c) the unique hydrostatic pressure conditions observed due to deep groundwater table

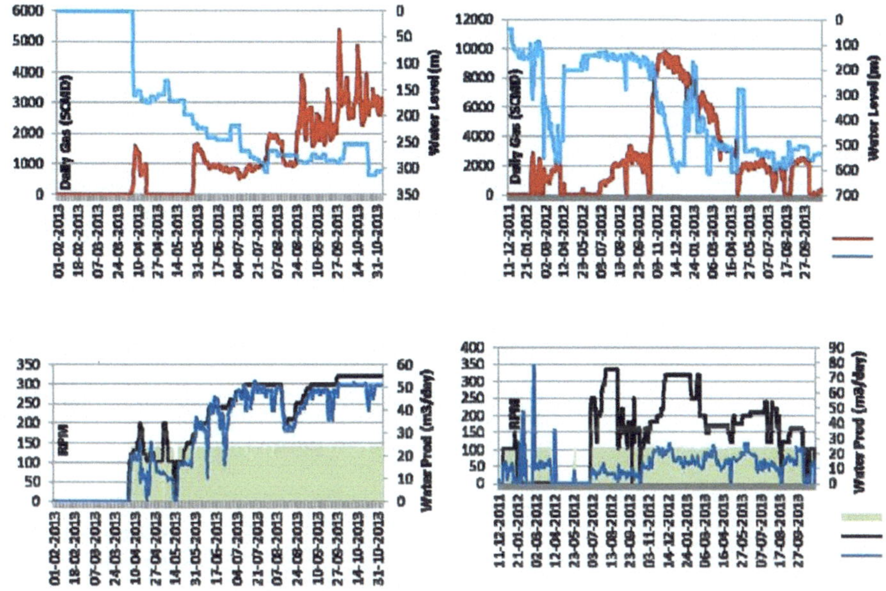

Fig. 2 Production profile of water and gas at Raniganj East (after Essar 2013)

occurrence creating dry conditions in zone of well influence as a result rate of water and gas production more or less similar, and (d) the similar rate of water and gas production indicating pore pressures responsible for release of gas controlled by hydrostatic pressure.

The results of analysis of CBM and mine-produced water samples such as pH, EC, turbidity and TDS values for CBM water vary from 8.260 to 8.720, 3090 to 4600 µs/cm, 0.600 to 2.360 NTU and 2070.300 to 3082.000 mg/L, respectively. The pH, EC, turbidity and TDS values for mine water range from 6.820 to 8.580, 623 to 1513 µs/cm, 0.740 to 2.300 NTU and 417.410 to 1013.710 mg/L, respectively. The mine water of is mildly acidic to alkaline in nature, while CBM-produced water is typically rich with concentrations of total dissolved solids than coal mine water.

Cations and anions concentration distribution for CBM and mine water is given in ternary diagram (Figs. 3 and 4). It is observed that CBM water contains wide distribution of Na^+ and HCO_3^-, while mine water contains SO_4^{2-} and HCO_3^-. Water produced from deeper coal formations can contain NO_3^-, Cl^-, metals and high levels of total dissolved solids, which makes it unsafe for drinking purposes (Jackson and Reddy 2007; Jamshidi and Jessen 2012; Mendhe et al. 2015). In water, sulphate is usually derived from the weathering of sulphide-bearing minerals like pyrite (FeS_2), or dissolution of gypsum ($CaSO_4{\cdot}2H_2O$) or anhydrite ($CaSO_4$). Pyrite (FeS_2) usually occurs as a secondary mineral sin the Gondwana coals and associated sediments.

Figure 5 signifies the concentrations of different ions of CBM, Mine water which denotes that CBM water have higher concentrations of Na^+ and HCO_3^- with

Fig. 3 Ternary diagram for cations of CBM and mine water

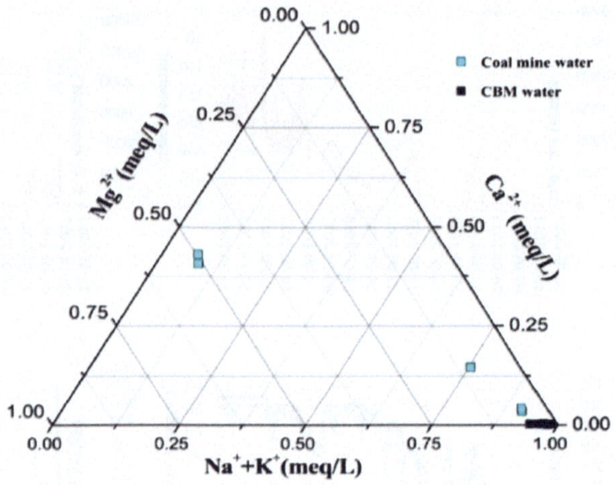

Fig. 4 Ternary diagram for anions of CBM and mine water

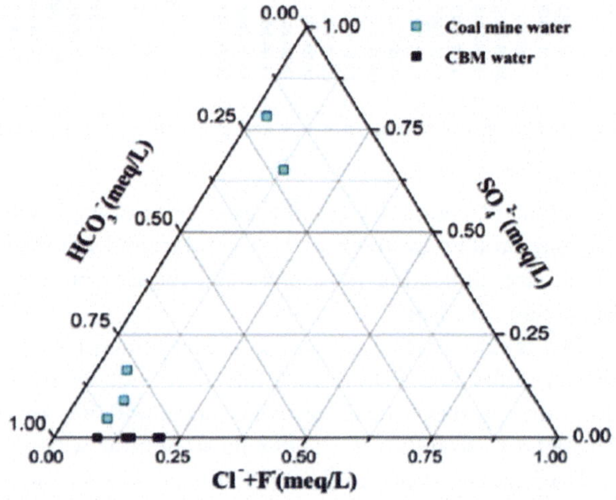

higher SAR (Sodium Adsorption Ratio), whereas mine water has higher concentration of SO_4^{2-} because of presence of pyrite in coal seams compared to river water.

$$\mathrm{SAR} = \mathrm{Na}^{+} / \sqrt{1/2\left(\mathrm{Ca}^{2+} + \mathrm{Mg}^{2+}\right)}$$

The surface disposal and agriculture use of CBM-produced water are restricted due to high values of SAR, which may cause infiltration, surface crusting and also reduces the permeability of soil (Van Voast 2003). The values of major ions and SAR shows that bicarbonate and sodium concentration in CBM water are relatively high, ranging from 2129.400 to 2771.300 and 349.800 to 976.100 mg/L, respectively, whereas for mine water it varies within values of 132.450–1023.950 and

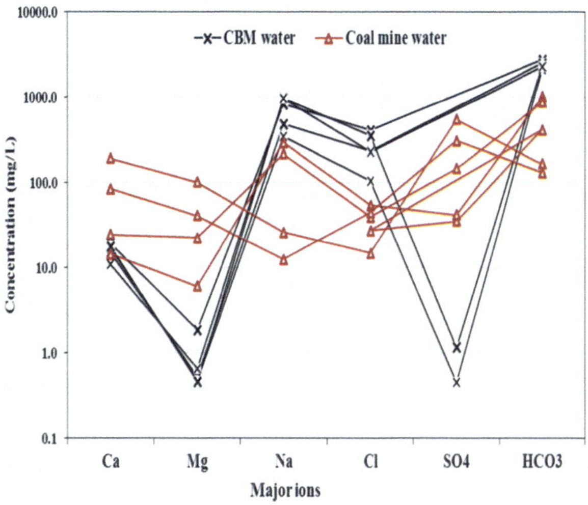

Fig. 5 Schollar plot for major ions

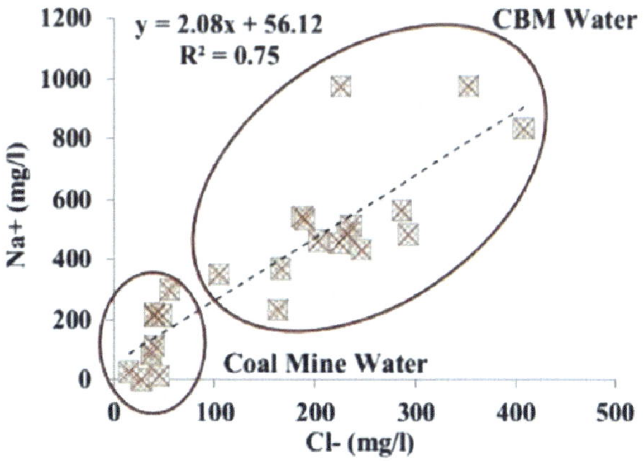

Fig. 6 Relationship between concentration of CT and Na^+ CBM and mine water

0.000–297.300 mg/L. Trace metals have similar range of distribution in both CBM and coal mine water, except manganese concentration is observed relatively high in mine water. The relationship between Cl^- and Na^+ is presented in Fig. 6. It displays a very good correlation for both CBM and mine water. SAR and Na^+ concentrations vary proportionately to each other (Fig. 7). The stiff plots of cations and anions of CBM and mine water are presented in Figs. 8 and 9. It is observed that CBM water contains wide distribution of Na^+ and HCO_3^-, while mine water contains SO_4^{2-}, HCO_3^-, Ca^{2+} and Mg^{2+}. In deep coal beds, dissolved bicarbonate is a direct product of carbonate dissolution by oxygenated recharge waters and of both the sulphate reduction and methanogenesis processes (Freeze and Cherry 1979). The CBM-produced waters are mostly devoid of divalent cations like calcium and magnesium but enriched in concentrations of sodium. The prime reason for the

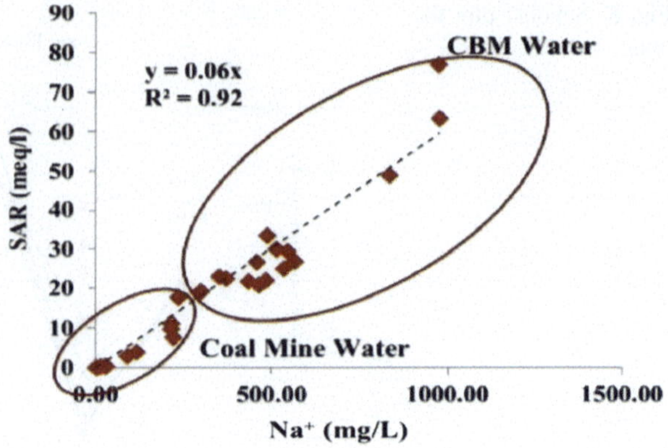

Fig. 7 Relationship between concentration of Na^+ and SAR in CBM and mine water

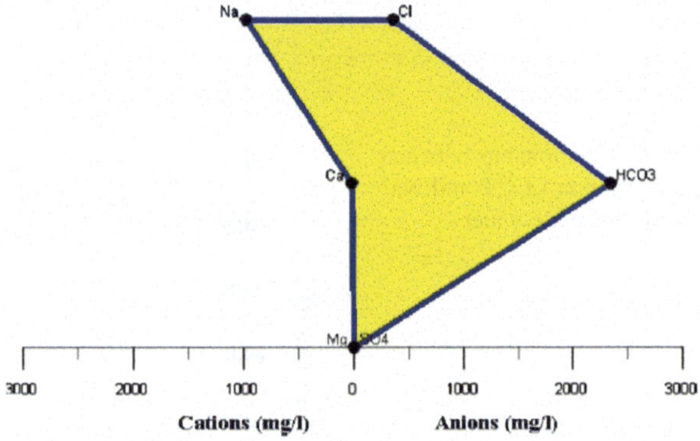

Fig. 8 Stiff diagram for CBM-produced water

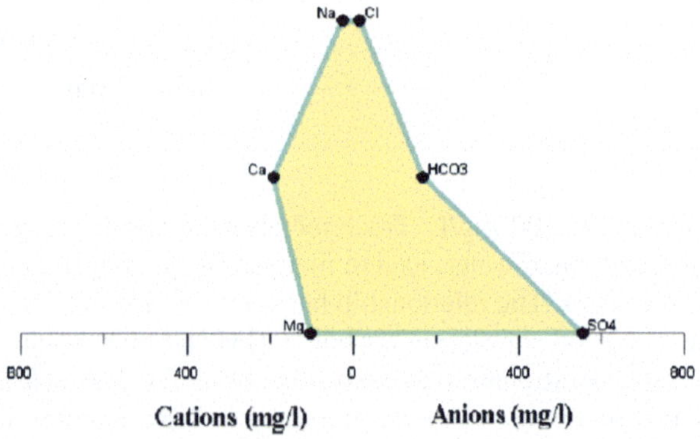

Fig. 9 Stiff diagram for coal mine produced water

depletion of calcium and magnesium is likely to be the precipitation of calcite and dolomite due to the presence of the elevated bicarbonate concentrations (Eaton 1935; Van Voast 2003; Pashin 2007).

After examining the quality and quantity of produced water, following options may be useful for appropriate use and disposal of CBM water in Damodar basin.

Irrigation may be a suitable option for CBM-produced water subsequent to desalinization and proper treatment. Irrigation has several critical aspects which need to be taken care for proper balance of soil quality and crops grown in the area. The CBM-produced water quality should be assessed with respect to salinity (EC) and sodicity (SAR) to provide an initial indication of the general suitability for irrigation. The trace metals should also be evaluated to find any potential impacts to irrigated soils and plants. The characteristics of produced water and irrigated soil, like drainage, texture and chemical variations should be considered with respect to local crops growing in the area. The lethal effects over soil due to elevated salinity and on plant productivity is well identified (Maas and Hoffman 1977). The high salinity in CBM-produced water may reduce the soil pore water required for plant use. Hence, plants need to devote more energy to extract water from the soil when raised concentrations of salts are added in the root zone (Western Fertilizer Handbook 1995; Barbour et al. 1998; Bauder and Brock 2001). The increase in energy required to extract water results in decreased plant productivity in soils with elevated concentrations of soluble salts. While all plants exhibit decreased productivity with increasing concentrations of soluble salts, the threshold and degree to which salinity affects crop yield varies between species (Maas and Hoffman 1977).

Shallow groundwater contained lower TDS, high sulphate concentrations and different hydrochemical characteristics in comparison to groundwater in coal bed methane aquifers (Hamawand et al. 2013). It is observed that CBM-produced water from Raniganj basin contains most of the parameters such as E.C, TDS, bicarbonate, alkalinity, sodium and fluoride are exceeding the permissible limits given by WHO (1997) and BIS (1991) IS: 10,500 for potable water or drinking water and irrigation purpose. For mine water bicarbonate, sodium and iron content in some areas exceeds the permissible limits. Impounding CBM water by pumping it into storage facilities, reservoirs and ponds has traditionally been a preferred water management option for CBM operators and may be one of the effective methods in Damodar basin. These impoundments are well known as infiltration ponds, evaporation ponds, or zero-discharge ponds. Drinking water availability is the major issue in Damodar basin; however the large quantity of water generated from CBM production wells can be potential freshwater sources for various applications, including potable consumption. These challenges include high treatment cost, potential chronic toxicity of the treated produced water and public acceptance. Because of the need of desalination and removal of a large number of chemical compounds, RO will most likely be used for potable reuse applications. The facility of reverse osmosis created by Essar Oil in their CBM block in Raniganj coalfield is shown in Fig. 10. It is emphasized that the main challenges present in CBM-produced water are desalination, degassing, suspended solids removal, organic compounds removal, heavy metal and others. Achieving the various treatment goals requires the use of multiple treatment technologies, including physical, chemical, and biological treatment processes (Ahmadun et al. 2009; Basumatary et al. 2010; Mendhe et al. 2015). Even if RO is capable of removing many organic compounds at high efficiency, the combined chronic toxicity of the organic compounds existing in mixture in the RO permeate needs to be carefully

Fig. 10 Facility for reverse osmosis (RO) of CBM-produced water at Raniganj (*Courtesy* Drinking water project-Essar Oil)

evaluated before direct reuse is implemented. It has been shown that the toxicity of dissolved organic compounds in produced water can be additive (Glickman 1998). The current drinking water standards were established based on the human health risk associated with individual contaminants. Although existing technologies have been demonstrated to meet current drinking water standards (Ahmadun et al. 2009; Xu et al. 2008), there are concerns on unknown toxic effects or unknown toxic compounds. Some of the technologies are removal of TDS by precipitation, electrochemical or photo-catalytic oxidation, nano-filtration or reverse osmosis, removal of metal through aeration, settling, sand filtration with suspended solids removal, coagulation/flocculation, sedimentation and filtration. The surface discharge and sub-surface injection of the produced water should be treated up to the requirements of the locals and state regulatory limitations for discharge and injection (Sadler and George 1995; Ganjegunte et al. 2011; Johnston et al. 2008). The overall objective of this study is to treat and manage the large volume of produce water with possibly alleviate operating costs and at the same time, provide high quality water for discharge that can be beneficially used downstream.

A holistic mine water treatment technology should be an optimized combination of the processes in a sequential way to fulfil WHO norms for potable water such as adsorption ion/exchange to reduce concentration of Ca, Mg, Na, K, Cl, NO_3, SO_4, F and PO_4; aeration to lower the values of Fe; filtration to get down concentration of cations, anion and bacteria; electro-coagulation to diminish TDS and cations and photo-catalysis ozonation for removal of organic material and bacteria. Figure 11 shows the facility developed for treatment of mine water by Bharat Coking Coal Limited (BCCL) at Dhanbad with the help of CIMFR Dhanbad, considering the application of new improved technology to increase coal production, reduce the possibility of contamination of other water resources, help in solving water crisis problem and also the cost involved in power and human resource should be utilized in a beneficial way.

Fig. 11 Facility for mine-produced water treatment at Dhanbad (*Courtesy* Drinking water project-CIMFR-BCCL)

Summary and Conclusion

Coalbed produced water is an inevitable and integral step in CBM and coal mines advancements. Where the fields are large and volume of water substantial, the challenges will be significant. Initial water purification and disposal create a primary problem that must overcome to establish profitable methane production. Water disposal is a deciding factor in developing marginally economical properties. The anticipated schedule of water production throughout the life of the project is needed for an accurate economic evaluation. The investigation carried out at Raniganj basin for CBM and underground coal mine produced water reveals the following summary and conclusions:

- The quality assessment of CBM- and mine-produced water is useful for evaluating water quality from different geological formations, which normally have distinctly different geochemical signatures. Coal mine water is relatively higher in dissolved calcium (Ca^{2+}), magnesium (Mg^{2+}), chloride (Cl^{-}) and sulphate (SO_4^{2-}), whereas water from the deep coal beds associated with adsorbed methane gas is comparatively higher in dissolved sodium (Na^{+}) and bicarbonate (HCO_3^{-}).
- The CBM water is categorized as Na–K type, Na–HCO_3 type and HCO_3 type, whereas the coal mine water may be categorized as the Ca–Mg–HCO_3, HCO_3–Cl–SO_4 and Na–HCO_3 type in Damodar basin.
- The treatment technologies for use of CBM and mine water for irrigation, drinking and impounding considering removal of pH, EC, TDS, alkalinity, bicarbonate, sodium, fluoride, metals content and SAR values such as precipitation, electrochemical or photo-catalytic oxidation, nano-filtration or reverse osmosis, removal of metal through aeration, settling, sand filtration with suspended solids removal, coagulation/flocculation, sedimentation and filtration.
- The surface discharge and sub-surface injection of the produced water should be treated up to the requirements of the locals and state regulatory limitations for discharge and injection. The effective management of CBM and coal mine water in Damodar basin required more specific scientific investigation before adoption of any disposal method.

References

Ahmadun FL, Pendashteh A, Abdullah LC, Biak DRA, Madaeni SS, Abidin ZZ (2009) Review of technologies for oil and gas produced water treatment. J Hazard Mater 170:530–551

APHA.AWWA.WPCF (1992) Standard methods for the examination of water and waste water, 16th edn. APHA, Washington D.C.

ALL Consulting (2003) Handbook on coal bed methane produced water: management and beneficial use alternatives. Prepared by ALL Consulting for the Ground Water Protection

Research Foundation, U.S. Department of Energy, and U.S. Bureau of Land Management, pp 1–322
Barbour MG, Burk JH, Pitts WD, Gillian FS, Schwartz MN (1998) Terrestrial plant ecology. Benjamin/Cummings, Menlo Park
Basumatary JK, Mandal D, Datta GC (2010) Physico-chemical investigations on produced water compositions of Jharia and Bokaro CBM blocks, Jharkhand, India—a tool for effective produced water management, Petrotech, New Delhi, India, 31 Oct–3 Nov 2010
Bauder JW, Brock TA (2001) Irrigation water quality, soil amendment, and crop effects on sodium leaching. Arid Land Res Manag 15:101–113
Eaton FA (1935) Changes in the composition of ground waters resultant to anaerobic sulfate decomposition and the attendant precipitation of calcium and magnesium. Transactions American Geophysical Union 16th Annual Meeting, Apr 1935, pp 512–516
Essar Group (2013) Coal bed methane-monetization challenges in India. International workshop on development of coal based nonconventional energy resources in India, 12–13 Nov 2013, Ranchi, India
Freeze RA, Cherry JA (1979) Groundwater. Englewood Cliffs, Prentice Hall, New Jersey, p 604
Ganjegunte GK, Trostle CL, Braun RJ (2011) Irrigation effects of cooling tower effluent on soil chemistry and Alfalfa in the Rio Grande river basin. Land Degrad Dev 22:410–424
Glickman AH (1998) Produced water toxicity: steps you can take to ensure permit compliance. Presented at the API Produced Water Management Technical Forum and Exhibition, Lafayette, LA, 17–18 Nov 1998
Hamawand I, Yusaf T, Hamawand SG (2013) Coal seam gas and associated water: a review paper. Renew Sustain Energy Rev 22:550–560
Jackson RE, Reddy KJ (2007) Geochemistry of coalbed natural gas (CBNG) produced water in Powder River Basin, Wyoming: salinity and sodicity. Water Air Soil Pollut 184:49–61
Jamshidi M, Jessen K (2012) Water production in enhanced coalbed methane operations. J Petrol Sci Eng 92–93:56–64
Johnston R, Vance GF, Ganjegunte GK (2008) Irrigation with coalbed natural gas coproduced water. Agric Water Manag 95:1243–1252
Maas EV, Hoffman GJ (1977) Crop salt tolerance ñ current assessment. J Irrig Drain Div 103 (IR2):115–134
Mendhe VA, Mishra P, Varade AM (2010) Coal seam reservoir characteristics for coalbed methane in North and South Karanpura Coalfields, Jharkhand, sedimentary basins of India: recent developments. Gondwana Geol Mag 12:141–152
Mendhe VA, Mishra S, Varma AK, Singh AP (2015) Assessment of coalbed methane produced water and its management options in Raniganj Basin, West Bengal, India. Appl Water Sci 1–9
Moore TA (2012) Coalbed methane: a review. Int J Coal Geol 101:36–81
Nuccio V (1997) Coalbed methane—an untapped energy resource and an environmental concern. USGS Fact Sheet FS-019–97, U.S. Geological Survey
Pashin JC (2007) Hydrodynamics of coalbed methane reservoirs in the BlackWarrior Basin: key to understanding reservoir performance and environmental issues. Appl Geochem 22:2257–2272
Rice CA, Nuccio V (2000) Water produced with coalbed methane. USGS Fact Sheet FS-156
Rice CA, Ellis MS, Bullock JH Jr (2000) Water co-produced with coalbed methane in the Powder River Basin, Wyoming: preliminary compositional data. U.S. Geological Survey Open File report 00-372, Denver, Colorado
Rice CA, Bartos TT, Ellis MS (2002) Chemical and isotopic composition of water in the Fort Union and Wasatch formations of the Powder River Basin, Wyoming and Montana: implications for coalbed methane development, The Rocky Mountain of Geologists, Coalbed methane of North America-II
Sadler LY, George O (1995) Concentration of saline produced water from coalbed methane gas-wells in multiple-effect evaporator using waste heat from the gas compressor and compressor drive engine. Desalination 101(2):76–169
Scharufnagel RA (1993) Coalbed methane production in hydrocarbons from coal. In: Law BE, Rice DD (eds) AAPG studies in geology #38, pp 341–359

Taulis M, Milke M (2013) Chemical variability of groundwater samples collected from a coal seam gas exploration well, Maramarua, New Zealand. Water Res 47:1021–1034

Van Voast WA (2003) Geochemical signature of formation waters associated with coalbed methane. AAPG Bull 87:667–676

Western Fertilizer Handbook (1995) Produced by the Soil Improvement Committee of the California Fertilizer Association. Interstate Publishers Inc., Sacramento, California

Xu P, Drewes JE, Heil D (2008) Beneficial use of co-produced water through membrane treatment: technical-economic assessment. Desalination 225(1–3):1–139

Water Supply–Demand Assessment in Ur River Watershed in Tikamgarh District

V. C. Goyal, T. Thomas, S. Goyal and R. V. Kale

Abstract The water resources management has of late become a very challenging task due to manifold challenges of multi-sectoral demands, decreasing availability coupled with the looming threats of climate change. The Bundelkhand region in central India is primarily dependent on rain-fed agriculture. In this semi-arid region, drought is becoming frequent, and there is an increasing pressure on scarce resources such as farm land, water and pastures. The growing water demand across competitive sectors, increasing severity of droughts, declining groundwater levels, and deteriorating water quality are some of the crucial problems faced by the stakeholders in the water sector of this region. Effective solutions to the water problems must tackle both the supply and demand, identify user's actual needs and proceed with appropriate technologies. Integrated Water Resources Management (IWRM) approach is considered to be particularly useful for the water resources management in such semi-arid regions of India. Estimation of the available water supplies and the demands is the first and foremost task in developing management strategies for water-scarce regions. This paper presents the results of a detailed water balance study carried out for Ur River watershed in Tikamgarh district of Madhya Pradesh (India) by quantifying the important hydrological components during 1999–00 to 2010–11. The spatial information pertaining to the topography, land use and soil type have been extracted using the ArcGIS 9.3 and crop water requirements computed by CROPWAT 8.0. It is observed that surface runoff of 301, 206 and 333 MCM was generated during the wet years 1999–00, 2003–04 and

V. C. Goyal (✉) · R. V. Kale
National Institute of Hydrology, Jalvigyan Bhavan, Roorkee, Uttaranchal, India
e-mail: vcgoyal@yahoo.com

R. V. Kale
e-mail: ravikale2610@gmail.com

T. Thomas
National Institute of Hydrology, Regional Centre, Bhopal, India
e-mail: thomas_nih@yahoo.com

S. Goyal
MP Resource Atlas Division, M.P. Council of Science & Technology, Bhopal, India
e-mail: sgoyalbsg@gmail.com

V. P. Singh et al. (eds.), *Water Resources Management*, Water Science and Technology Library 78, https://doi.org/10.1007/978-981-10-5711-3_21

2008–09 for an annual rainfall of 1212, 1035 and 1196 MCM, respectively. The water budget helps to understand the overall water availability and demand scenario, particularly during periods of droughts, so that effective water resources management schemes can be devised.

Keywords Water balance · Supply–demand · Bundelkhand · Water resources management · IWRM

Introduction

The Intergovernmental Panel on Climate Change (IPCC) states that the availability and distribution of freshwater resources will be greatly affected by climate change and the vulnerability to water scarcity could increase (Parry 2007). Also the human population is increasing at a steady rate throughout the world (USCB 2003). This creates a problem since many areas all over the world are struggling with declining fresh water supplies. Agricultural water use is expected to increase by 18% in the next 30 years (Kundzewicz and Parry 2001). Also the industrial water use in Asia is expected to increase even faster, possibly tripling in the next 30 years (Kundzewicz and Parry 2001). Therefore, it is certain that there will be additional pressures on the water resources systems under the multiple challenges of population growth, climate change and delay in implementing water-efficient technologies in the agriculture and industrial sectors.

Many environmental problems are caused by changes in aspects of the hydrological cycle. Any change in any of the components of the water balance is ought to affect the other components of the water balance. The supply–demand scenario of a region can be understood by conducting a detailed water balance study. However, understanding of the hydrological cycle is very important before undertaking the task of water balance. Water balance helps to quantify the relationships between precipitation, surface and groundwater runoff, evaporation, transpiration, and aquifer drafts, and provide a framework for future planning of sustainable exploitation of the available water resources. Water balance modelling combined with field experiments can give us a better understanding of the components of the hydrological cycle from which appropriate management options can be developed. A water balance model can be considered as a system of equations designed to represent some aspects of the hydrological cycle. The specific methods for water budget analysis will change according to the scale of the study area and operative temporal variations. The variations in spatial scale include regional, watershed-wide, sub-watershed level or site scale whereas the temporal variations for a water budget on a watershed to site scale could vary from hours, days, months, seasonal or annual. For shorter time-steps, soil moisture accounting becomes more important in terms of antecedent soil moisture storage and infiltration in order to partition surface runoff from infiltration.

Without accurate water balance information, it is not possible to manage the available water resources of any country/region. When working on the water balance, it has to be clearly understood that the available water is a highly dynamic and variable process, both spatially and temporarily. Therefore, the methodology which is directly dependent on the time unit and is a function of the measured hydro-meteorological and hydrological data is the most significant element in such studies. Due to the human influence, change of the water demands and climatic variations and/or changes, the water balance of an area cannot be considered as final (unchanging). Therefore the water balance components must be constantly monitored, controlled and updated. The major aim of any water balance attempt is for the long-term sustainable management of water resources.

Monthly water balance models were first developed in the 1940s by Thornthwaite (1948) and later revised by Thornthwaite and Mather (1955, 1957). These models have since been adopted, modified, and applied to a wide spectrum of hydrological problems (Gabos and Gasparri 1983; Alley 1984, 1985; Vandewiele et al. 1992; Xu and Vandewiele 1995). Recently, they have been employed to explore the impact of climatic change (Schaake and Liu 1989; Arnall 1992; Xu and Halldin 1996). They also have been utilized for long range stream flow forecasting (Alley 1985; Xu and Vandewiele 1995). Toebes (1961) studied the water balance of natural catchments wherein he assumed that the evapotranspiration decreases with decrease in soil moisture. Tuffuor and Labadie (1973) used a nonlinear time-variant model for modelling monthly or seasonal data for the Todzie River in Ghana. Hughes (1982) modified the daily model developed by Roberts (1978) and converted it to a monthly model. The model uses two parameters to divide precipitation into direct runoff, infiltration, and evapotranspiration, with another two parameters used for regulation of the underground reservoir. Gleick (1986) reviewed various approaches for evaluating the regional hydrologic impacts of global climatic change and presented a series of criteria for choosing among different methods. He concluded that the use of monthly water balance models appears to offer significant advantages over other methods in accuracy, flexibility, and ease of use. Gleick (1987) developed and tested a monthly water balance model for climatic impact assessment for the Sacramento basin. The model works well under conditions of stationary climate and has capabilities to incorporate changes in climatic variables. The results suggest that the application of such models may provide considerable information on regional hydrologic effects of climate change than is currently available.

A water balance model was developed by Schaake et al. (1996) based on a statistical model with a small number of parameters and of intermediate complexity, by averaging the main hydrologic processes. Vandewiele et al. (1992) proposed a series of models which are variants of the monthly water balance model developed by Van der Beken and Byloos (1977). Xu and Singh (1998) discuss the relevance of various aspects of the practical application of water balance models. Kumar (2001) studied the water balance of Krishnai River basin according to Thornthwaite's concept of potential evapotranspiration. Zhang et al. (2001) developed a simple water balance model that requires only vegetation, annual total stream flow and rainfall data with the intention of assessing impacts of land use changes on mean annual water yield. Singh and Prasad (2004) used the water balance techniques for computing seasonal and geographical patterns of water availability to facilitate better management of available water resources. Anatoliki et al. (2005) estimated the annual water balance for Anthemountas River basin using the annual precipitation and water consumption. Hsin et al. (2007) applied a water balance concept with two models in the Ching-Shui watershed to describe the groundwater recharge. Tekleab et al. (2011) analysed the water balances of twenty catchments in the Upper Blue Nile basin using a top-down modelling approach based on Budyko's hypotheses for better understanding the water balance dynamics of upper Blue Nile catchments on annual and monthly temporal time-scales and on a meso-scale to large spatial scales.

Study Area

The Ur River basin, a tributary of the Dhasan River located in Tikamgarh district of Madhya Pradesh has been selected for carrying out the assessment of water availability under various alternate climate scenarios. The study area represents the typical topography and geology of the Bundelkhand region and is one of the most vulnerable areas in respect of climate change and drought-related indicators. Ur River basin lies on the Bundelkhand plateau and extends between latitudes 24° 35′ 00″N and 25° 05′ 00″N and between 78° 50′ 00″E and 79° 10′ 00″E longitudes with a total geographical area of 990.37 km^2. The total geographical area of the watershed is 990.37 km^2. The maximum length of the watershed is about 119 km from north to south with an average width of about 80 km. The Ur River flows in a south to north-east direction. Ur River watershed is bounded by Chhatarpur district to the east, Lalitpur district to the west; Jhansi district to the north and Sagar district to its south. The location map of the study area is given in Fig. 1. The climate of study area is semi-arid with four seasons. The winter season extends from December to February followed by the summer season from March to mid-June; rainy season from mid-June to September and the transition season spans from October to November. After February, the temperature rises progressively. May is generally the hottest month with mean daily maximum temperature at about 43 °C which may sometimes rise up to even 47 °C on individual days. The relative humidity is high during the monsoon season being generally above 70%, whereas air is comparatively dry during the rest of the year. The summer season is dry with relative humidity of less than 20%.

The topography of the watershed is undulating and comprises of very high hills along the ridge line with the elevation varying between 200 and 400 m above mean sea level (msl). The elevation gradually decreases from the southern part of the watershed towards the north. The Ur River flows in a north-eastern direction till its confluence with Dhasan River. The map showing the topography of the region is given in Fig. 2.

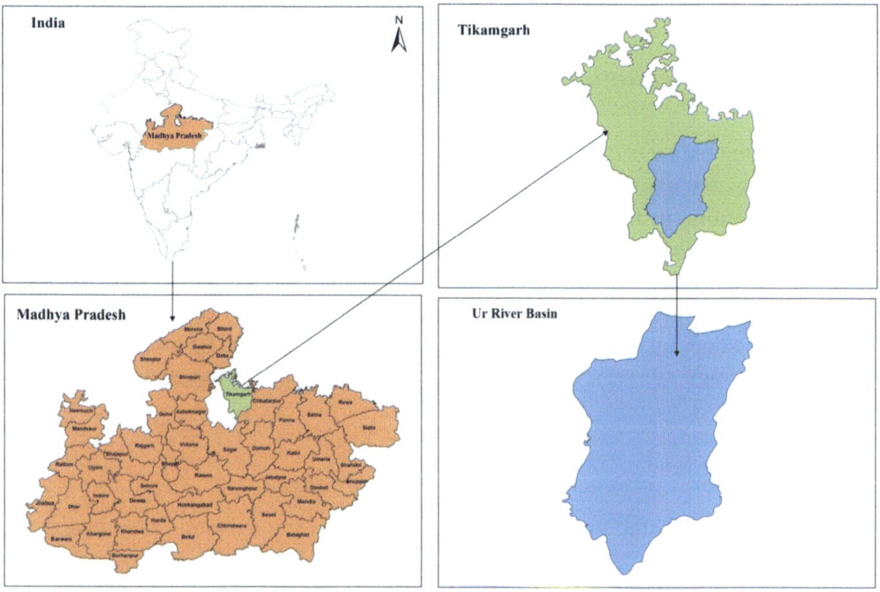

Fig. 1 Index map of the Ur basin

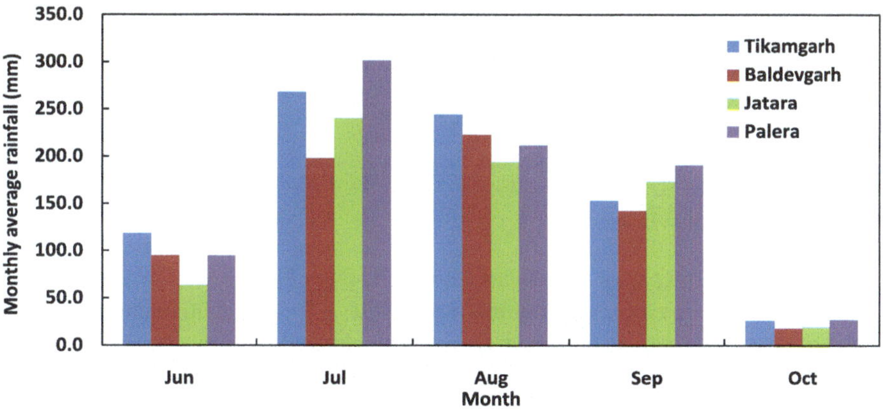

Fig. 2 Monthly average rainfall (long-term) at influencing rain gauge stations

The area covered by the settlement is 19.98 km^2, whereas dense forest occupies an area of 44.56 km^2. The river and water bodies cover the area of 34.47 km^2, fallow land covers 69.25 km^2 and barren land covers 110.54 km^2 of the watershed. The major portion of the watershed is covered by sandy loam soil which is about 635.46 km^2 and covers 68.05% of the total watershed area.

Methodology

Water Balance Equation

With the water balance data, it is possible to compare the individual sources of water in the system over different periods of time and to establish the degree of their effect on variations in the water regime. Also the water balance enables the evaluation of one unknown component of the water balance from all other known components. A water balance equation is a mathematical expression of water entering and leaving a specified volume over a given time period. A specified area can be a watershed, an aquifer, a lake, etc. The time period can be a season, year, or can be an average of yearly values. The basic concept of the water balance is expressed in Eq. 1,

$$I = O + \Delta S \tag{1}$$

where, I = inflows to the system; O = outflows from the system; ΔS = change in storages.

If inflow = outflow then ΔS is zero; if inflow > outflow then ΔS is positive and if inflow < outflow then ΔS is negative. The broad components of the water balance may be divided more precisely into four groups namely inflows, water use, outflows, and storages.

Water budgeting analysis for the Ur River watershed has been conducted in order to decide strategies for development and management of water resources in the watershed on seasonal basis using the water balance equation of the following form:

$$P + \mathrm{GW}_{\mathrm{in}} - Q_{\mathrm{dsro}} - Q_{\mathrm{b}} - E_{\mathrm{t}} - \mathrm{ET}_{\mathrm{f}} - \mathrm{ET}_{\mathrm{c}} - D_{\mathrm{dom}} - D_{\mathrm{liv}} - \mathrm{GW}_{\mathrm{out}} = \Delta S_{\mathrm{s}} + \Delta S_{\mathrm{g}} + U_{\mathrm{w}} \tag{2}$$

where, P = rainfall; Q_{dsro} = direct surface runoff; Q_{b} = base flow; E_{t} = evaporation from tanks; ET_{f} = evapotranspiration from forest; ET_{c} = evapotranspiration from cropped areas; D_{d} = domestic usage; D_{l} = livestock needs; ΔS_{s} = change in surface water storages; ΔS_{g} = change in groundwater storage; U_{w} = unaccounted water.

The various components of the water balance and their estimation methods including the inflows, outflows, and storage components are given in the following section.

System Inflows

Inflows represent the water inflows and it includes (a) precipitation, (b) surface water inflows, and (c) groundwater inflows. The precipitation includes rainfall, and all other forms like snow, dew, hail, etc. Except in areas with regular snowfall, this is practically equal to the rainfall. Groundwater and surface inflows are water flowing into the study area from surrounding areas. Surface water and groundwater inflows into the study area from surrounding areas are highly dependent on the developmental activities of the surrounding areas. If watershed-based development is undertaken in the surrounding areas, water will be used there itself, which can therefore reduce these inflows substantially. Only if it is sure that some of those inflows will continue to be available on a long-term basis it should be included as dependable inflows. Since the present study area corresponds to a classical watershed bounded by ridges and with a single exit point, surface inflows are eliminated and groundwater inflows are also considered to be negligible but still estimated based on the prevailing water table gradients. Therefore, the rainfall as well as the groundwater inflow is the inflow terms being considered in the water balance equation. The mean areal rainfall for the watershed has been estimated using the Thiessen polygon method, wherein the representative weights for each of the four influencing rain gauge stations were obtained. The mean areal seasonal rainfall has been estimated using Eq. 3,

$$P_{\text{seas}} = \frac{A_1P_1 + A_2P_2 + A_3P_3 + A_4P_4}{A_1 + A_2 + A_3 + A_4} \tag{3}$$

System Water Use

The water use inside the system includes evaporative losses from water bodies such as ponds, tanks, and reservoirs inside the study area, evapotranspiration by crops and trees, water use for domestic purposes and livestock needs and water demands for non-agricultural purposes like industries, water sports, etc. In the Ur River watershed, the major water use demands include domestic demands, livestock demands, and evapotranspiration from plants and crops. As there are no industries and water recreational activities, the industrial and recreational demands have not been considered. Presently there is no information pertaining to the storage details including the area-elevation-capacity tables/curves for the few tanks present in the watershed. Lack of this information makes the task of estimation of precise reservoir evaporation rather difficult. Therefore this component has been estimated indirectly and is subjective. Subsequently after conducting the bathymetric survey of some of the major ponds, the information can be used for computing the reservoir evaporation component as well as change in storage in the reservoirs. The methodology adopted in the estimation of water use from the system is given below.

Evaporation from Water Bodies

Some of the water stored in the water bodies like ponds, lakes, tanks and reservoirs will evaporate and be lost from the system. This evaporation takes place from the surface storages as well as from saturated soil surfaces. Evaporation loss from surface storages is obvious, but the other evaporative loss may need some explanation. Whenever there is a heavy spell of rain water tends to accumulate in some areas. If such areas are large and rain accumulates for considerable time in these areas after a heavy spell, it results either in standing water or in a saturated soil surface; the evaporative loss from these areas may have to be taken into account. This component also needs to be taken into account when we plan for watershed development. This component of water balance represents a pure loss, since this quantity of water returns to nature. It can be considered as the unproductive expenditure of water from the system. So attempts should be initiated to reduce this component as much as is possible.

The actual evaporation from the tanks is estimated by multiplying the reservoir area with pan coefficient as given in Eq. 4.

$$\mathrm{EV}_{\mathrm{tank}} = \mathrm{EV}_{\mathrm{pan}} * P_{\mathrm{c}} \tag{4}$$

where, $\mathrm{EV}_{\mathrm{tank}}$ = Actual evaporation from tank (mm); $\mathrm{EV}_{\mathrm{pan}}$ = Evaporation from pan (mm); P_{c} = Pan coefficient.

The actual evaporation volume for a particular month is then computed by Eq. 5 as given below:

$$\mathrm{VEV}_{\mathrm{tank}} = \frac{1}{2}(\mathrm{WSA}_{\mathrm{sm}} + \mathrm{WSA}_{\mathrm{em}}) D_{\mathrm{w}} \mathrm{EV}_{\mathrm{tank}} \tag{5}$$

where, $\mathrm{VEV}_{\mathrm{tank}}$ = Volume of actual evaporation from tank (m^3); $\mathrm{WSA}_{\mathrm{sm}}$ = Water spread area at the start of the month; $\mathrm{WSA}_{\mathrm{em}}$ = Water spread area at the end of the month; D_{w} = Average depth of water in the tank.

Evapotranspiration

Evapotranspiration includes water utilized by all types of plants in the study area, including that used by crops, grasses and trees, and also includes both the rainfall and the externally applied water used by plants. Evapotranspiration component represents the productive expenditure of water. This can be sub-divided into two groups namely,

(a) Evapotranspiration from cropped areas
(b) Evapotranspiration from non-cropped areas.

A further break up for different cropped areas and species is required in estimating as to how much water is being utilized how effectively and in deciding on how to get the best out of our productive water use. Also a track of the externally applied water and its use is to be kept which includes the record of the quantity of water being applied for each of the crops and amount which is being actually utilized by the crops. The conversion of rainfall or externally applied water into evapotranspiration depends on the capacity of the soil to store moisture in its root zone. Only that portion of rainfall or applied water which is stored in the root zone becomes available to the plant. Even with the same rainfall and the same amount of applied water, if the soil is capable of storing much more water, then abundant water becomes available to the plant.

Evapotranspiration from Cropped Areas

The evapotranspiration from cropped areas have been computed by the Penman–Monteith equation which is the sole standard method of determining crop evapotranspiration as suggested by FAO. It is a method with strong likelihood of correctly predicting ET_o in a wide range of locations and climates and has provision for application in data-short situations. The FAO Penman–Monteith method to estimate ET_o is given in Eq. 6,

$$ET_o = \frac{0.408\Delta(R_n - G) + \gamma \frac{900}{T+273} u_2 (e_s - e_a)}{\Delta + \gamma(1 + 0.34u_2)} \quad (6)$$

where ET_o = reference evapotranspiration (mm day^{-1}); R_n = net radiation at the crop surface (MJ m^{-2} day^{-1}); G = soil heat flux density (MJ m^{-2} day^{-1}); T = mean daily air temperature at 2 m height (°C); u_2 = wind speed at 2 m height (m s^{-1}); e_s = saturation vapour pressure (kPa); e_a = actual vapour pressure (kPa); $e_s - e_a$ = saturation vapour pressure deficit (kPa); Δ = slope vapour pressure curve (kPa °C^{-1}); γ = psychometric constant (kPa °C^{-1}).

The equation uses standard climatological records of solar radiation, air temperature, humidity and wind speed. The crop evapotranspiration has been worked out for 10 days on daily basis considering four growth stages of the crop, namely initial stage, development stage, growth stage and the final stage. The crop coefficients for the various crops which vary during the different stages of the crop growth have been used along with the Penman–Monteith (FAO) reference crop evapotranspiration values to arrive the water crop evapotranspiration for each of the ten-day period. The crop coefficient is maximum during the mid-stage or growth stage. The total crop evapotranspiration is arrived at by summing up the ten-day crop evapotranspiration over the entire crop period. The village-wise crop evapotranspiration is then added to get the total crop evapotranspiration for the study area for a particular crop.

The crop water requirement for the various crops grown in the watershed has been estimated using CROPWAT 8.0. It is a decision support system developed by the Land and Water Development Division of FAO for planning and management of irrigation. CROPWAT initially computes the reference crop evapotranspiration, the potential rate based on the climatological data as mention above. The actual water requirement is thereafter computed on a 10 days on daily basis using the reference evapotranspiration, rainfall, crop information including the crop coefficients and the soil properties. The information on cropping pattern consists of the crop planting date, crop coefficient data files including K_c values, stage in days, root depth, and depletion fraction along with the area planted (0–100% of the total area).

Evapotranspiration from Forested Areas

The evapotranspiration needs of the unirrigated non-cropped area in open and dense forests in the watershed are met by the monsoon rains and groundwater. The potential evapotranspiration for the deep-rooted trees has been estimated based on the potential rates obtained by Penman–Monteith method. The evapotranspiration from the forested areas is assumed to take place at the potential rate as the forests comprise deep-rooted trees and the water table is shallow in the study area during most parts of the year. The actual evapotranspiration from non-cropped forested areas has been estimated using the following Eq. 7,

$$\mathrm{VET_f} = A_\mathrm{f}\mathrm{ET_f} \tag{7}$$

where $\mathrm{VET_f}$ = Volume of evapotranspiration from forested areas (m^3); A_f = Area under forests (m^2).

Water Use for Domestic and Livestock Use

Water is an essential human need for drinking, cooking, washing, cleaning, and sanitation and for livestock. Water used for these purposes is not a productive expenditure like that expended in evapotranspiration but a useful and essential expenditure. A separate account of water used for these purposes is to be kept. This is a first priority need as well as a higher quality need and so working out its magnitude as a component of the water balance is imperative. The drinking and sanitation water needs require special attention and they should be seen as part of the overall problem of improving water resource availability in an area. Four blocks, namely Tikamgarh, Baldevgarh, Jatara and Palera, fall inside the Ur watershed and the water demands for the domestic use have been computed based on the per capita demand of fresh water and the population based on the Census of

2011. Similarly, the livestock water demands have also been computed based on the per capita demand for livestock and their Census of 2007.

Water Use for Non-agricultural Production

There may be other non-agricultural production activities which require water like industries, power plants, fisheries or water sports. The water use on this account needs to be estimated and a record is to be maintained on the sources tapped and the amount of water utilized by each of these users. This is a productive expenditure of water and such use should be encouraged if the problems of pollution and waste disposal are taken care and water use on this account does not cut into other priority needs and lastly if it creates a sufficiently high rate of income generated per unit of water utilized. However, for Ur watershed there is no such requirement of water for any of these uses.

System Outflows

The outflows pertain to the amount of water flowing out of the study area and mainly comprise two components namely, surface water outflow and groundwater outflow. This represents the unutilized flows, but not necessarily losses. In the watershed development plans efforts are made to minimize these outflows as much as possible. These outflows often have an important role in the downstream environment and perennial flows in the stream. The outflows from the system include the surface water outflows and groundwater outflows.

Surface Water Outflow

The surface water outflow has been estimated by the Soil Conservation Service Curve Number (SCS-CN) model at the outlet of the watershed. The SCS-CN model is based on the single parameter Curve Number (CN) which depends on the land use, land cover, soil type and the antecedent moisture conditions prevailing in the watershed. The composite curve number (CCN) for the watershed is estimated as 78 using hydrologic soil group and land use for the AMC-II condition (average). The direct surface runoff has been estimated using the SCS-CN model given in Eqs. 8–10.

$$Q = \frac{(P - I_a)^2}{(P - I_a + S)} \quad \text{for } P > I_a \tag{8}$$

$$Q = 0, \quad \text{for } P \leq I_a \tag{9}$$

$$S = \frac{25{,}400}{\text{CN}} - 254 \tag{10}$$

where Q = direct surface runoff (mm); S = potential retention (mm); CN = curve number; I_a = initial abstraction = 0.2S for general soils; 0.3S for AMC-I and black soils; and 0.1S for AMC-III.

Groundwater Outflow

The topography of the watershed being highly undulating, there is a significant elevation difference between the watershed outlet and the stream origination point. Moreover the groundwater which gets replenished also gets lower due to the groundwater outflow from the watershed towards the downstream areas. The contours of the groundwater levels have been drawn in ArcGIS 9.3 and the watershed boundary has been divided into different reach lengths and groundwater levels on both sides of the reach studied. Depending on the hydraulic gradient the groundwater outflow has been determined using Eq. 11,

$$Q = TiL \tag{11}$$

where T = transmissivity (m^2/day); i = hydraulic gradient; L = length of reach; Q = rate of flow (m^3/day); The hydraulic gradient of the groundwater flow is given by Eq. 12,

$$i = h/L \tag{12}$$

where h = difference in groundwater levels between the observation well inside the watershed and observation well downstream for the particular reach (m); L = reach length (m).

System Storages

The storages being accumulated in the system can be considered to comprise three main components, namely (a) surface storages, (b) groundwater storages, and (c) root zone storages. The surface storages include all the water stored on the surface in ponds, tanks and behind check dams. Groundwater storage includes all water stored as free water below the root zone of plants. Root zone storage includes the moisture held in the soil in the root zone of plants. This component is not often considered as a storage element in the water balance, but it has an important role to play. These storage components represent the savings and are important in tiding over bad years in which rainfall may not be adequate.

So as a part of the watershed development process efforts should be made to replenish the storages in good years as these storages are of utmost importance and act as cushion against the variations in rainfall and its pattern. The increases or decreases in these storages help to identify whether the demands are being met from the annual water inflows or the storages are also being exploited alongside for satisfying the demands. The important condition for sustainability of the water resource is that the storages should not decrease in average years but should increase sufficiently in good years to compensate for the additional withdrawals from the storages during the bad years. Generally, at the end of summer season the surface storages are practically empty and soil moisture in the root zone may also be negligible. If the water balance is started and ended as per the water year from June to May then both these storage terms remain unaffected. So only the change in groundwater storage needs to be accounted for. The change in groundwater storage have been computed based on the change in groundwater levels between each season, specific yield of the aquifers and the area of aquifers as given in Eq. 13.

$$\Delta \mathrm{GW_s} = S_\mathrm{y} A \Delta \mathrm{GWL}_{(\mathrm{before-after})} \quad (13)$$

Unaccounted Water

The unaccounted water is lumped together and worked as a residual term which practically means that all components in the water balance equations are computed and the difference in the water balance and unaccounted components are taken as other outgoings.

Establishment of Water Balance

The water balance for the Ur River watershed has been established after estimation of the water balance components that play a major role in the hydrology of the watershed. The values of all these components have been substituted in the water balance equation. The unaccounted water in the monsoon season gives an idea of the accuracy with which the water balance components have been evaluated. It also contains the components which could not be evaluated due to lack of data.

Analysis and Results

Inflows to the System

Computation of Mean Areal Rainfall

The long-term daily average rainfall series (normal) at each of the four stations have been obtained by taking the average daily rainfall values spanning 1999–2010. The monthly average rainfall at the influencing rain gauge station is given in Fig. 2. The mean areal rainfall in the watershed has been computed based on the Thiessen polygon method. The rain gauge stations at Tikamgarh, Jatara, Baldevgarh and Palera influence the rainfall pattern in the watershed. It has been observed that the monsoon season rainfall varies between 344.27 MCM in 2007–08 which was a widespread drought year in Bundelkhand and 1211.66 MCM in 1999–00 with the average basin rainfall of 792.77 MCM.

Water Use in the System

Evapotranspiration from Cropped Areas

The reference evapotranspiration has been computed using the CROPWAT8.0 developed by FAO using the climatic data, viz. maximum and minimum temperature, relative humidity, wind speed and sunshine radiation in the Penman–Monteith equation. The average daily reference evapotranspiration at Tikamgarh is 4.30 mm/day. The variation of the monthly ET_o is given in Fig. 3. The effective rainfall has been computed based on monthly rainfall by using the rain module of CROPWAT8.0 with the USDA method of effective rainfall computation. Subsequently, the crop module of CROPWAT8.0 has been used to assign the crop parameters during the various crop growth stages namely, initial stage, development stage, growth stage and the final stage. After finalizing the crop parameters the soil module of CROPWAT 8.0 has been used to assign the soil parameters namely, total available moisture which is the difference between the field capacity and wilting point of the soil, maximum infiltration rate, maximum rooting depth, and initial available soil moisture. Thereafter the CWR module of CROPWAT 8.0 has been used to compute the crop water requirement for 10 days on a daily basis.

The crops have been considered to be sown during different dates based on the actual field conditions as it is not possible to sow the entire area under a particular crop on a single day due to many operational constraints like availability of machinery for ploughing and sowing and many other spatially and temporally varying factors that come into play. The total crop evapotranspiration is estimated by adding the ten-day crop water requirement over the entire crop period.

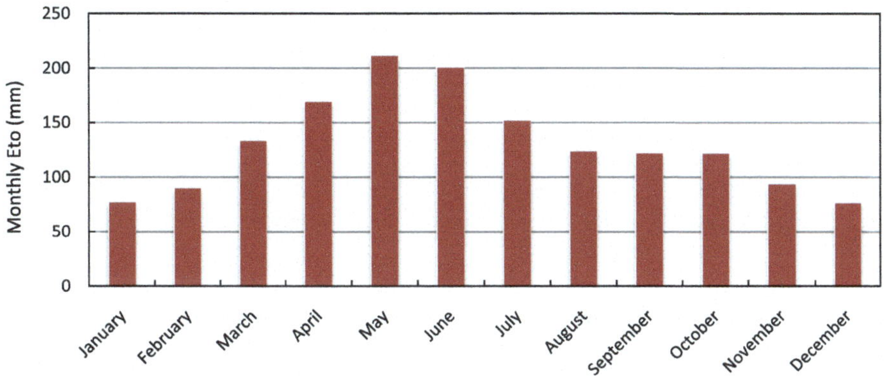

Fig. 3 Monthly variation of ET_o at Tikamgarh

Soyabean is the principal crop grown during the kharrif season whereas wheat is the predominant crop in rabi season. The computation of crop water requirement of wheat grown during the rabi season of 2007–08, which was a drought year too, is given in Table 6.6. The total water requirement for wheat without considering the effective rainfall is 444.30 mm whereas after deducting the effective rainfall of 0.60 mm, the actual crop water requirement for wheat is 443.70 mm. The total amount of water required at the field for the wheat crop during 2007–08 is 443.70 mm. However, the actual amount of irrigation water required for the wheat crop will be substantially more considering the conveyance losses and application losses in the process of water reaching the field after being released from the dams/ storage structures. The total water requirement for soyabean after considering the effective rainfall of 267.30 mm is 170.0 mm during 2007–08. Therefore only supplemental irrigation requirements have to be planned for the kharrif crop as much of its water requirements are met by the monsoon rainfall. During normal and high rainfall years, the water requirements are further reduced and varied between 27.60 mm in 1999–00 to 82.90 mm in 2008–09. The water demands of the major crops of both seasons have been summed up to arrive at the seasonal crop water demand in the watershed. For some of the crops sown during the kharrif season and extending further into the rabi season, the total crop water requirement during each season has been computed separately. The details of the seasonal crop water requirement for the various blocks of Tikamgarh falling in Ur River watershed is given in Table 1. The crop water requirement for the kharrif (monsoon) crop varies between 7.99 MCM during 1999–00 which had been a wet year to 44.73 MCM during 2007–08 which had been a drought year in the Ur watershed. However the water requirement for the rabi (non-monsoon) crop varies between 218.96 MCM in 1999–00 to 127.62 MCM in 2005–06.

Table 1 Seasonal crop water requirement in Ur River watershed

S. No.	Year	Kharrif (monsoon) (MCM)	Rabi (non-monsoon) (MCM)
1.	1999–00	7.99	218.96
2.	2000–01	31.70	216.02
3.	2001–02	48.14	155.39
4.	2002–03	43.90	160.24
5.	2003–04	14.55	140.28
6.	2004–05	26.62	135.72
7.	2005–06	31.78	127.62
8.	2006–07	44.58	129.00
9.	2007–08	44.73	141.73
10.	2008–09	28.32	134.83
11.	2009–10	12.48	120.61

Table 2 Seasonal water requirement for deep-rooted trees (forests)

S. No.	Forest type	Kharrif season (MCM)	Rabi season (MCM)
1.	Dense forest	32.80	38.80
2.	Open forest	93.75	110.84
Total		126.58	149.64

Evapotranspiration from Forested Areas

The forested area has been classified into two classes, viz. dense forests comprising of thick healthy vegetation with good canopy cover and open forests which comprise sparse forests with less canopy cover. Substantial area in the Ur watershed is comprised of dense forests (45.57 km^2) and open forests (130.18 km^2) and is therefore one of the important land use categories in the watershed. The dense and open forests are located all along the hills on the ridge line and also sparsely on the flat topped hills inside the watershed and there is not much change in the forested area for the period under analysis. The seasonal evapotranspiration from the forested area which includes both dense and open forests in the watershed have been computed and given in Table 2. The water requirement by forests has been computed to be 126.58 MCM in monsoon season and 149.64 MCM in non-monsoon season.

Water Use for Domestic Purposes and Livestock Needs

The domestic water consumption has been estimated based on the population of villages of four blocks namely Jatara, Tikamgarh, Palera and Baldevgarh which lies inside the watershed. The demand has been assumed at the rate of 135 l/capita/day

Table 3 Domestic water demands

S. No.	Year	Monsoon (MCM)	Non-monsoon (MCM)
1.	2001–02	6.16	8.53
2.	2002–03	6.19	8.58
3.	2003–04	6.29	8.71
4.	2004–05	6.38	8.84
5.	2005–06	6.48	8.97
6.	2006–07	6.57	9.10
7.	2007–08	6.66	9.23
8.	2008–09	6.76	9.36
9.	2009–10	6.85	9.50

as it is considered that there shall be no distinction between the water demand for urban and rural population. There are 197 villages located in the four blocks falling in the Ur River watershed and the total population inside the watershed is 3,660,000 as per the 2011 census records. The population in the villages falling inside the study area in Jatara, Palera, Tikamgarh and Baldevgarh blocks are 195,488, 10,993, 59,498 and 100,021, respectively, based on the 2011 census. Also the population data pertaining to the study area is available based on the 2001 census. A linear rate of increase in population has been assumed between 2001 and 2011 to estimate the population of the intermediate years. The domestic water demand in the Ur River watershed is given in Table 3. A rate of 135 l/capita/day is taken into account for the improved living standards in future when piped water supply systems and flushing toilets will ultimately be available in rural household too. The main source of livelihood of the local population other than agriculture is dairy farming and therefore considerable livestock population exists in the watershed. These livestock too requires water for drinking and its upkeep. The livestock water use in the villages inside the study area has been calculated at the rate of 40 l/capita/day. The livestock population in the villages falling inside the study area in Jatara, Palera, Tikamgarh and Baldevgarh blocks are 121,602, 12,590, 103,775 and 113,262 respectively based on the 2007 census. The total water required for livestock varies between 2.14 MCM in monsoon season and 2.96 MCM in non-monsoon season.

System Outflows

Surface Runoff

The surface water outflow from the system has been estimated by the Soil Conservation Curve Number (SCS-CN) method at the outlet of the watershed. The SCS model is based on the single parameter curve number which depends on the land use, land cover, soil type and the antecedent moisture conditions prevailing in the watershed. The land use information of the watershed has been extracted from the satellite digital data. IRS LISS-III and Aster satellite data covering the entire watershed area has been used for the land use classification. The digital image processing of the satellite data was performed using the Maximum Likelihood Classification (MLC) to prepare the land use and topography map of the watershed. Ten land use classes have been identified in the watershed based on the digital classification verified thereafter by ground truth survey. The land use classes include agriculture, barren, open forest, dense forests, water bodies and built-up areas. The surface runoff by the SCS-CN method has been computed and given in Table 3.

Groundwater Outflow/Inflow

The groundwater levels are being monitored from a network of observation wells located inside and surrounding the watershed. For computation purpose, the boundary of the watershed has been divided into a number of small reaches so that each reach has an observation well within and outside the watershed. The groundwater outflow from the watershed was then determined for each reach by using the gradient of water table and transmissivity of the aquifer system. The reach-wise outflows were then added to arrive at the total groundwater outflow and inflow taking place from the boundaries of the watershed as given in Table 4. The groundwater outflow takes mostly from the downstream portions of the watershed both in the monsoon as well as the non-monsoon season. The groundwater outflow from the watershed has been computed as 0.034 MCM during the monsoon season and 0.045 MCM during the non-monsoon season. At the upstream reaches of the watershed based on the water table gradients, the groundwater inflow takes place in the watershed both during the monsoon season (0.068 MCM) and the non-monsoon season (0.081 MCM).

Table 4 Seasonal runoff from Ur River watershed

S. No.	Year	Monsoon (MCM)	Non-monsoon (MCM)
1.	1999–00	300.55	0.00
2.	2000–01	44.65	0.00
3.	2001–02	20.53	0.00
4.	2002–03	108.52	0.00
5.	2003–04	205.90	0.00
6.	2004–05	76.65	0.00
7.	2005–06	63.24	0.00
8.	2006–07	22.01	0.00
9.	2007–08	0.04	0.00
10.	2008–09	313.02	0.00
11.	2009–10	76.55	0.00

Change in Storage in the System

The change in storages from the system includes the change in surface water storage and change in groundwater storage. The change in surface storages includes the difference in storages between at the start and end of the particular season in all the ponds and tanks located in the watershed. This component could not be directly evaluated due to the lack of data pertaining to the water levels and elevation-area-capacity of these tanks. The change in groundwater storage includes difference between the groundwater storages at the start and end of the particular season. The change in groundwater storage has been evaluated by considering the difference between the groundwater levels at the start and end of the season and specific yield for the various types of formations in the watershed. The change in groundwater storage in the monsoon season has been considered by taking the difference in groundwater level between monsoon season groundwater level and preceding non-monsoon season groundwater level and multiplying with the specific yield and area of the watershed. Similarly the groundwater storage in the non-monsoon season has been considered by taking the difference in groundwater level between monsoon season groundwater level and following year pre-monsoon groundwater level and multiplying it with the specific yield and area of the watershed. The change in groundwater storage during the period of analysis from 1999–00 to 2009–10 is given in Table 5. The change in groundwater storage varies between 461.04 MCM in 1999–00 to 49.53 MCM in 2007–08 during the monsoon season whereas it varies between −400.23 MCM in 1999–00 to −92.76 MCM during the non-monsoon season.

Water Budget

The water budget of the Ur River watershed has been carried out on a seasonal time scale for two seasons namely, monsoon season (June–October) and non-monsoon season (November–May). The water budget during the monsoon season yields an estimate of the major components and also helps to identify those components which can be utilized more effectively to conserve the precious water resources within the watershed. The water budget computations for 2002–03 are given in Table 6. It is observed from the analysis that during the period of analysis, there have been many below normal rainfall years, viz. 2000–01, 2001–02, 2002–03, 2004–05, 2005–06, 2006–07, 2007–08, 2009–10. This indicates that during the

Table 5 Groundwater outflows and inflows (2005–06)

S. No.	Outflows/inflows	Monsoon (MCM)	Non-monsoon (MCM)
1.	Groundwater outflow	0.034	0.045
2.	Groundwater inflow	(0.068)	(0.081)

Table 6 Change in groundwater storage

S. No.	Year	Rainfall (MCM)	Monsoon (MCM)	Non-monsoon (MCM)
1.	1999–00	1211.65	461.04	−400.231
2.	2000–01	633.36	250.18	−324.89
3.	2001–02	541.34	383.743	−424.310
4.	2002–03	627.01	289.24	−255.07
5.	2003–04	1034.72	352.19	−314.47
6.	2004–05	701.39	305.3	−333.2
7.	2005–06	623.66	295.9	−318.6
8.	2006–07	520.23	179.88	−192.99
9.	2007–08	344.26	49.53	−92.76
10.	2008–09	1195.75	304.53	−319.85
11.	2009–10	668.48	298.51	−315.97

Table 7 Water budget of Ur River watershed (2002–03)

Water balance components	Monsoon	Non-monsoon
Inflows		
Rainfall	627.01	25.54
Groundwater inflow	10.326	13.847
Outflows		
Domestic demands	6.192	8.711
Livestock demands	1.736	2.405
Agriculture demands	43.90	160.24
Surface runoff	108.52	0.00
Forested areas	126.58	149.64
Evaporation from tanks	1.05	1.57
Groundwater outflow	3.705	5.12
Change in storage		
Change in storage (GW)	289.24	−255.07
Unaccounted water	56.40	−33.24
Percentage error	9.00	

11 years of analysis only 3 years namely, 1999–00, 2003–04 and 2008–09 had rainfall above its normal value. This shows that the watershed has now become prone to receiving less rainfall than it used to in previous times due to many reasons including changing climate, change in topography, and large-scale unsustainable exploitation of natural resources among others (Table 7).

It is observed that sufficient runoff is generated from the basin during above normal rainfall years. The surface runoff of 300.55 MCM, 205.90 MCM and 333.02 MCM is generated for the wet years having rainfall of 1211.65 MCM, 1034.72 MCM and 1195.75 MCM during 1999–00, 2003–04 and 2008–09, respectively. The comparison of the seasonal rainfall and seasonal runoff at the outlet of the basin is given in Fig. 4. Here it is mentioned that the surface runoff has

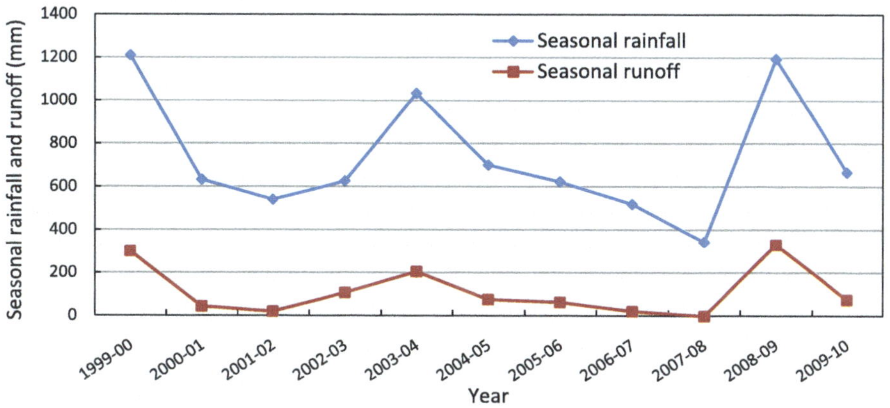

Fig. 4 Comparison of seasonal rainfall and runoff in Ur River watershed

been computed at the catchment outlet but as there are few tanks located within the basin, and some of the runoff generated within the catchment gets stored inside these water bodies, but could not be accounted for lack of data pertaining to elevation-area-capacity relationships and lake water levels as most of these water retaining structures have been constructed during the Bundela and Chandela era. As such, the surface runoff of obtained by the SCS-CN method at the outlet of the watershed also includes the quantity of water which would otherwise have been stored in the tanks. Therefore the actual runoff available at the catchment outlet will be comparatively less than the indicated figures. However, the actual amount of runoff flowing down the streams is not known as there is no discharge measurement site anywhere in the Ur River watershed and therefore the results could not be validated. Also due the lack of data pertaining to tank water levels, the seasonal change in surface water storage in these tanks could not be estimated, which is a vital component in the watershed owing to the large number of tanks, some of which are very big tanks storing considerable volume of water. Some of the big tanks do not dry up even if the monsoon in the subsequent months is late by even up to two months. The percentage error in the water balance computations is indicated by the ratio of the unaccounted water to the seasonal rainfall during the monsoon season. The error in computation includes the errors in computations of individual inflow and outflow components of the water balance equation and also includes some of the components which have not been estimated due to lack of data, viz. change in surface water storage, base flow component in the surface runoff computations. The error in computation is limited between −5.65% in 2001–02 to 32.05% in 2003–04, which can be considered reasonable, considering the complexities in the estimation of all the major water balance components.

The results can be improved subject to the estimation of change in surface water storage by installing effective monitoring mechanisms of lake water levels in the tanks along with bathymetric survey for understanding the storage characteristics of the water bodies. Similarly the base flow component in the stream flow can also be estimated by monitoring the discharges in the river systems and thereafter developing relationships for estimation of base flows.

As most of the years are below normal rainfall years, the quantification of the vital hydrological components during this period is critical for understanding the overall hydrological regime and water availability scenario in the basin. Also the detailed water balance will be beneficial for planning purposes as the computed demands are highest and available supplies are lowest during dry/low flow years. This is the worst case scenario for which appropriate development and alleviation mechanisms can be planned and implemented. Appropriate measures should be initiated to arrest the surface runoff from the watershed for multiple uses, including increasing the area under agriculture through enhanced irrigational facilities, adopting changed cropping pattern depending on the constraints on water availability, land availability, fertilizers, etc. and use of water-efficient technologies. The agricultural demands in the watershed is also limited as compared to the total rainfall availability even during the below normal rainfall years considered in the analysis. Therefore efforts should be made to increase the area under crops particularly in the fallow lands, and land without scrubs depending on the soil properties and water availability scenario.

Conclusions

The comprehensive water budgeting study performed in the Ur River watershed gives an insight into the various significant components of the hydrological cycle. The estimation of the various water budget components have been carried out by collecting relevant data based on field investigations through established techniques. The errors in the overall water balance depicts the accuracy with which the individual components have been estimated and includes some of the important components that could not be quantified due to non-availability of requisite data. However the range of the errors in computation varying between −5.65 to 32.05% can be considered to be reasonable as some of the important components including the change in surface water storage and base flow have not been estimated which otherwise would have reduced the unaccounted water component thereby reducing the error. The study reveals that rainfall which is the prime driver for all the hydrological process in the basin has been below normal in 8 of the 11 years for which the study has been conducted. The surface runoff is the one of the major outflow components along with the evapotranspiration demands of the cropped areas and forested areas. It has been observed that sufficient runoff is generated during the normal rainfall as well as wet years, but the water availability scenario changes during dry years with widespread water scarcity in the study area.

Looking into the scenario of frequent droughts in the study area, it is felt that integrated water resources management (IWRM) approach needs to be adopted for the optimal use of the natural resources in the basin. A suitable IWRM plan can be prepared to tap the surface water at suitable sites within the watershed so as to provide the much required irrigation water for bringing more areas under agriculture and creation of fisheries and agro-forestry as other sources of livelihood. The available surface water should be tapped at suitable locations within the watershed for creating additional command areas for increased agricultural production. Few sites for creating additional water storage structures can been identified in the watershed depending on the topography and possible command area. Efforts can also be made to use the surface water for creating grazing lands for the livestock population of the villages which is otherwise creating pressure on the forest resources in the watershed. Artificial recharge of groundwater can be given more thrust for recharging the confined aquifers by identifying the weathered granitic zones, for alternative source of water supply during droughts.

References

Alley WM (1984) On the treatment of evapotranspiration, soil moisture accounting and aquifer recharge in monthly water balance models. Water Resour Res 20(8):1137–1149

Alley WM (1985) Water balance models in one month ahead stream flow forecasting. Water Resour Res 21(4):597–606

Anatoliki SA, Fikos I, Ziankas G, Rizapoulou A, Famellos S (2005) Water balance estimation in the Anthemountas river basin and correlation with underground water level. Global NEST J 7 (3):354–359

Arnall NW (1992) Factors controlling the effects of climate change on river flow regimes in humid temperate environment. J Hydrol 132:321–342

Gabos A, Gasparri L (1983) Monthly runoff model for regional planning. Water Int 8:42–45

Gleick PH (1987) The development and testing of a water balance model for climate impact assessment: modelling the Sacramento basin. Water Resour Res 23(6):1049–1061

Gleik PH (1986) Methods for evaluating the regional hydrologic impacts of global climatic changes. J Hydrol 88:97–116

Hsin F, Yeh H, Shing T, Chang MH, Hsu KC, Lee CH (2007) GIS and SBF for estimating groundwater recharge of a mountainous basin in the Wu River watershed, Taiwan

Hughes DA (1982) Conceptual catchment model parameter transfer studies using monthly data from the Southern Cape Coastal lakes Region, Report 1/82, Hydrological Research Unit, Rhodes University, Graham's town

Kumar SR (2001) Water balance study of Krishnai river basin according to Thornthwaite's concept of potential evapotranspiration, CS/AR-2/2000-01

Kundzewicz ZW, Parry ML (2001) Chapter 13. In: McCarthy JJ, Canziani OF, Leary NA, Dokken DJ, White KS (eds) Climate change 2001. Impacts, adaptation, and vulnerability. Contribution of Working Group II to the third assessment report of the Intergovernmental Panel on Climate Change. Cambridge University Press, Cambridge, pp 641–692

Parry ML (2007) Climate change 2007: impacts, adaptation and vulnerability: contribution of Working Group II to the fourth assessment report of the Intergovernmental Panel on Climate Change. Cambridge University Press, UK

Roberts PJT (1978) A comparison of the performance of selected conceptual models of the rainfall runoff process in semi-arid catchments near Grahams town, Report 1/78, Hydrological Research Unit, Rhodes University, Grahams town

Schaake JKV, Duan Y (1996) Simple water balance model for estimating runoff at different spatial and temporal scale. Geophys Recharge 101(D3):7461–7475

Schaake JC, Liu C (1989) Development and application of simple water balance models to understand the relationship between climate and water resources. In: Kavvas ML (ed) New directions for surface water modelling. Proceedings of the Baltimore symposium, May 1989, IAHS Publ. No. 181

Singh RK, Prasad VH (2004) Remote sensing and GIS approach for assessment of the water balance of a watershed. Hydrol Sci J 49(1)

Tekleab S, Uhlenbrook S, Mohamed Y, Savenije HHG, Temesgen M, Wenninger J (2011) Water balance modelling of upper Blue Nile catchments using a top-down approach. Hydrol Earth Syst Sci 15:2179–2193

Thornthwaite CW (1948) An approach toward a rational classification of climate. Geogr Rev 38 (1):55–94

Thornthwaite CW, Mather JR (1955) The water balance, vol 8(8). Publications in Climatology. Laboratory of Climatology. Drexel Institute of Technology, pp 1–104

Thornthwaite CW, Mather JR (1957) Instructions and tables for computing potential evapotranspiration and the water balance, vol 10(3). Publications in Climatology, Drexel Institute of Technology, pp 185–311

Toebes C (1961) The phi-index for infiltration rates. Handbook of Hydrology, Proceedings No. 1. SCRCC, Wellington

Tuffuor S, Labadie JW (1973) A nonlinear time variant rainfall-runoff model for augmenting monthly data. Water Resour Res 19(6):1161–1166

USCB, Robert J, Galvin MD, Governor MPH (2003) Commission estimated populations in Connecticut as of 1 July 2003, State of Connecticut, Department of Public Health, Health Care Quality, Statistics, Analysis and Reporting

Van der Beken A, Byloos J (1977) A monthly water balance model including deep infiltration and canal losses. Hydrol Sci Bull 22(3):341–351

Vandewiele GL, Xu CY, NiLar W (1992) Methodology and comparative study of monthly water balance models in Belgium, China and Burma. J Hydrol 134:315–347

Xu CY, Halldin S (1996) The effect of climate change in river flow and snow cover inthe NOPEX area simulated by a simple water balance model. In: Proceedings of Nordic hydrological conference, Alkureyri, Iceland, vol 1, pp 436–445

Xu CY, Singh VP (1998) A review on monthly water balance models for water resources investigations. Water Res Manag 12(1):20–50

Xu CY, Vandewiele GL (1995) Parsimonious monthly rainfall runoff models for humid basins with different input requirements. Adv Water Resour 18:39–48

Zhang L, Dawes WR, Walker GR (2001) Response of mean annual evapotranspiration to vegetation changes at catchment scale. Water Resour Res 37(3):701–708

Analysis of Spring Discharge in the Lesser Himalayas: A Case Study of Mathamali Spring, Aglar Watershed, Uttarakhand

Vikram Kumar and Sumit Sen

Abstract In a hilly terrain, water from spring is one of the main sources of domestic water supply. There is a huge concern in relation to spring discharge, i.e. drying up or becoming seasonal in this region. For better understanding of spring discharge behaviour, this study applied various methods of recession curve and time series. Specific objectives of this study are (i) to compare recession curves using simple exponential, hyperbola and least square methods, (ii) to develop the autocorrelation of spring discharge and rainfall series and (iii) to analyse the flow duration curve using mean daily spring discharge data. A fracture- and contact-type spring located in the Aglar watershed of the Yamuna River basin has been instrumented for collection of continuous discharge data (Feb 2014–July 2015). Recession curve analysis using the method of least square found to be more accurate than exponential and hyperbola for six selected events with NSE (0.82–0.94) and RMSE (0.14–1.08). Flow duration curve analysis reveals that 90% of measured data is less than 23 lpm which can be taken as the characteristic value for minimum spring flow. Results of this study will be useful for conserving the spring discharge during the monsoon period and for various watershed management practices, such as developing a sustainable irrigation system.

Keywords Spring discharge · Recession curve analysis · Flow duration curve

Introduction

Water is a foremost need for survival of human and ecosystems, therefore sustainable management of water resources becomes major challenge in arid and semi-arid regions, which are 30% of the total earth's area (Middleton and Thomas 1997). With increasing pressure of population and climate change there is an urgent need to understand the variability and available water with their source.

V. Kumar (✉) · S. Sen
Department of Hydrology, Indian Institute of Technology Roorkee, Roorkee 247667, India
e-mail: 25.vikram@gmail.com

V. P. Singh et al. (eds.), *Water Resources Management*, Water Science and Technology Library 78, https://doi.org/10.1007/978-981-10-5711-3_22

Quantification of distribution of discharge in streams and springs has major role in water resources planning and development (Hajkowicz and Collins 2007). Especially in the data-scarce region like the Himalayas, hydrological models could be beneficial for better understanding of hydrological processes and management of water resources. In lesser Himalayas, the rocks are highly weathered and fractured which regulate the groundwater path for spring systems. These are one of the important sources of domestic water supply (Waltham 1971; Fiorillo 2009). Discharge of these springs are either drying up or reducing by deforestation in hill slope (Valdiya and Bartarya 1991). There are still limited studies in this region to understand the impact of land use/land cover change or climate variability on spring discharge. Spring discharge is varying in nature due to varying storage and recharge behaviour (Valdiya and Bartarya 1989, 1991; Negi and Joshi 1996, 2004). Spring discharge time series are often used for the understanding the hydrological process and for the characterization of the aquifer systems (Smart and Hobbs 1986; Kiraly 2002).

Discharge hydrograph is representations of the flow rate with time, which comprises rising limb, peak flow and falling limb, respectively. Recession curve is the portion of discharge hydrograph in the falling limb that originates after the peak flow. Analysing the behaviour of hydrograph during recession time of springs flow could provide information related to hydrogeology especially in fracture-type or conduit flows. Studying the hydrograph by recession curve approach is favoured over other geophysical and geological techniques (Dreiss 1982; Bakalowicz 2005).

There are factors which influence the shape of hydrograph and are classified into two categories, external and internal. Physical features, climate variability and land use/land cover which primarily control the net recharge are considered as external factors, whereas hydraulic properties of aquifer system such as hydraulic conductivity, transmissivity and aquifer thickness are internal factors. Besides the above factors rainfall characteristics such as intensity, spatio-temporal distribution and duration also influence the shape of discharge hydrograph and peak discharge. On assessment of different spring hydrograph recession curve for different springs, it is noted that the behaviour of recession curves, such as gradient and shape (i.e. recession coefficients), is mainly influenced by the rainfall intensity and geometry of fracture system (Kovacs et al. 2005). Chaulagain (2006) and UNFCCC (2007) pointed out the change in climatic conditions in terms of reduction in number of rainy days, and decrease in rainfall amount could cause severe water scarcity in Himalayas.

Discharge from springs characterizes combination of the number of hydrogeological processes that decide total amount of recharge and storage in an aquifer system (Kresic 1997). Spring discharge hydrograph analysis makes it possible to gain information about hydraulic stresses, characterization of aquifer flow, and hydraulic properties of system (Baedke and Krothe 2001; Pinault et al. 2001; Bonacci 1993). A wide number of different techniques such as graphical method, time series analysis, and spectral analysis techniques have been applied and are reviewed by White (1988) and Ford and Williams (1989) for characterizes of flow behaviour. Kresic (1997) described step-by-step review of the recession analysis technique. Discharge hydrograph of spring directly describes the physical processes occurring in an aquifer system, and spring discharge having distinct aquifer types shows alteration in its discharge hydrographs (Barnes 1939; Brutsaert and Nieber 1977; Mangin 1984; Amit et al. 2002; Malvicini et al. 2005). Therefore analysis of discharge hydrograph behaviour can be treated as a main signature for the aquifer features. Even though most of the analysis of a spring discharge hydrograph is based on Darcian theory, and the same concept is being used by Baedke and Krothe (2001), Shevenell (1996), Padilla et al. (1994), Sauter (1992), Milanovic (1981). The method of characterizing spring flow is based on the equation below, whereby the recession curves of spring hydrographs are analysed to calculate the value of α, the recession slope:

$$Q_t = Q_o \times e^{\propto \times (t - t_0)} \tag{1}$$

where

- t is any time since the beginning of the recession for which discharge is calculated,
- t_0 is the time at the beginning of the recession, usually set equal to zero,
- Q_t is spring discharge at time t,
- Q_0 is spring discharge at the start of the recession (t_0), and
- α defines the slope, or recession constant, that expresses both the storage and transmissivity properties of the aquifer.

The behaviour of recession curves obtained from spring discharge hydrograph typically reveals the features of flow rate of spring. The fast flow in an aquifer is governed by conduit inside aquifer systems along with distribution of rainfall over it whereas the low flow is largely governed by low hydraulic conductivity (Padilla et al., 1994). The model suggested by Maillet is extensively used (Mangin 1975; Soulios 1991; Vogel and Kroll 1992; Sugiyama 1996; Lastennet and Mudry 1997; Vasileva and Komatina 1997; Long and Derickson 1999), but according to (Boussinesq 1903, 1904) the general behaviour of discharge is nonlinear from aquifer systems. Extensively used exponential equation proposed by Maillet (1905) is an approximate solution; however, the equation solved by Boussinesq (1903, 1904) is in the form of quadratic and is an exact solution that describes the flow behaviour. According to Dewandel et al. (2003) solution of equation derived by Boussinesq (1903, 1904) can give quantitative information of aquifer characteristics. The solution obtained from quadratic form mostly fits the entire recession behaviour, whereas the solution by Maillet approach overestimates the duration of the 'influenced' depth of flow therefore it underestimates the storage volume of an aquifer system.

Continuous discharge series is also useful in understanding the behaviour of aquifer by plotting the flow duration curve. Estimation of flow duration curve at gauge, partial gauge and ungauged stations in Illinois are developed by Mitchell (1957). Same concept was being also applied in different other gauging locations to understand the flow behaviour (Saville et al. 1933; Cross and Bernhang 1949;

Singh 1971; Dingman 1978; Vogel 1990, etc.). With the advancement of computer technology (Fienberg 1979) found that there has been decline in relative use of this duration curve for displaying statistical information. Application of flow duration curve has been used in irrigation planning and other water resources planning (Searcy 1959; Chow 1964; Male and Ogawa 1984; Warnick 1984; Alaouze 1989). Nowadays flow duration curve have been used to validate the results of hydrologic models (Hansen et al. 1996; Ye et al. 1997). There is a hypothesis is that spring flow duration curve gives the signature of aquifer which reflects the properties of the head water, therefore it can be considered for regional studies (Yu et al. 2002).

Analysis of time series data provides an insight about hydrologic processes and quick response of aquifer systems (Duffy and Gelhar 1986). In an aquifer system, to assess the 'memory effect' (Kresic and Stevanovic 2010), autocorrelation function of the discharge time series is generally used to judge the interdependence of discharge values. Cross-correlation approach is commonly adopted to understand the behaviour of discharge and its relation with rainfall (Mangin 1984; Padilla and Pulido-Bosch 1995; Panagopoulos and Lambrakis 2006; Kovacic 2010). Application of autocorrelation and cross-correlation among daily spring discharge and daily rainfall are presented in many hydrological studies (Angelini 1997; Fiorillo and Doglioni 2010). In this study, analyses of recession curves by different approach were applied to eleven selected major events which include the event from monsoon and non-monsoon period. Flow duration curve and autocorrelation analyses of time series were applied to mean daily spring discharge and rainfall which is being monitored for a hilly terrain spring system located in Aglar watershed.

Study Area

Uttarakhand is a mountainous state of India. The northern part of these mountains is known for its snow covered peaks, rivers and valley, highly rich flora and fauna, various ecosystems and highly varied topography with few coverings of dusty plains in the southern part of the state. Uttarakhand is well gifted with water resources and forest. High snowfall and perennial rivers in the catchments act as the lifeline for the complete hydrological cycle. In this state more than 60% of an inhabitant totally depends on agriculture for their living and their source of income is predominantly dependent on this mountain agriculture. A large part of the land in the State (90%) is hilly with steep gradients and loose soils that lead to frequent erosion of soil during the high intensity precipitation and particularly during the monsoon seasons.

The Mathamali spring is complex, which lies in the Valley of Aglar watershed, western lesser Himalayas, Uttarakhand, and it is the only perennial spring available in the vicinity for the domestic water supply. Aglar is sub-catchment for Yamuna basin and lies in the geo-coordinates N38° 25′ to N30° 25′ and E77° 58′ to E78° 18′, (Fig. 1). According to the observed mean daily discharge from

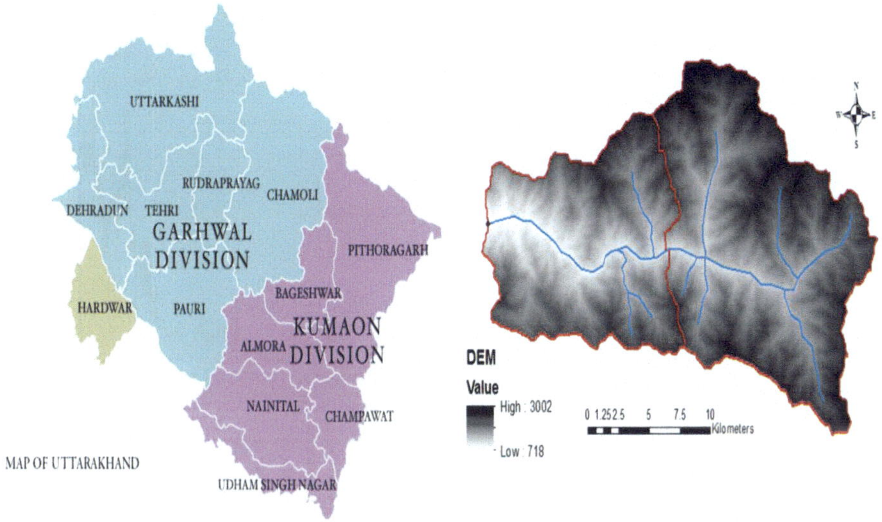

Fig. 1 Location and DEM of Aglar watershed

1 February 2014 to 18 August 2015, the Mathamali springs had daily mean average of 12.74 lpm, a maximum discharge of 45.2 lpm (2 September 2014), and a minimum of 4.7 lpm (28 March 2014). It is believed that rainfall is only primary source of recharge to the aquifer. Climate of the area is sub-tropical to temperate with the total rainfall is 1901.7 mm based on the records from 1 February 2014 to 18 August 2015. The maximum rainfall in this area occured in last 100 years is 1097 mm in September 1995.

Small catchment and steep hill slope river valley are the primary physiographic feature of the Aglar watershed, and surrounding areas of the watershed consists of mixed (forest, agriculture and barren) hilly terrain whose altitudes ranges from 450 to 3022 m above mean sea level. Agriculture field is found in small patches and is only source of income of 60% of population. The strata consist of different types of rocks such as quartzite, shale, lime stone, slate and phyllite (IIRS 1989). It is an assumption that the spring in this area arises because of unconformity in geology, where permeable strata low meet ground surface (Hao et al. 2012).

Methods of Analysis

Data Characteristics

Daily spring discharge data from February 2014 to August 2015 was collected from Mathamali spring location (Fig. 2). According to this figure, spring discharge fluctuates regularly therefore discharge from the Mathamali spring is

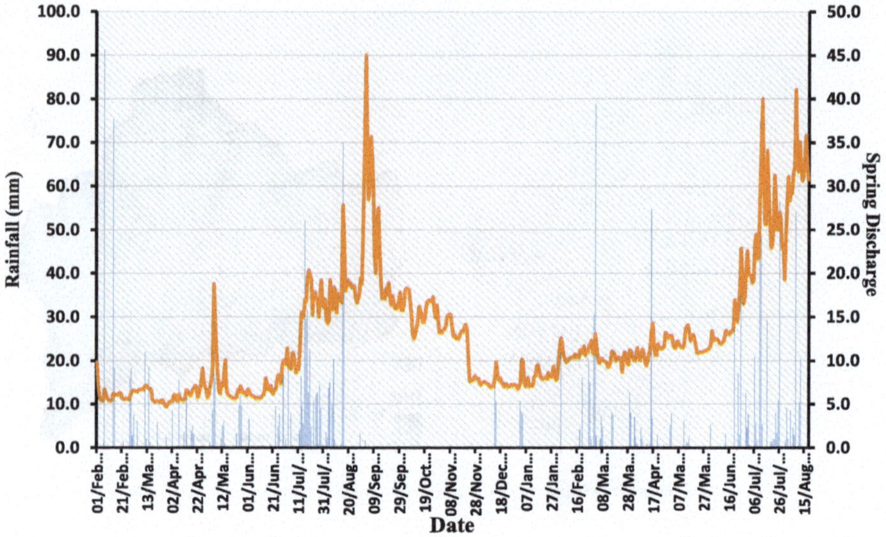

Fig. 2 Spring discharge and rainfall variation with time

non-stationary. The rainfall data was used from a tipping bucket rain gauge, which is located about 200 m from the location of spring discharge whereas spring discharge level fluctuations were measured using capacitance-based water leveler with data logger equipped with HS flume. Average annual rainfall from 2014 to 2015 is 1290 mm in this region.

Data Analysis

A number of methods have been used in past for calculating base flow recession (Toebes and Strang 1994; Hall 1968; Gilman 1977a, b; Stichler and Herrmann 1982; Youngs 1985a, b). Some of methods that have applied in this study are discuss below here.

Simple Exponential

The widely used equation for recession is perhaps exponential forms in different types (Barnes 1939; Laurenson 1961; Kunkle 1962; Knisel 1963; Toebes et al. 1969; Hall 1968; Venetis 1969; Singh and Stall 1971; Yates and Snyder 1975; Anderson and Burt 1980):

$$Q_t = Q_0 \times \exp\left(-\frac{t}{a}\right) \tag{2}$$

where Q_0 is flow at a selected time or at $t = 0$, Q is flow at unit time which is often taken as 1 day, α is constant and t is time. Above equation has been derived by Boussinesq (1877) by obtaining linear solutions and applied by Maillet (1905).

A parameter k, called recession constant or depletion factor is calculated by

$$k = \exp\left(-\frac{1}{a}\right) \tag{3}$$

The value of recession constant is always less than unity (usually more than 0.9). The constant α is known as storage delay factors and has the dimension of time. For example the constant α denotes the time required for the flow to decrease by a factor equal to e or one natural cycle as shown in below Fig. 3.

Least Squares Method

Least squares method was firstly used by James and Thompson (1970) to determine the recession constants. Alternatively Eq. (2) can be written in the form of

$$Q_t = kQ_{t-1} \tag{4}$$

k can be estimated by minimizing

$$R = \sum_{t=1}^{N} [Q(t) - kQ(t-1)]^2 \tag{5}$$

Fig. 3 Typical recession curve

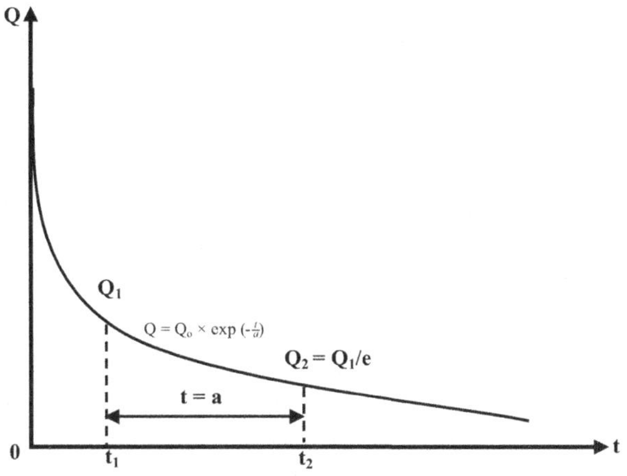

Least square estimate of k can be obtained by setting the derivative of R with respect to k equal to zero and solving for the value of k. This yields

$$k = \frac{\sum_{t=1}^{N} Q(t)Q(t-1)}{\sum_{t=1}^{N} Q(t-1)^2} \tag{6}$$

When we substitute the value $N = 1$ in Eq. (6), result is same as given by Eq. (4). If the data series are free from errors then this method is a convenient method comparatively. James and Thompson (1970) further noted that k would be more reliable if the flow sequence is longer and contained larger flows.

Hyperbola

An equation developed by Boussinesq (1904) for the horizontal impermeable lower boundary and having water table in curvilinear shape with zero water level is in the form of

$$Q = \frac{Q_0}{(1+at)^2} \tag{7}$$

where a is constant. In Europe for spring discharge Eq. (7) has been widely used (Maillet 1905; Hall 1968; Toebes et al. 1969). This equation is more suitable where if no further rainfall occurred until flow ceased. Another general equation derived by replacing 2 in Eq. (7) by an exponent, for characterizing water table recession in uniform soils drained by parallel drain lines (Young 1985a, b).

The parameter a, is calculated by

$$a = \frac{1}{t}\left[\left(\frac{Q_0}{Q}\right)^{0.5} - 1\right] \tag{8}$$

However, the value of a in this approach will be influenced by on the choice of Q.

Autocorrelation

The linear dependency of the series of continuous dataset over a given time duration is quantify by autocorrelation functions (Larocque et al. 1998) whereas the cross-correlation function quantifies the dependency of the series between the two or more continuous data series. Autocorrelation function describes the correlation of a time series with its own past and future data. Positive autocorrelation might be considered a specific form of 'persistence', a tendency for a system to remain in the same state from one observation to the next. In autocorrelation instead of using two distinct time series such as rainfall series and discharge series, the correlation can be computed for discharge series and rainfall series and the respective time series can be lagged by one or more time units. In the autocorrelation of first order, the lag will be of one time step. The first-order autocorrelation coefficient is the simple correlation coefficient of the first $N - 1$ observations, x_t, $t = 1, 2, \ldots, N - 1$ and for the next $N - 1$ observations, x_t, $t = 2, 3, \ldots, N$. The correlation between x_t and x_{t+1} is given by

$$r_t = \frac{\sum_{t=1}^{N-1}\left(x_t - \bar{x}_{(1)}\right)\left(x_{t+1} - \bar{x}_{(2)}\right)}{\sqrt{\left[\sum_{t=1}^{N-1}\left(x_t - \bar{x}_{(1)}\right)^2\sum_{t=2}^{N}\left(x_t - \bar{x}_{(2)}\right)^2\right]}} \tag{9}$$

where $\bar{x}_{(1)}$ is the mean of the first $N - 1$ observations and $\bar{x}_{(2)}$ is the mean of the last $N - 1$ observations.

Flow Duration Curve

Flow duration curve (FDC) is commonly used tools in water resources, which gives the graphical representation of frequency distribution of total flow from the watershed. It represents the percentage of duration in which discharge is equaled or exceeded for a given region. The FDC can be easily constructed for a desired gauging location at daily or monthly time series. Information from FDC can be adopted for water resource assessment which includes design of irrigation system, water supply, design of hydropower, evaluation of flow for ecological balance and many more. The spring flow duration curve is constructed using daily observed spring discharge of 564 days.

Microsoft excel is used for programming the Weibull equation to determine the spring discharge at different percentages

$$P_m = P\left[Q \geq q\left(m\right)\right] = \frac{m}{n+1} \tag{10}$$

whereas p_m is probability of exceedance of $q(m)$, $q(m)$ denotes daily observed spring discharge, m is the rank that is obtained after arranging all the discharge data in decreasing order, and n is the number of data. The X-axis of FDC represents the percentage of time that a certain discharge value is equaled or exceeded whereas Y-axis signifies the flow quantity associated with the given duration. In FDC, on X-axis zero indicates the maximum discharge value in the observed record which is condition of flooding and 100 correspond to the minimum discharge value which is condition of drought. For example, a flow duration interval of 25% related with a spring discharge of 16 lpm suggests that 25% of all observed daily average spring discharge values equal or exceed 16 lpm.

Performance Evaluation

The RMSE many times also referred by the root mean square deviation, RMSD, which is a commonly used for performance evaluation of model, measures the difference between value obtained through model and the values actually observed. The Nash–Sutcliffe model efficiency coefficient (NSE) is frequently used to assess the efficiency of hydrological models. Reliability on estimated discharge data by different approaches was calculated using mean squared error (RMSE) and Nash–Sutcliffe model efficiency (NSE) for all six events using the following equations:

$$\mathrm{RMSE} = \sqrt{\frac{\sum_{i=1}^{n} \left(X_{\mathrm{obs},i} - X_{\mathrm{model},i}\right)^2}{n}} \tag{11}$$

$$\mathrm{NSE} = 1 - \frac{\sum_{i=1}^{n} \left(X_{\mathrm{obs},i} - X_{\mathrm{model}}\right)^2}{\sum_{i=1}^{n} \left(X_{\mathrm{obs},i} - \overline{X_{\mathrm{obs}}}\right)^2} \tag{12}$$

where X_{obs} is observed values and X_{model} is modelled values at time i.

Essentially, the closer the model efficiency is to 1, the more accurate the model is.

Results and Discussions

One of the main factor which is required for the estimation of flow behaviour into the aquifer system is the recession coefficient. Therefore to understand the recession behaviour of Mathamali spring, from the observed discharge series, major six events were selected on the basis of a 10-day antecedent rainfall condition. It has been mentioned that antecedent rainfall influences the hydrograph as well as plays an important role in aquifer recharge (Fiorillo and Wilson 2004). The 10-day antecedent rainfall for event 1, 2 and 3 was 30.1, 63.0 and 104.3 mm respectively, whereas for events 4, 5 and 6 it was 19.4, 18.9 and 9 mm. For this study, the events were classified as dry, if the 10-day antecedent rainfall is less than 30 mm. Similarly, if the 10-day antecedent rainfall is greater than 30 mm, the events were in wet condition. Influence of these rainfall conditions on the recession timing and decay of spring discharge yield is summarized in (Table 1).

The behaviour of each recession curve is varying with the different approach adopted for analysing the spring discharge. The difference in these recession curve pattern for each event depends on the time gap between two rainfall events. Every recession curve analysis approach has its own applicability and sometimes is site specific. Widely adopted methods for analysing recession curve such as simple exponential, least square and hyperbola approach has applied for understanding the spring discharge behaviour. RMSE and NSE efficiency criteria were applied for all

Table 1 Summary of recession events and rainfall observed at Mathamali spring

Event	Recession duration (days)	10-day antecedent rainfall (mm)	Max. daily rainfall (mm)	Recession curve	
				Start (lpm)	Finish (lpm)
1	14–20 Mar 2014 (7)	30.1	21.9	6.9	5.4
2	14–22 May 2014 (9)	63	24.7	10.2	5.7
3	16–27 Aug 2014 (12)	104.3	70.1	23.9	17.2
4	14 Dec 2014–01 Jan 2015 (19)	19.4	19.4	10	7.1
5	04–12 Jan 2015 (9)	18.9	10.7	10	7.3
6	15–26 May 2015 (12)	9	6.4	13.2	11.1

Table 2 Comparison of model performance

Event No.	Exponential		Hyperbola		Least square method	
	RMSE	NSE	RMSE	NSE	RMSE	NSE
1	0.44	0.81	1.12	-0.24	0.19	0.9
2	0.43	0.83	1.05	-0.03	1.08	0.93
3	0.89	0.65	2.49	-1.7	0.64	0.82
4	0.37	0.52	0.67	-0.19	0.14	0.94
5	0.31	0.77	0.94	-1.01	0.28	0.82
6	0.2	0.91	1.23	-1.03	0.22	0.94

three approaches to assess the performance of the models. Performance of modelled spring discharge with observed spring discharge is summarized in Table 2.

Results indicate that the least square method, in which the recession constant k is related to the preceding flow magnitudes, provides a considerably better fit to the observed spring discharge data than that of the simple exponential and hyperbola method. However, approach other than the least square did not simulate spring discharge recession flow satisfactorily. Initial recession behaviour was overestimated whereas lower part of recession curve was underestimated.

The spring discharge at Mathamali varies from 5 to 45 lpm with a mean discharge of 13 lpm, which is almost same as that of a spring discharge measured at Pauri 12.79 lpm, (Tarafdar 2011). However, the variation between the minimum and maximum is large comparatively. The behaviour of spring discharge hydrograph is similar to the rainfall pattern (Fig. 2). The quick recession of spring discharge at Mathamali suggests that the spring might have restricted storage capacity or the aquifer system is having large discontinuities with more permeability (Stevanovic 2010). The geography of spring location is intemperate climatic conditions, where in monsoon period frequent rainfall hinders continuous recession curve of longer duration. The change in recession coefficient of spring hydrograph with different events may be due to drainage resistance changes with respect to time within the source aquifer (Van de Griend et al. 2002). On the basis of observed hydrograph shape during the recession period the hydraulic behaviour of the aquifer can be estimated during dry periods and droughts (Fiorillo et al. 2007).

Autocorrelation

For understanding the behaviour of site rainfall and spring discharge, autocorrelation of these parameters were analysed. Data of the year 2014–15 were used for the autocorrelation analysis. Figure 4 shows the autocorrelation function of the daily spring discharge and rainfall. The autocorrelation coefficient of daily spring discharge decay gradually and have a long time lag which shows a strong linear interrelationship and annual repetition behaviour of spring discharge. Behaviour of daily rainfall is entirely different from daily spring discharge. Autocorrelation coefficient of daily rainfall reaches to zero quickly, which indicates that there is a lack of dependency and have uncorrelated behaviour (Angelini 1997).

Estimated lag in the springs discharge was high with a value of more than 155 days and the uniformity in the slope indicates that spring discharge was in the base flow during the period (Russo et al. 2014). Whereas the slopes are very less in daily rainfall as compared to spring discharge and changes frequently indicates less memory effects. The autocorrelation function of Mathamali spring was very similar to that of Scirca spring in Central Italy. The correlation coefficient of Mathamali

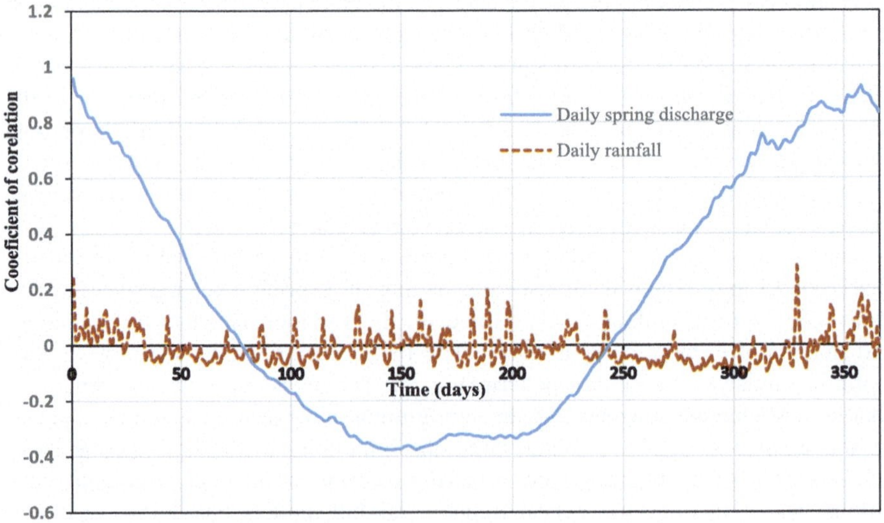

Fig. 4 Auto-correlation function of daily spring discharge and daily rainfall

spring reaches a value of 0.2 in around 60 days and zero value in 80 days which resembles with the Scirca spring having values 70 days and 89 days for correlation coefficient 0.2 and 0, respectively (Angelini 1997).

Flow Duration Curve (FDC)

Flow duration curve is constructed using daily spring discharge data which offers comprehensive way of observing flow duration characteristics of a spring or stream. An FDC is constructed by arranging the flow values in the decreasing order of magnitude, and assigning rank to each flow values. Alternative way is to plot the all ranked discharge against corresponding assigned ranks which is again expressed in percentage. Log probability plot of the discharge is common way of representing FDC (Smakhtin 2001). Figure 5 shows the flow duration curve of Mathamali spring.

The rate of change of flow behaviour can be interpreted from the slope of duration curve where flat slope at the tail of Fig. 5 indicates a slow response of the catchment to rainfall in this region. The largest volume from the spring is available during the monsoon period followed by non-monsoon period. Further analysis of the monthly cumulative spring discharge during the monsoon period (July, August and September) is about 73% of the total spring discharge. This clearly shows a reasonable seasonality in the spring discharge, which is common in such region where rainfall has significant role.

If we look at Fig. 5, the daily discharge value at 20% exceedance, its 17.1 lpm, which does not mean that discharge is 17.1 for 20% of time, but it mean that the discharge is equalled or exceeded for 20% of the time. Discharge is often expressed

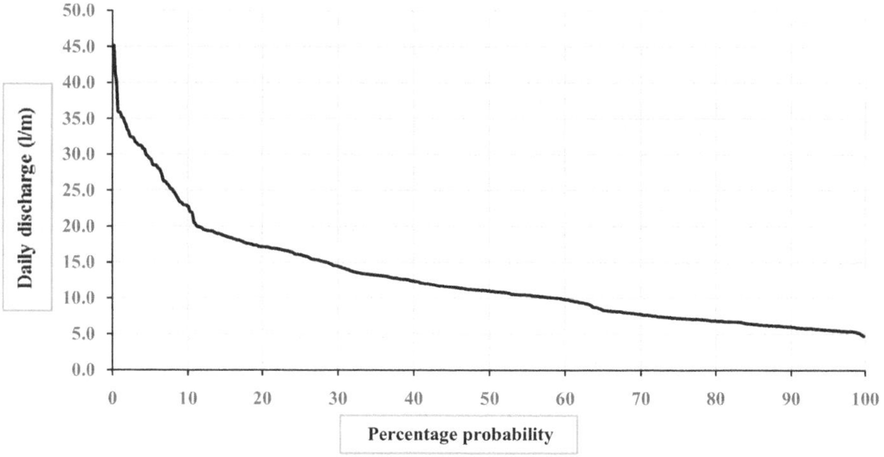

Fig. 5 Flow duration curve of Mathamali spring

as Q and the exceedance value as a subscript, so Q_{20} mean the discharge equalled or exceeded for 20% of the time. The duration curve often used to describe the nature of discharge (flashy or steady). A full year discharge data must be used to make sure that a wet winter period included without the corresponding dryer summer to balance the resulting FDC. Discharge between Q_1 and Q_{10} are considered high flow in which flow correspond to Q_1 would be extreme flood. Discharge between Q_{10} and $Q_{70\ \text{would}}$ be medium range of flow and discharge from Q_{70} to Q_{100} are the low flow and important for drought studies. From Fig. (5), it is very clear that Q_{90} of observed discharge data is less than 23 lpm which can be taken as the characteristic value for minimum spring flow. This minimum flow data can be considered while framing policy and guideline to address the drought situations for this type of aquifer system.

Conclusion

The study reveals the better suitability of available approaches for recession curve analysis for Mathamali spring discharge located in lesser Himalayan Aglar watershed. Least square method for recession curve analysis better fits the observed discharge data with NSE (0.82–0.94) and RMSE (0.14–1.08), than simple exponential and hyperbola approach. The varying recession behaviour of spring discharge may be due to the soil moisture movement, hydraulic conductivity and storage of an aquifer. The autocorrelation coefficient of Mathamali spring decays gradually and has a time lag of 155 days which shows a strong linear interrelationship and annual repetition behaviour of spring discharge. Analysed monthly cumulative spring discharge during the monsoon period (July, August and September) is about 73% of the total spring discharge. Analyses of flow duration curve reveals that 90 percentile of observed discharge data is less than 23 lpm which can be taken as the characteristic value for minimum spring flow. This study can be used for short and long-term planning to address the water resources issues in this watershed and it can also be used for comparing the spring characteristics at other mountainous watershed.

Acknowledgements The authors would like to acknowledge the Indian Institute of Technology, Roorkee, for funding under grant no. HYD/FIG/100582 towards field visits and instrumentations. The authors also wish to thank Mr. Sumer for the field support.

References

Alaouze CM (1989) Reservoir releases to uses with different reliability requirements. Water Resour Bull 25(6):1163–1168

Amit H, Lyakhovsky V, Katz A, Starinsky A, Burg A (2002) Interpretation of spring recession curves. Ground Water 40:543–551

Anderson MG, Burt TP (1980) Interpretation of recession flow. J Hydro 46:89–101
Angelini P (1997) Correlation and spectral analysis of two hydrogeological systems in Central Italy. Hydrol Sci J 42:425–438
Baedke SJ, Krothe NC (2001) Derivation of effective hydraulic parameters of a karst aquifer from discharge hydrograph analysis. Water Resour Res 37(1):13–19
Bakalowicz M (2005) Karst groundwater: a challenge for new resources. Hydrogeol J 13:148–160
Barnes BS (1939) The structure of baseflow recession curve. Trans Am Geophys Union 20: 721–725
Bonacci O (1993) Karst spring hydrographs as indicators of karst aquifers. Hydrol Sci J des Sci Hydrol 38(l):51–62
Boussinesq J (1904) Recherches theoriques sur lecoulement des nappes deau infiltrees dans le sol et sur le debit des sources. J Math Pure Appl 10:5–78
Boussinesq J (1877) Essai sur la throrie des eaux courantes. *M&noires prrsentrs par divers savants a lAcademie des Sciences de lInstitut National de France*, Tome XXIII, No. 1
Boussinesq J (1903) Surun mode simple decoulement des nappes deau dinfiltration a lit horizontal, avec rebord vertical tout autour lorsquune partie de ce rebord est enlevee depuis lasurface jusquaufond. C R Acad Sci 137:5–11
Brutsaert W, Nieber JL (1977) Regionalized drought flow hydrographs from a mature glaciated plateau. Water Resour Res 34:233–240
Chaulagain NP (2006) Impacts of climate change on water resources of Nepal: the physical and socioeconomic dimensions. Unpublished Ph.D. Thesis, International Institute of Management, Energy and Environment Management, University of Flensburg, Germany
Chow VT (1964) Handbook of applied hydrology. Mc-Graw Hill Book Co., New York
Cross WP, Bernhagen RJ (1949) Flow duration. Ohio streamflow characteristics, Bull. 10, part 1, Ohio Dept. of Natural Resour., Div. of Water
Dewandel B, Lachassagneb P, Bakalowicz M, Wengb P, Al-Malki A (2003) Evaluation of aquifer thickness by analysing recession hydrographs. Application to the evaluation Oman ophiolite hardrock aquifer. J Hydrol 274:248–269. doi:10.1016/S0022-1694(02)00418-3
Dingman SL (1978) Synthesis of flow-duration curves for unregulated streams in New Hampshire. Water Resour Bull 14(6):1481–1502
Dreiss SJ (1982) Linear kernel for karst aquifer. Water Resour Res 18:865–876
Duffy CJ, Gelhar LW (1986) A frequency domain analysis of groundwater quality fluctuations: interpretation of field data. Water Resour Res 22(7):1115–1128
Fennessey NM, Vogel RM (1990) Regional flow duration curves for ungaged sites in Massachusetts. Water Resour PIng Mgmt ASCE 116(4):530–549
Fienberg SE (1979) Graphical methods in statistics. Am Statistician 33(4):165–178
Fiorillo F, Doglioni A (2010) The relation between karst spring discharge and rainfall by cross-correlation analysis (Campania, southern Italy). Hydrogeol J 18(8):1881–1895. doi:10.1007/s10040-010-0666-1
Fiorillo F, Esposito L, Guadagno FM (2007) Analyses and forecast of the water resource in an ultra-centenarian spring discharge series from Serino (Southern Italy). J Hydrol 36:125–138
Fiorillo F, Wilson RC (2004) Rainfall induced debris flows in pyroclastic deposits, Campania (southern Italy). Eng Geol 75(3):263–289
Ford Derek, Williams PW (1989) Karst Geomorphology and Hydrology. Unwin Hyman, London, p 601
Foster HA (1934) Duration curves. Trans ASCE 99:1213–1267
Francesco Fiorillo (2009) Spring hydrographs as indicators of droughts in a karst environment. J Hydrol 373:290–301
Gilman K (1977a) Dilution gauging on the recession limb: 1. Constant rate injection method. Hydrol Sci Bull 22:353–369
Gilman K (1977b) Dilution gauging on the recession limb: 2. The integration method. Hydrol Sci Bull 22(4):469–481
Hajkowicz S, Collins K (2007) A review of multiple criteria analysis for water resource planning and management. Water Resour Manag 21:1553–1566

Hall FR (1968) Base flow recessions—a review. Water Resour Res 4(5):973–983
Hansen DP, Ye W, Jakeman AJ, Cooke R, Sharma P (1996) Analysis of the effect of rainfall and streamflow data quality and catchment dynamics on streamflow prediction using the rainfall-runoff model IHACRES. Environ Softw 11(1–3):193–202
Hao Y, Liu G, Li H, Li Z, Zhao J, Yeh TJ (2012) Investigation of karstic hydrological processes of Niangziguan Springs (North China) using wavelet analysis. Hydrol Process 26(20):3062–3069
IIRS (1989) Geohydromorphology, soil, and vegetation land use mapping of Aglar watershed (U. P.) using remote sensing techniques
James LD, Thompson WO (1970) Least squares estimation of constants in a linear recession model. Water Resour Res 6(4):1062–1069
Kiraly L (2002) Karstification and Groundwater Flow. In: Gabrovsek F (ed) Proceedings of the Conference on Evolution of Karst: from Prekarst to Cessation. Postojna-Ljubljana, pp 155–190
Knisel WG (1963) Baseflow recession analysis for comparison of drainage basin and geology. J Geophys Res 68:3649–3653
Kovacic G (2010) Hydrogeological study of the Malenscica karst spring (SW Slovenia) by means of a time series analysis. Acta Carsologica 39(2):201–215
Kovacs A, Perrochet P, Kiraly L, Jeannin P (2005) A quantitative method for characterisation of karst aquifers based on the spring hydrograph analysis. J Hydrol 303:152–164
Kresic N, Stevanovic Z (2010) Groundwater hydrology of springs: engineering, theory, management, and sustainability. Elsevier
Kresic N (1997) Quantitative solutions in hydrogeology and groundwater modeling: Boca Raton. CRC Lewis Publishers, 461 p
Kunkle GR (1962) The baseflow duration curve, a technique for the study of groundwater discharge from a drainage basin. J Geophys Res 67:1543–1554
Larocque M, Mangin A, Razack M, Banton O (1998) Contribution of correlation and spectral analyses to the regional study of a karst aquifer (Charente, France). J Hydrol 205:217–231
Lastennet R, Mudry J (1997) Role of karstification and rainfall in the behaviour of a heterogeneous karst system. Environ Geol 32:114–123
Laurenson BE (1961) A study of hydrograph recession curves of an experimental catchment. J Inst Eng Aust 33:253–258
Long AJ, Derickson RG (1999) Linear system analysis in a karst aquifer. J Hydrol 219:206–217
Maillet E (1905) Essai d'hydraulique souterraine et fluviale. Libraire Sci. A. Herman, Paris
Male JW, Ogawa H (1984) Tradeoffs in water quality management. Water Resour Plng Mgmt 110 (4):434–444
Malvicini CF, Tammo SS, Walter MT, Parlange J, Walter MF (2005) Evaluation of spring flow in the uplands of Matalom, Leyte, Philippines. Adv Water Resour 28:1083–1090. doi:10.1016/j.advwatres.2004.12.006
Mangin A (1984) Pour une meilleure connaissance des systemes hydrologiques a partir des analyses correlatoire et spectrale. J Hydrol 67:25–43. doi:10.1016/0022-1694(84)90230-0
Mangin A (1975) Contribution a letude hydrodynamique des aquiferes karstiques. Ph.D. These, Universitéde Dijon, p 124
Middleton N, Thomas D (eds) (1997) World Atlas of desertification, 2nd edn. Arnold, London
Milanovic PT (1981) Karst hydrogeology. Water Resources Publications, Littleton, p 434
Mitchell WD (1957) Flow duration curves of Illinois streams. Illinois Department of Public Works and Buildings, Division of Waterways
Negi GCS, Joshi V (2004) Rainfall and spring discharge patterns in two small drainage catchments in the Western Himalayan Mountains, India. Environmentalist 24:19–28
Negi GCS, Joshi V (1996) Geo-hydrology of springs in a mountain watershed: the need for problem solving research. Curr Sci 71(10):772–776
Padilla A, Pulido-Bosch A (1995) Study of hydrographs of karstic aquifers by means of correlation and cross-spectral analysis. J Hydrol 168:73–89
Padilla A, Pulido-Bosh A, Mangin A (1994) Relative importance of base flow and quick flow from hydrographs of karst spring. Ground Water 32:267–277

Panagopoulos G, Lambrakis N (2006) The contribution of time series analysis to the study of the hydrodynamic characteristics of the karst systems: application on two typical karst aquifers of Greece (Trifilia, Almyros Crete). J Hydrol 329:368–376. doi:10.1016/j.jhydrol.2006.02.023
Pinault JL, Plagnes V, Aquilina L, Bakalowicz M (2001) Inverse modeling of the hydrological and hydrochemical behaviour of hydrosystems: characterization of karst system functioning. Water Resour Res 37(8):2191–2204
Russo SL et al. (2014) Recession hydrographs and time series analysis of springs monitoring data: application on porous and shallow aquifers in mountain areas (Aosta Valley). Environ Earth Sci 73(11):7415–7434 (2014)
Sauter M (1992) Assessment of hydraulic conductivity in a karst aquifer at local and regional scale. Ground Water Manage 10:39–57
Saville T, Watson JD (1933) An investigation of flow-duration characteristics of North Carolina streams. Am Geophys Union Trans, pp 406–425
Searcy JK (1959) Flow duration curves. *Water supply paper* 1542-A, U.S. Geological Survey, Reston, Virginia
Shevenell L (1996) Analysis of well hydrographs in a karst aquifer: estimates of specific yields and continuum transmissivities. J Hydrol 174:331–355
Singh KP (1971) Model flow duration and streamflow variability. Water Resour Res 7(4):1031–1036
Singh KP, Stall JB (1971) Derivation of base flow recession curves and parameters. Water Resour Res 7(2):292–303
Smakhtin VU (2001) Low flow hydrology: a review. J Hydrol 240(3):147–186
Smart PL, Hobbs SL (1986) Characterization of carbonate aquifers: a conceptual base. In: Proceedings of the environmental problems in Karst Terranes and their solutions conference, Bowling Green, KY, pp 1–14
Soulios G (1991) Contribution à lètude des coubes de recession des souces karstiques: example du pays hellènique. J Hydrol 124:29–42
Stevanovic Z (2010) Utilization and regulation of springs. In: Kresic N, Stevanovic Z (eds) Groundwater hydrology of springs: engineering, theory, management, and sustainability. Elsevier, Amsterdam, pp 339–388
Stitchler W, Hermann A (1982) Environmental isotopes as tracers in water balance studies of mountainous watersheds. In: Proceedings Berne Symposium on Hydrological Research Basins, vol 2, pp 357–368
Sugiyama H (1996) Analysis and extraction of low flow recession characteristics. Water Resour Bull 32:491–497
Tarafdar S (2011) Understanding the dynamics of high and low spring flow: a key to managing the water resources in a small urbanized hill slope of Lesser Himalaya, India. Environ Earth Sci 70:2107–2114. doi:10.1007/s12665-011-1493-y
Toebes C, Morrissey WB, Shorter R, Hendy M (1969) Base-flow-recession curves. In: Handbook of hydrological procedures. Proc. 8, Ministry of Works, Wellington, New Zealand
Toebes C, Strang DD (1994) On recession curves, 1. J Hydrol, NZ, pp 2–15
UNFCCC (2007) Climate change: impacts, vulnerabilities and adaptations in developing countries. Climate Change Secretariat (UNFCCC), Bonn, Germany
Valdiya KS, Bartarya SK (1989) Diminishing discharges of mountain springs in a part of Kumaun Himalaya. Current Sci 58(8):417–426. ISSN 0011-3891
Valdiya KS, Bartarya SK (1991) Hydrological studies of springs in the catchment of the Gaula River, Kumaun Lesser Himalaya, India. Mt Res Dev 11(3):239–258
Van de Griend AA, De Vries J, Seyhan E (2002) Groundwater discharge from areas with variable specific drainage resistance. J Hydrol 259:203–220
Vasileva D, Komatina M (1997) A contribution of the alpha recession coefficient investigation in karts terrains. Theor Appl Karstol 10:45–54
Venetis C (1969) A study of the recession of unconfined aquifers. Bull Int Assoc Sci Hydrol 14 (4):119–125

Vogel RM, Kroll CN (1992) Regional geohydrologic-geomorphic relationships for the estimation of low-flow statistics. Water Resour Res 28:2451–2458

Waltham AC (1971) Caving in Himalaya. Himalayan J 31:3

Warnick CC (1984) Hydropower engineering. Prentice-Hall Inc., Englewood Cliffs, New Jersey, pp 59–73

White WB (1988) Geomorphology and Hydrology of karst terrains. New York

Yates P, Snyder WM (1975) Predicting recessions through convolution. Water Resour Res 11 (3):418–422

Ye W, Hansen DP, Jakeman AJ, Sharma P, Cooke R (1997) Assessing the natural variability of runoff: Clarence Basin catchments, NSW, Australia. Math Comput Simul 43(3–6):251–260

Youngs EG (1985a) Characterization of hydrograph recession of land drains. J Hyd 82:17–25

Youngs EG (1985b) A simple equation for predicting water-table drawdowns. J Agr Rngg Res 31:28–321

Yu P, Yang T, Liu CA (2002) Regional model of low flow for Southern Taiwan. Hydrol Processes 16(10):2017–2034

Submerged Geotextile Sand Containers for Coastal Defence

Kiran G. Shirlal and Ramesh Reddy Mallidi

Abstract Geotextile sand containers (GSCs) used as submerged reefs are increasingly being incorporated into coastal defence solutions because of they are and eco-friendly and cost effective. These cause wave-breaking and reduce the wave action on the lee side. In the present experimental study, trapezoidal submerged reefs of 0.25 m height are constructed with GSCs weighing about 485 g placed in different alignments, like Perpendicular, Parallel and Flemish, at a uniform slope of 1V:2H. These models are tested in water depth of 0.3 m with wave heights of 0.1 and 0.12 m and different wave periods varying from 1.5 to 2.5 s. During the investigation, damage of the submerged breakwater and wave transmission are measured. After the tests, it is found that though the structure is damaged, it is still effective in inducing wave-breaking and energy dissipation resulting in effective damping of waves. The wave transmission coefficient K_t varied between 0.65 and 0.9.

Introduction

A breakwater, with its crest at or below SWL, is a submerged breakwater and is used for creation of a protected area of water or for protection of coastal structures and beaches from damage of erosion caused by wave action.

Since the crest of the structure is at lower elevation, wave overtopping and/or breaking takes place due to which armour units are subjected to relatively smaller wave force thus the structure can be economically constructed with relatively smaller sized armour. Johnson et al. (1951) studied the wave damping action of submerged breakwater and recognized that wave steepness (H/L), crest width (B), structure height (h) and water depth (d) are important parameters. They concluded that for a given height of the structure, steeper waves are effectively damped than

K. G. Shirlal (✉) · R. R. Mallidi
Department of Applied Mechanics and Hydraulics, National Institute of Technology Karnataka, Surathkal, Mangaluru 575025, India
e-mail: kshirlal@gmail.com

V. P. Singh et al. (eds.), *Water Resources Management*, Water Science and Technology Library 78, https://doi.org/10.1007/978-981-10-5711-3_23

flat waves due to breaking of waves. Submerged breakwater induces wave breaking causing great turbulence on lee side. Current and turbulence together on lee side of submerged breakwater have a strong power of erosion on a sandy bottom and can thus prevent siltation. They also offer resistance through friction and turbulence created by breakwater interference in wave field causing maximum wave damping and energy dissipation, minimum wave reflection and bottom scour, and maximum sand trapping efficiency and are used for coastal protection (Baba 1985; Pilarczyk and Zeidler 1996).

Apart from conventional breakwaters other types of obstructions such as vertical/inclined barriers have also been investigated for managing steep waves and reduce their influence on the leeside (Rao et al. 2009). The authors conducted 1:30 scale model studies on submerged thin plate barriers inclined at various angles for regular wave heights ranging from 0.05 to 0.15 m of periods 1–2.2 s and water depths varying from 0.3 to 0.5 m. It was concluded that inclined plates at 600 angle were effective enough to provide the value of $K_t < 0.6$ for an entire range of wave parameters studied.

Fridman et al. (2010) authors discuss an experimental study on the impact of single and double submerged barriers on the propagation of a solitary wave package with the characteristic horizontal scale exceeding the water layer depth by a factor of 10–30. The study was undertaken in a 5 m long and 0.105 m deep basin with single and double vertical barriers. The height of the barriers and the distance between them were varied during the experiments, while keeping the position of the first barrier fixed with respect to the wave generator (1.02 m). The ratio of the barrier height (h) to the water depth (d) was varied from 0.3 to 1.2. It was concluded that showed that double barriers were more efficient than single barriers of the same size in reducing the tsunami runup, a minimum runup existed for a particular distance between two submerged barriers and as the barrier height increased, the relative amplitude of the minimum runup decreased.

Kriebel and Bollmann (1996) describe the small-scale experiments on wave transmission at vertical wave barriers with varying wall penetration w/d of 0.4–0.7 where, w is the depth of vertical wall penetration in a water depth d. The results were compared with three theories. The theories for predicting regular wave transmission past vertical wave barriers are power transmission theory, modified power transmission theory that includes effects of wave reflection and the Eigen function expansion theory. The tests were conducted in a wave tank 36.6 m long, 2.43 and 1.52 m deep. A thin rigid wall 0.05 m thick was placed about 18.3 m from the wave maker. The wave periods varied from 0.9 to 2.5 s and wave heights ranged from 0.025 to 0.23 m. The wave steepness, H/L, varied between 0.01 and 0.06 with most tests being in the range of 0.02–0.04. It was observed that transmission coefficient K_t decreased with increase in wall penetration w/d. None of the theories considered accurately predict K_t.

Geotextile sand container (GSC) structures have clear advantages compared to other breakwater structures that environmentally damaging quarrying and transporting of rock are not required and this structure can be easily dismantled in the case of adverse and unforeseen situations. With the expected worsening of coastal erosion, a soft, and inexpensive solution like GSC structures will be a strong contender to replace a more conventional hard engineering solution as coastal communities which are highly developed will be severely affected. Several types of materials and container systems have been developed specifically for the designing of coastal erosion protection systems (Pilarczyk 1996) and for beneficial uses of dredged material. Such applications have made engineering construction eco-friendly, highly durable and economical while allowing the speedier construction. The ability to use local material and labour makes construction easier and faster and possible in isolated areas. The use of GSC structures allows for a more flexible and adaptive approach that can be more easily modified if the desired outcome is not satisfactory or if the design conditions change. GSC structures can even be easily removed in the event that the GSC structure does not perform.

Geotextiles have been widely used in coastal applications (Oh and Shin 2006). The concept of employing sand-filled containers of various sizes and shapes for erosion control has been in existence for centuries. The initial evolution of sand-filled containers for coastal erosion control has occurred over the past 400 years, since the Dutch first employed small sand-filled containers to shore up their dike and dam structures in the seventeenth century. The myriad of enormous dikes surrounding Holland provide continuous protection to a nation which is sited largely below sea level (Harris and Sample 2009).

Sand-filled containers have been utilized to construct a variety of traditional coastal erosion control structures, including both shore-perpendicular and shore-parallel structures. These systems have included jetties, groins, vertical seawalls, sloped revetments, breakwaters and sill structures with varying degrees of success. Due to the lower costs and less need for heavy equipment, sand-filled containers have been used in developing countries (Harris and Zadikoff 1999). Geotextiles have several applications those could be especially important for coastal management programs. They are environmental restoration, flood prevention and erosion control (CPRA 2012). Extensive studies have been undertaken on the hydraulic stability of single geotextile tubes (Pilarczyk 2000; Recio and Oumeraci 2009; Dassanayake and Oumeraci 2013) and these have been used in coastal protection works (Kudale 2013, 2015; Sundar 2013; Sundar and Sannasiraj 2013). However, current understanding of hydraulic stability and wave transmission of stacked GSC structures is limited and calls for further investigations (Ramesh 2014). The performance study of such a GSC structure like a submerged reef could be both an innovative and low-cost solution for various coastal projects is the real motivator for the present study.

Objectives

The objectives of the present physical model study are to examine GSC armour stability and wave transmission of submerged reef constructed with GSC armour arranged in different alignments like Parallel, Perpendicular and Flemish.

Experimental Details

Wave Flume

The wave flume of Marine Structures Laboratory of the Department of Applied Mechanics and Hydraulics, National Institute of Technology Karnataka, Surathkal which generates monochromatic waves is used to test the physical models of the GSC reefs. Figure 1 gives a schematic diagram of the experimental setup.

The wave flume is 50 m long, 0.71 m wide and 1.1 m deep. About 15 m length of the flume is provided with a smooth bed and side walls and also glass panels on one side to facilitate photography of test models. This reduces wave reflection significantly. It has a 41.5 m long smooth concrete bed. Gradual transition is provided between normal bed level of the channel and that of wave generating chamber by a ramp. The flume has a 6.3 m long, 1.5 m wide and 1.4 m deep chamber with a the bottom hinged flap at one end which generates waves. The wave filter consists of a series of vertical asbestos cement sheets spaced at about 0.1 m centre to centre and parallel to length of the flume. A fly-wheel and bar chain link the motor with the flap. By changing the eccentricity of bar chain on the fly-wheel, the wave height can be varied for a particular wave period.

The changing of frequency through inverter, waves of desired wave period can be generated. The flap is controlled by an induction motor of 11 kW power at 1450 rpm. This motor is regulated by an inventor drive (0–50 Hz), rotating in a speed range of 0–155 rpm. In this flume, monochromatic waves of 0.08–0.24 m heights and periods of 0.8–4.0 s in a maximum water depth of 0.5 m can be generated. Wave reflection from the structure does not interfere with freshly generated incident waves. In all tests, a series of five waves of uniform height were generated and measurements were limited to exclude any effects of wave reflection from the structure, beach or from the wave maker. Therefore, any kind of wave reflection was not considered in this study.

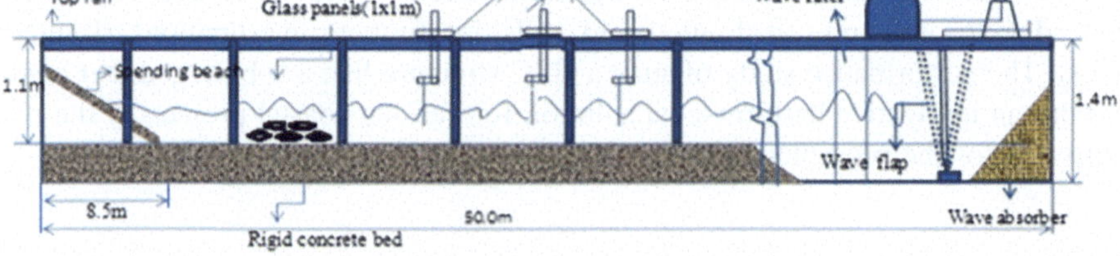

Fig. 1 Schematic diagram of experimental set up with GSC reef model

Instrumentation

The capacitance type wave probes along with amplification units are used for data acquisition. Three such probes are used during the experimental work for acquiring incident wave heights (*H*) as shown in Fig. 1. The spacing between probes is adjusted approximately to one-third of the wave length to ensure accuracy (Isaacson 1991). During experimentation, signals from wave probes are verified online through a software using the earlier calibration data stored and recorded by the computer through the data acquisition system. These are then processed to extract the incident waves using another software.

Test Models

The GSCs were made with a non-woven polypropylene polymer material. The properties of this material are conforming to PIANC (2011), and are not mentioned here according to the supplier's request. The sand used as fill material in the GSCs is beach sand from NITK beach at Surathkal with a median grain size of 0.45 mm. Each GSC unit was 0.16 m long, 0.08 m wide and 0.035 m high and weighed about 485 g when completely filled with sand. 1:30 scale models of the submerged reef were constructed manually with GSCs at a distance of 3 m from the spending beach on the flume bed. See Fig. 1. The GSCs were arranged in trapezoidal manner, layer wise with a slope of 1V to 2H to a height (*h*) 0.25 m and varying crest widths (*B*). In the models, GSCs were aligned in four different alignments with respect to direction of wave propagation namely Parallel (single layer), Parallel with two layers, Flemish and Perpendicular as shown in Fig. 2.

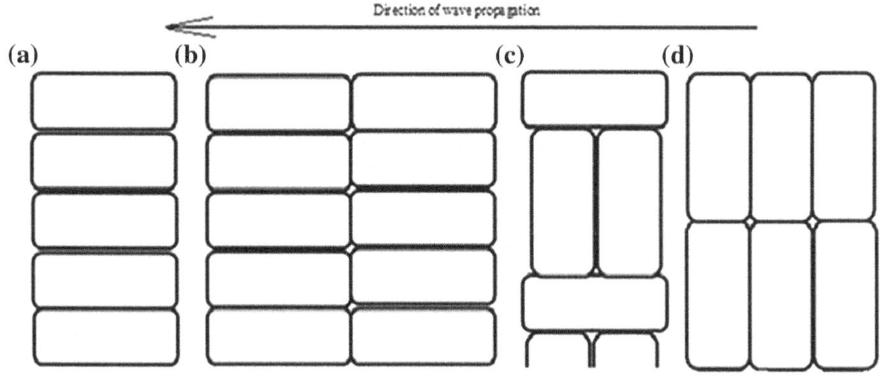

Fig. 2 GSC alignments **a** parallel (single layer), **b** parallel double layer, **c** Flemish, **d** perpendicular

Test Procedure

The models designed as submerged GSC reefs are tested for the stability of their armour (GSC) units and wave transmission while subjected to varying wave height and wave period in a water depth of 0.3 m in a two-dimensional wave flume. The incident wave height, wave transmission and movement of GSC units, i.e. damage of the reef of constant reef crest height are recorded along with number waves passing during physical model investigation. The test methodology is shown in Fig. 3.

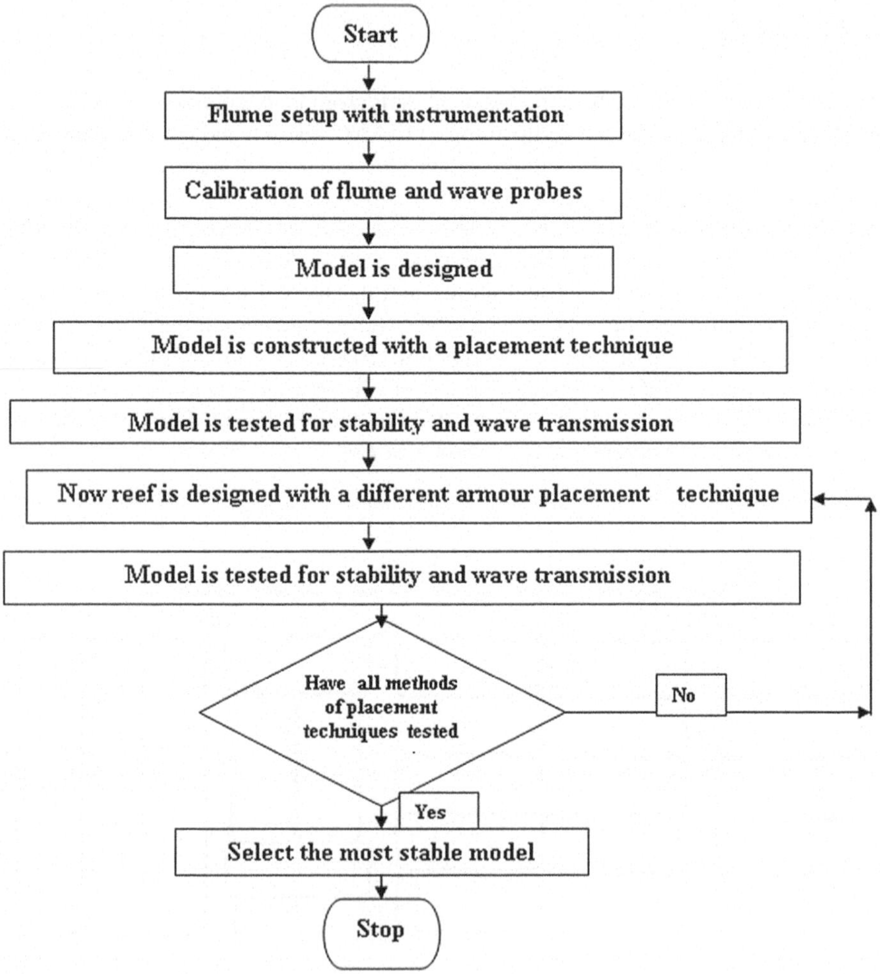

Fig. 3 Flow chart of the methodology

The wave flume is filled with potable water to the required depth. Before starting the experiment, the flume (with all its constituents) was calibrated to produce the incident waves of different combinations of wave height and wave periods. Combinations that produced the secondary waves in the flume were not considered for the experiments. The wave probes were calibrated at the beginning and at the end of the test runs. These models were tested for stability in a water depth (d) of 0.3 m (then the freeboard $F = d - h = 0.05$ m) with varying waves of heights (H) of 0.1 and 0.12 m, and wave periods (T) of 1.5–2.5 s. The wave data acquired by the probes were recorded by a computer and analysed using software. Occasionally, the wave heights were measured manually to crosscheck the instrumental data.

Results

In this section, the influence of wave steepness parameter and wave activity (as number of waves) on reef damage and wave transmission is analysed for a constant relative reef height (h/d) of 0.83 (i.e. $h/F = 5$).

Damage

All the GSCs present in the crest layer of different reef models are considered to be critical. Percentage damage is computed as number of GSCs detached against those in the critical layer and plotted against deep water wave steepness parameter (H_0/gT^2) for wave activity of 100, 500, 1000, and 1500 waves (actually the damage was measured up to 1800 waves) as shown in Fig. 4 for a constant relative reef height (h/d) of 0.83 (where H_0 is the deep water wave height). The GSCs with parallel double layer alignment suffered no damage throughout the test regardless of wave steepness parameter and storm duration.

The maximum detachment of GSCs for the parallel alignment (single layer) was 55–100% for H_0/gT^2 up to 0.0019 (i.e. gentle to moderately steep waves), while the reef was stable and the damage was least, i.e. about 0–5%, for the steepest waves (i.e. H_0/gT^2 of 0.0043) for a wave activity up to 1500 waves. The other GSC alignments suffered increased detachment with the wave activity with maximum being about 90–100%. The damage was almost constant after 1500 waves and the reef profile stabilized. However, it was observed that damage of a submerged GSC reef exhibited inconsistent behaviour with H_0/gT^2.

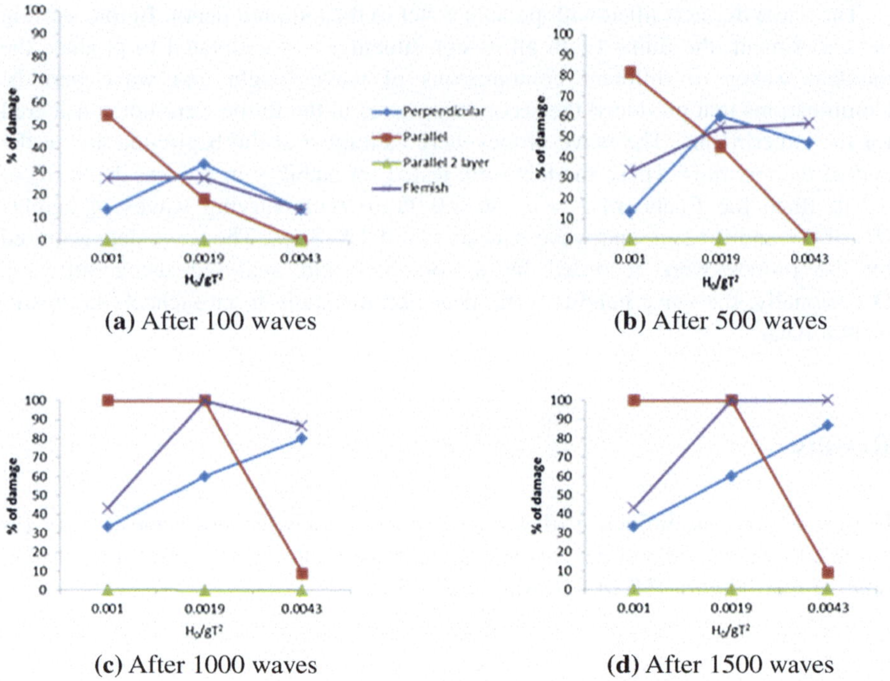

Fig. 4 Plot of percentage damage against H_0/gT^2 for various GSC alignments and storm durations

Wave Transmission

The submerged breakwater successfully trips the steeper waves and dissipates a major portion of wave energy. The submerged breakwater has two main energy dissipation mechanisms that attenuate wave height. First, energy is dissipated when the wave breaks due to abrupt change in water depth at the submerged breakwater; secondly, some energy dissipation may occur due to surface friction offered by GSC layer of the submerged reef (but no value for surface friction has been assumed in this study). The effectiveness of reef in damping of waves increases with an increase in wave steepness for all the alignments of GSC armour. This is clearly illustrated in Fig. 5 for initial condition and storm durations of 100, 500 and 1000 waves for a constant relative reef height (h/d) of 0.83. The K_t was found reducing in the order of parallel (single row), Flemish, parallel with two layers and perpendicular alignments up to about 100 waves changed the pattern later.

The wave transmission coefficients K_t at the initial condition, i.e. immediately after construction and after passing 100 waves are comparable except for the case of perpendicular alignment as shown in Fig. 5a, b. It can be said that the damage caused by a longer wave duration has not changed the reef's wave transmission character even after the reef suffering damages up to 55%. The wave transmission

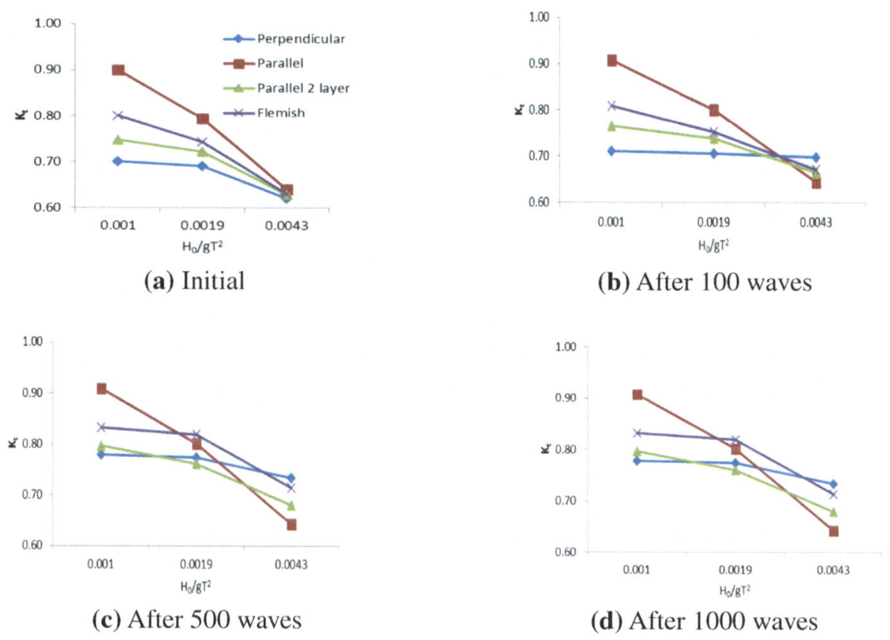

Fig. 5 Plot of K_t against H_0/gT^2 for various GSC alignments and storm durations

coefficient K_t continues to follow more or less the same pattern after suffering damage due to passing 500 waves. The K_t varied between 0.65 and 0.9 for various GSC alignments. After passing 500 waves while the damages to the reef increased up to 80%, K_t increased by about 5–12% compared to the previous case for all the GSC alignments. K_t for parallel (single row) remained almost constant throughout the experiment; while parallel alignment with two layers of GSCs exhibited constant trend after the passage of 100 waves. Though the damage of the reef increased up to 100% after passing 1000 waves, the K_t for all GSC alignments remained almost constant after about 500 waves. Parallel alignment with two layers of GSCs in the crest, though was stable with zero damage, and exhibited 6.6% increase in K_t. It is found that though the structure is damaged, it is still effective in inducing wave breaking and energy dissipation resulting in effective damping of waves. This is because, though the damage displaced the armour units from their initial position, these units interfered with the wave field and offered the resistance and created turbulence to the passing wave thus inducing wave breaking and damping the waves.

Conclusions

From this study following conclusions are drawn:

Reef Damage

Damage of a submerged GSC reef exhibits inconsistent behaviour with H_0/gT^2, but increases with increase in storm duration. A reef model should be tested for minimum storm duration of 1000–1500 waves to define its damage status completely. Parallel alignment with two layers of GSCs in the crest is the only alignment that is stable and the reef damage is zero for the complete range of test conditions.

Wave Transmission

The wave transmission is influenced by GSC alignment and reef damage. The effectiveness of reef in damping of waves increases with an increase in wave steepness for all the alignments of GSC units. The K_t varied between 0.65 and 0.9. The wave transmission coefficients K_t at the initial condition, i.e. immediately after construction and after passing 100 waves (i.e. even after the reef suffering damages up to 55%.) are almost similar except for the case of perpendicular GSC alignment. After passing 500 waves while the damages to the reef increased up to 80%, K_t increased by about 5–12% compared to the previous case (i.e. after passing 100 waves). Though the damage increased with the number of waves passed, K_t almost stabilized after passing 500 waves. Parallel alignment with two layers of GSCs in the crest, though was stable with zero damage and exhibited 6.6% increase in K_t. It is found that though the structure is damaged, it is still effective in inducing wave breaking and energy dissipation resulting in effective damping of waves.

The above conclusions are derived from a small scale physical model study conducted within a limited wave conditions. However, a large flume with large waves may give a different number of waves to test stability. Also different wave steepness would give different results.

References

Baba M (1985) Design of submerged breakwater for coastal protection and estimation of wave transmission coefficient. In: Proceeding of 1st national conference on dock and harbour engineering, vol 1. IIT Bombay, India, pp B 13–B 26

CPRA (2012) Louisiana's comprehensive master plan for a sustainable coast: committed to our coast. Coastal protection & restoration authority, Louisiana, Baton Rouge, LA

Dassanayake DT, Oumeraci H (2013) Geotextile sand container structures for shore protection—hydraulic stability formulae and nomograms. In: Proceedings 2nd international workshop on geosynthetics and modern materials in coastal protection and related applications, IIT Madras, India, pp 68–81
Fridman AM, Alperovich LS, Shemer L, Pustil'nik L, Shtivelman D, Marchuk An G, Liberzon D (2010) Tsunami wave suppression using submarine barriers. Physics—Uspekhi Russian Acad Sci 53(8):809–816
Harris LE, Zadikoff G (1999) Concrete tetrahedrons and sand-filled geotextile containers: shoreline stabilization and other uses in developing countries. In: Proceedings 5th international conference coastal & port engineering in developing countries, Cape Town, South Africa, COPEDEC, pp 2245–2253
Harris LE, Sample JW (2009) The evolution of multi-celled sand-filled geo-synthetic systems for coastal protection and surfing enhancement. Reef J 1(1)
Isaacson M (1991) Measurement of regular wave reflection. J Waterways Port Coast Ocean Eng ASCE 117:553–569
Johnson JW, Fuches RA, Morison JB (1951) The damping action of submerged breakwater. Trans AGU 32(5):704–718
Kriebel DL, Bollmann CA (1996) Wave transmission past vertical wave barriers. Coastal Eng, pp 2470–2481
Kudale MD (2013) Use of gabions and geotubes in coastal protection works-Indian scenario. In: Proceedings 2nd international workshop on geosynthetics and modern materials in coastal protection and related applications, IIT Madras, India, pp 37–48
Kudale MD (2015) Use of sand filled geotextile tubes for sustainable coastal protection. ISH News, 24 (II), January, pp 3–8
Mallidi RR (2014) Physical modeling of submerged inclined serrated rough plate breakwater. M-Tech Thesis. National Institute of Technology, Karnataka, Surathkal, India
Oh YI, Shin EC (2006) Using submerged geotextile tubes in the protection of the East Korean shore. J Coast Eng 53:879–895
PIANC (2011) The application of geosynthetics in waterfront areas. The World Association for Waterborne Transport Infrastructure, Brussels, Belgium, p 111
Pilarczyk KW (1996) Geotextile systems in coastal engineering—an overview. In: Proceedings 25th ICCE, ASCE, pp 2114–2127
Pilarczyk K (2000) Geosynthetics and geosystems in hydraulic and coastal engineering. In: Balkema, A (ed) Rotterdam, The Netherlands
Pilarczyk KW, Zeidler RB (1996) Offshore breakwaters and shore evolution control. In: Balkema AA (ed), Rotterdam, The Netherlands
Rao Subba, Shirlal KG, Varghese Roobin V, Govindaraja KR (2009) Physical model studies on wave transmission at the submerged inclined plate breakwater. Ocean Eng 36:1199–1207
Recio J, Oumeraci H (2009) Processes affecting the hydraulic stability of coastal revetments made of geotextile sand containers. J Coast Eng 56(3):260–284
Sundar V (2013) Coastal protection measures along few stretches of Indian coast. In: Proceedings 2nd international workshop on geosynthetics and modern materials in coastal protection and related applications, IIT Madras, India, pp 147–158
Sundar V, Sannasiraj SA (2013) Coastal protection using geosynthetic at Shankarpur, West Bengal, India. In: Proceedings 2nd international workshop on geosynthetics and modern materials in coastal protection and related applications, IIT Madras, India, pp 159–164

The manufacturer's authorised representative in the EU is Springer Nature Customer Service Centre GmbH, Europaplatz 3, 69115 Heidelberg, Germany. If you have any concerns regarding our products, please contact ProductSafety@springernature.com

Printed and bound by CPI Group (UK) Ltd, Croydon, CR0 4YY
30/09/2026
02225440-0012